AF344315

Eco-Friendly Synthesis of Fine Chemicals

RSC Green Chemistry Book Series

Series Editors: James H Clark, *Department of Chemistry, University of York, York, UK*
George A Kraus, *Department of Chemistry, Iowa State University, Iowa, USA*

Green Chemistry is one of the most important and rapidly growing concepts in modern chemistry. Through national awards and funding programmes, national and international courses, networks and conferences, and a dedicated journal, Green Chemistry is now widely recognised as being important in all of the chemical sciences and technologies, and in industry as well as in education and research. The RSC Green Chemistry book series is a timely and unique venture aimed at providing high level research books at the cutting edge of Green Chemistry.

Titles in the Series:

Alternative Solvents for Green Chemistry
By Francesca M. Kerton, *Department of Chemistry, Memorial University of Newfoundland, St. John's, NL, Canada*

Eco-Friendly Synthesis of Fine Chemicals
Edited by Roberto Ballini, *Department of Chemical Sciences, University of Camerino, Camerino, Italy*

The Future of Glycerol: New Uses of a Versatile Raw Material
By Mario Pagliaro, *CNR, Instiute of Nanostructured Materials and Institute for Scientific Methodology, Palermo, Italy and Michele Rossi, Department of Inorganic Chemistry, University of Milan, Milan, Italy*

Visit our website on www.rsc.org/books

For further information please contact:
Sales and Customer Care, Royal Society of Chemistry, Thomas Graham House, Science Park, Milton Road, Cambridge, CB4 0WF, UK
Telephone: +44 (0)1223 432360, Fax: +44 (0)1223 426017, Email: sales@rsc.org

Eco-Friendly Synthesis of Fine Chemicals

Edited by

Roberto Ballini
Department of Chemical Sciences, University of Camerino, Camerino, Italy

RSCPublishing

ISBN: 978-1-84755-908-1
ISSN: 1757-7039

A catalogue record for this book is available from the British Library

© Royal Society of Chemistry 2009

Published by The Royal Society of Chemistry,
Thomas Graham House, Science Park, Milton Road,
Cambridge CB4 0WF, UK

Registered Charity Number 207890

For further information see our web site at www.rsc.org

Preface

The late 20th and early 21st centuries have seen a phenomenal growth of the global economy and a continuous improvement of the standard of living in industrialized countries. Sustainable development has consequently become an ideal goal in recent years. In the early 1990s the concept of "green chemistry" was launched in the USA as a new paradigm, and since 1993 it has been promoted by the National Science Foundation (NSF) and the Environmental Protection Agency (EPA).

The success of the pharmaceutical industry is, in large part, due to the towering achievement of organic chemistry – a mature science that emerged as a distinct discipline well over 150 years ago. This history is both a blessing and a curse. Many of our most reliable strategies for assembling target molecules employ reactions that are 50 to 100 years old and that are often named in honour of their discoverers. During these early years, the chronic toxicological properties of chemicals were often completely unknown and many, unwittingly, became indispensable tools of the trade.

Early pioneers in green chemistry included Trost (who developed the atom economy principle) and Sheldon (who developed the E-Factor). These measures were introduced to encourage the use of more sustainable chemistry and to provide some benchmarking data to encourage scientists to aspire to more benign synthesis.

The present book is intended to provide an important overview of various processes and procedures devoted to the eco-sustainable syntheses of fine chemicals.

The contributions of catalyses as well as of photochemistry, high pressure and microwave irradiation are thoroughly examined. Nevertheless, the key role of solvents has also been considered. In addition, a chapter has been dedicated to the application of a simple reaction to the synthesis of complex molecules.

RSC Green Chemistry Book Series
Eco-Friendly Synthesis of Fine Chemicals
Edited by Roberto Ballini
© Royal Society of Chemistry 2009
Published by the Royal Society of Chemistry, www.rsc.org

Thus, I believe that this book represents an important contribution to eco-sustainable chemistry and should be of interest for both young and senior researchers involved in this field.

I would like to thank all the distinguished scientists and their co-authors for their rewarding, timely and well-referenced contributions. Grateful acknowledgements are also offered to the RCS editorial staff.

Roberto Ballini

Contents

RSC Green Chemistry Book Series
Eco-Friendly Synthesis of Fine Chemicals
Edited by Roberto Ballini
© Royal Society of Chemistry 2009
Published by the Royal Society of Chemistry, www.rsc.org

Catalysis in Non-conventional Reaction Media

MARCO LOMBARDO AND CLAUDIO TROMBINI

Università di Bologna, Dipartimento di Chimica "G. Ciamician", Via Selmi 2, I-40126 Bologna, Italy

1.1 Introduction

1.1.1 The Context

The concept of green or sustainable chemistry was born around 1990 thanks to a small group of chemists who, ahead of the times, clearly saw that the need for more environmentally acceptable processes in chemical industry had to become a top priority in R&D activities. In 1990, the Pollution Prevention Act was the spark that ignited awareness of the need for innovative chemical technologies that accomplished pollution prevention in a scientifically sound manner. In 1991 Paul Anastas coined the term and defined the field of "Green Chemistry". In the same year the first "Green Chemistry" program, the "Alternative Synthetic Pathways" research program, was launched. From a theoretical viewpoint, concepts like "atom economy" proposed by Trost[1] and the "E factor" introduced by Sheldon,[2] gave impetus to the creation of a new way of thinking about chemistry and to the development of a green metric able to provide quantitative support to compare the "greenness" of alternative products and processes.[3] From a functional point of view, several endeavours aimed to promote green chemistry activities got under way. In 1995 the Presidential Green Chemistry Challenge Award, proposed by Anastas to the White House,

RSC Green Chemistry Book Series
Eco-Friendly Synthesis of Fine Chemicals
Edited by Roberto Ballini
© Royal Society of Chemistry 2009
Published by the Royal Society of Chemistry, www.rsc.org

was approved. Since then, every year the Presidential Green Chemistry Challenge Awards highlight successes in research, development and industrial implementation of technologies that prevent pollution at source while contributing to the competitiveness of the innovators.[4] In 1997 Anastas co-founded the Green Chemistry Institute, which worked closely with industries and universities on environmental issues, and expanded its international network to consortia in 27 nations. Other initiatives in the green chemistry field rapidly spread throughout the world, *e.g.* in Italy, Canada, UK, Australia and Japan. The *Green Chemistry* journal was launched in 1999 by the Royal Society of Chemistry, and it was accompanied in 2008 by *ChemSusChem* published by Wiley-VCH. Less than two decades after its beginnings, green chemistry issues are providing an enormous number of challenges and are the central concern of those who practice chemistry in industry, education and research. In Europe the SusChem Technology Platform[5] and in the US the Technology Vision 2020 of the Chemical Industry[6] are turning their chemistry research agenda into new goals, jobs and business through innovation directed towards sustainability and environmental stewardship.

1.1.2   The 12 Principles of Green Chemistry

In 1998, Anastas and Warner teamed up to write the best selling and most cited book in the field of Green Chemistry, *Green Chemistry: Theory and Practice*, which explicated the 12 principles of Green Chemistry.[7] The principles, intended as guidelines for practical chemistry, provided a road map for chemists both of academia and industry to promote pollution prevention through environmentally conscious design of chemical products and processes. The 12 principles are:

1. It is better to prevent waste than to treat or clean up waste after it is formed.
2. Synthetic methods should be designed to maximise the incorporation of all materials used in the process into the final product.
3. Wherever practicable, synthetic methodologies should be designed to use and generate substances that possess little or no toxicity to human health and the environment.
4. Chemical products should be designed to preserve efficacy of function while reducing toxicity.
5. The use of auxiliary substances, *e.g.* solvents, separation agents, should be made unnecessary wherever possible and innocuous when used.
6. Energy requirements should be recognised for their environmental and economic impacts and should be minimized. Synthetic methods should be conducted at ambient temperatures and pressure.
7. A raw material feedstock should be renewable rather than depleting, whenever technically and economically practical.
8. Unnecessary derivatization (blocking group, protection/deprotection, temporary modification of physical/chemical processes) should be avoided wherever possible.

9. Catalytic reagents (as selective as possible) are superior to stoichiometric reagents.
10. Chemical products should be designed so that at the end of their function they do not persist in the environment and break down into innocuous degradation products.
11. Analytical methodologies need to be further developed to allow for real-time in-process monitoring and control prior to the formation of hazardous substances.
12. Substances and the form of a substance used in a chemical process should be chosen so as to minimize the potential for chemical accidents, including releases, explosions and fires.

The 12 guidelines essentially fall into four groups: efficient use of energy, hazard reduction, waste minimization and the use of renewable resources. Although each heading clearly indicates a proficient solution toward sustainability, the more guidelines or green technologies a master plan meets the greater is its intrinsic value. An example could be offered by a process involving the addition reaction (100% atom economy) of two starting materials derived from renewable feedstocks (rule no. 7) taking place in a perfect stereoselective way (waste minimization) at room temperature (energy efficient) in water (a prototypical green solvent) under the action of a low amount of a recyclable catalyst (rule no. 9).

The present chapter presents a few snapshots from the recent literature on the integration of the theme of catalysis with the topic of green solvents. The examples discussed witness how molecular design makes its own contribution to the breakthrough approach to innovative problem-solving, providing sustainable solutions to the chemical industry. Part of these cutting-edge results, suitably integrated by the state of the art process engineering, have the chance to evolve into low cost, robust, efficient and easy to operate sustainable technologies.

1.1.3 Catalysis

Adopting a K. B. Sharpless statement,[8] catalysis is the engine that drives the development of chemistry. Everybody can easily recognize that top achievements in applied chemistry are focused on industrial applications of catalysis, rational design, serendipitous discovery or combinatorial identification of new ligands, catalysts, new solid supports (organic, inorganic, amorphous or mesoporous silica phases, metal organic frameworks, *etc*).

An ideal catalyst should approach 100% selectivity while reaching high levels of productivity. Selectivity refers first of all to (i) chemoselectivity, which means the catalyst must be able to select preferred reactants from complex mixtures, (ii) regioselectivity, which means selection of preferred sites of the reacting substrate and (iii) stereoselectivity, which means preferred formation of a single stereoisomer.

At the beginning of this decade Gladysz defined what should be an ideal catalyst.[9] He proposed the following features: the rapid production (turnover frequency, TOF) of an infinite amount of product (turnover number, TON), preferentially at room temperature and under atmospheric pressure, which implies no deactivation and poisoning under the reaction conditions. This "ideal" catalyst does not require an inert atmosphere to operate, is insensitive to reactant impurities and affords product yields of 100%. Gladysz clearly noted that these unattainable limits can never be realized but help to focus attention on what we should strive for. The "infinite TON" limit, for example, would make catalyst recovery efforts unnecessary, a quest that is unrealistic. The design of recoverable catalysts has become, indeed, a central field of catalysis research.[10] In fact, catalyst decomposition associated with leaching of the active species and decomposition itself have to be taken into account. In practice, design efforts for effective recoverable catalysts must address the removal of these catalyst impurities from solution.

Catalysis is generally hypothesis driven: the chemist uses experience to envisage candidate catalysts, which are then tested, investigated and optimized, including the use of high throughput and library technologies. There is an increased need for new paradigms of how catalysis research and rational design has to be done. But there is an even more urgent need: that of addressing in R&D activities both classic economy-linked problems (cost, yield, selectivity, time, resistance to poisoning and deactivation, minimization of product contamination by catalyst residues),[11] and sustainability concerns. They focus on minimizing waste production, energy consumption and use of toxic and eco-toxic chemicals. An impressive example from the golden era of petrochemistry is the SOHIO ammoxidation of propane with ammonia, which replaced the IG FARBEN process based on acetylene and hydrogen cyanide.[12] Particularly in the case of costly and highly specialized catalysts the optimization of catalyst recovery and reuse is a fundamental demand, too, for economic and environmental reasons. Moreover, from a practical viewpoint, an optimum catalytic process should be wide in scope, easy to perform and insensitive to oxygen and water, which could become the solvent of choice. This is what characterizes, for example, the Cu-catalysed azide-alkyne cycloaddition, a reliable and practical reaction, which proved very useful, for example, for accelerating the drug discovery process.[13] The reaction, work-up and purification should use benign solvents, avoiding chromatography or time-consuming distillation or crystallization unit operations.

Catalysis is traditionally divided into heterogeneous and homogeneous catalysis. In classic solid/liquid or solid/gas heterogeneous catalysis, the catalyst provides a surface on which the reactants are temporarily adsorbed. Bonds in the substrate are weakened sufficiently by adsorption or chemisorption for new bonds to be created. Syngas conversion, hydrogenation and oxidation processes are by far the most important industrial applications. Catalyst synthesis technology is applied to the manufacture of high surface area metal species, including nanoparticles, and metal oxides, usually supported in another metal oxide such as alumina, silica and zeolites. The chemical industry often favours heterogeneous catalysis for the easy recovery and good stability of the catalyst, the practical downstream separation processes involved, and the easier recycling of the catalyst. However, the role played by phase partitioning and transport phenomena make the kinetic interpretation of a catalytic process a tough task. Analogous difficulties are encountered in the rationalization of heterogeneous catalysis in terms of a molecular phenomenon with well-defined surface organometallic intermediates and/or transition states. An example of paradigms developed to support a rational catalyst design is offered by the seven principles, or "seven pillars", proposed by Grasselli for tailoring new selective oxidation catalysts. The lattice oxygen of the metal oxide, metal–oxygen bond strength, host structure, redox properties, multifunctionality of active sites, site isolation and phase cooperation are all essential aspects to be controlled for a successful design.[14]

Homogeneous catalysis, on the other hand, has reached astonishing advanced levels in terms of molecular understanding of the elementary steps characterizing the catalytic cycle, bringing molecular insight to the design of new catalysts and even allowing the discovery of new reactions; examples are offered by Ziegler–Natta depolymerization[15] and alkane metathesis.[16] The pros of homogeneous catalysis are often the activity (TONs, TOFs) and selectivity, while cons are generally the poor stability, troublesome separation of the catalyst from the product, which is often contaminated by it, the quite difficult separation and reuse of the catalyst. Homogeneous asymmetric catalysis provides a powerful tool for the synthesis of optically active molecules that serve, for example, as Active Pharmaceutical Ingredients (APIs). Although numerous highly selective chiral catalysts have been developed over the past three decades, their practical applications in industrial processes are hindered by their high costs as well as difficulties in removing trace amounts of toxic metals from the organic product. In this case costly and time-consuming unit operations, such as crystallization, chromatography or distillation, are necessary to both purify the product and to recover and eventually reuse the catalyst.

To overcome these problems many different approaches have been employed to generate heterogenized asymmetric catalysts.[17] The most frequently used approach makes use of solid/liquid biphasic chemistry; in this context the term "supported reagent" is currently used to describe a wide range of materials involving an inorganic or organic support onto which a "catalyst" species has been chemically or physically adsorbed.[15,16]

Some of the key advantages of supported reagents compared to the unsupported homogeneous catalyst are:

- good dispersion of active sites, and the concentration of sites within small pores can lead to significant improvements in activity,
- the presence of molecular-sized pores and the adsorption of reactant molecules on the material surface can lead to improvements in reaction selectivity,
- easier and safer storage, work-up and handling,
- the opportunity to replace batch processes, as well as time- and solvent-consuming purifications (*e.g.* chromatography) with flow techniques exploiting a suitable immobilization strategy.

The heterogenized catalyst is, however, often less effective than their homogeneous counterparts. Thus there exists a need to develop new, innovative approaches toward the design of recoverable and reusable asymmetric catalysts with the aim of combining the advantages of heterogeneous and homogeneous catalysis.

Liquid/liquid biphasic, and sometimes gas/liquid/liquid triphasic heterogeneous conditions,[18] provide an alternative way of achieving phase separation, the catalyst being eventually confined in an immiscible liquid solvent.[19] In the ideal catalytic system operating under biphasic conditions the catalyst is immobilized in one solvent, which is immiscible with a second one in which the substrates/products are dissolved. Thus, the product is easily removed without being contaminated by the catalyst, and the catalyst can be simply recovered and reused. With metal-based catalysts, the efficient immobilization of the catalysts in a separate phase from that which contains the products is often performed by tailoring suitable ligands that confer the correct partition coefficient to the catalyst. One strategy adopted when the solvent of choice is water or an ionic liquid is based on the introduction of an ionic tag into ligands, transition metal complexes or organocatalysts, that confers the species of interest the right solubility in the reaction medium.[20] A beautiful, and historically one of the first examples, is the Ruhrchemie/Rhône-Poulenc propene hydroformylation process, where sulfonated triarylphosphine ligands make rhodium catalysts water-soluble, allowing for an easy separation of the catalyst from the butanal product by decantation.[21]

1.1.4   Green Solvents

Many of the traditional solvents that are favoured by organic chemists have been blacklisted by international regulations, for instance chlorinated hydrocarbon solvents have been severely curtailed. Thus, the question of the solvent requires a major rethink in terms of green chemistry issues; this need is driving the search for alternative reaction media. Such media are the basis of many of the cleaner chemical technologies that have reached commercial development. Typical non-conventional reaction media are supercritical carbon dioxide ($31.1\,°C$, $73\,atm$)[22] and supercritical water ($374\,°C$, $218\,atm$),[23] water under P/T standard conditions,[24] room-temperature ionic liquids,[25] up to and including solvent-free conditions.[26]

With supercritical fluids, a large degree of control over product selectivity and yield is possible by adjusting the temperature and pressure of the reactor. In many cases, supercritical fluids allow easy recovery of the product and separation of the catalyst. An example of industrial application of supercritical water is the synthesis of terephthalic acid. The traditional synthesis uses *p*-xylene and air with acetic acid as the solvent and a manganese/cobalt catalyst system at $190\,°C$ and $20\,atm$. The process is highly selective but it is also energy intensive. Poliakoff and co-workers developed a route that uses hydrogen peroxide as the oxidant and $MnBr_2$ as the catalyst. This reaction is the first selective oxidation to be carried out as a continuous process in supercritical water, affording terephthalic acid in more than 70% yield and with > 90% selectivity.[27]

Water under standard conditions, since the seminal contributions in the 1980s of Breslow in Diels–Alder reactions,[28] and Kuntz for hydroformylation processes,[21,29] has been enjoying an upsurge in applications, not only in enhancing reaction rates, but also performing new transformations that are impossible in conventional solvents. The combination of a small size and a three-dimensional hydrogen-bonded network results in a large cohesive energy density (2200 MPa), a high surface tension and a high heat capacity. These three characteristics give rise to the special properties known as hydrophobic effect, which also plays a critical role in the folding of biological macro-molecules, in the formation and stabilization of membranes and micelles, and in molecular recognition. Several of the most outstanding achievements with water have been carried out in the field of heterogeneous aqueous liquid/liquid biphasic systems. The introduction of separate phases in the context of "homogeneous" catalysis seems a contradiction. Actually liquid/liquid biphasic catalysis with organic/aqueous (but also organic/organic, fluorous/organic, ionic liquid/organic solvent pairs, including supercritical fluids, and systems with soluble polymers)[17a] generally takes place in a well-defined phase with all the activity/selectivity benefits of homogeneous catalysis. At the same time the decisive advantage of heterogeneous catalysis is attained, namely the immediate separation of catalyst and substrates/products without additional post-reaction separation steps. With organic/aqueous systems the catalyst and products can be forced to stay in different phases if they have sufficiently different polarities. Products are easily separated from the catalyst phase, which may be recycled in a suitable manner into the reactor.

Ionic liquids (ILs) have forcefully entered the field of synthetic and applied chemistry in this century. Examples of reactions and processes range from hydrogenation to biocatalytic reactions. ILs typically consist of nitrogen- or phosphorus-containing organic cations coupled with inorganic anions. An important property of ILs is that they have essentially no vapour pressure, which makes them optimal replacements for volatile organic compounds tra-ditionally used as industrial solvents. There is an almost endless number of possible ILs that can be made by combining various anions with a wide range of cations. In parallel, many homogeneous reactions have been transposed into ILs. Most of them are performed on a bench scale and, to date, just a few examples of pilot-scale or commercial-scale applications of ILs have been described. Cost, potential toxicity with very few data available, thermal stability, environmental impact, physical features such as the high viscosity, the difficult purification and difficult stirring are examples of limitations that are restraining the applications of ILs; notably, though, these limitations will be less important if ionic liquids are recycled. In most of the applications of ILs in catalysis, the focus is especially on the possibility of recycling the catalyst, limiting the loss of activity/selectivity and minimizing any loss of catalyst. This is the most important task when high-cost ligands and metals are used. Enantioselective catalytic hydrogenations applied to the production of APIs combine the two needs: (i) recycling the expensive catalyst without deterioration of its performance in terms of activity and enantioselectivity and (ii) separating the organic products free from trace amounts of metal contaminants.[30]

Owing to their unique physico-chemical properties (solvation, polarity, structure), ILs have been proved to be more than "physical" solvents, providing an unusual coulombic environment where kinetic constants, activation energies and thermodynamic parameters may be strongly modified with respect to their values in traditional molecular solvents. Indeed, tailored ILs with supplementary functionalities, usually referred to as task-specific ionic liquids (TSILs), can chemically interact with the catalyst, for example acting as an activator, a ligand, *etc.*

1.2 Organocatalysis in Green Solvents

Since 2000, the field of organocatalysis has been developing at an exponential rate, increasing from 10 papers in 2000 to 500 in 2007. The major benefit arising from the use of organocatalysts is the absence of metal impurities in the final product, particularly if an intermediate for the synthesis of an API is involved. However, technologically, there are still several drawbacks that are limiting industrial applications of organocatalysis. Typical problems are the high catalyst loading, difficult recycling of the catalyst, slow reaction rates, and solvent limitations. As outlined in two recent articles by Dondoni and Massi and by Barbas,[31] the most prominent goals in organocatalysis research are directed to the optimization of chemical efficiency and the search for new applications, *e.g.* in multicomponent one-pot processes and in key steps in total syntheses of target natural products and known compounds of biological and pharmaceutical relevance.

We will present here a selection of the best and most recent contributions to the field of organocatalysis in non-conventional solvents; attention is focused in particular on asymmetric carbon–carbon bond forming processes.

1.2.1 Asymmetric Cross Aldol Reactions

One of the most studied processes is the direct intermolecular asymmetric aldol condensation catalysed by proline[32] and primary amines,[33] which generally uses DMSO as solvent. The same reaction has been demonstrated to also occur using mechanochemical techniques, under solvent-free ball-milling conditions.[34] This chemistry is generally referred to as "enamine catalysis", since the electrophilic substitution reactions in the α-position of carbonyl compounds occur *via* enamine intermediates, as outlined in the catalytic cycle shown in Scheme 1.1. A ketone or an α-branched aldehyde, the donor carbonyl compound, is the enamine precursor and an aromatic aldehyde, the acceptor carbonyl compound, acts as the electrophile. Scheme 1.1 shows the TS for the rate-determining enamine addition step, which is critical for the achievement of enantiocontrol, as calculated by Houk.[35]

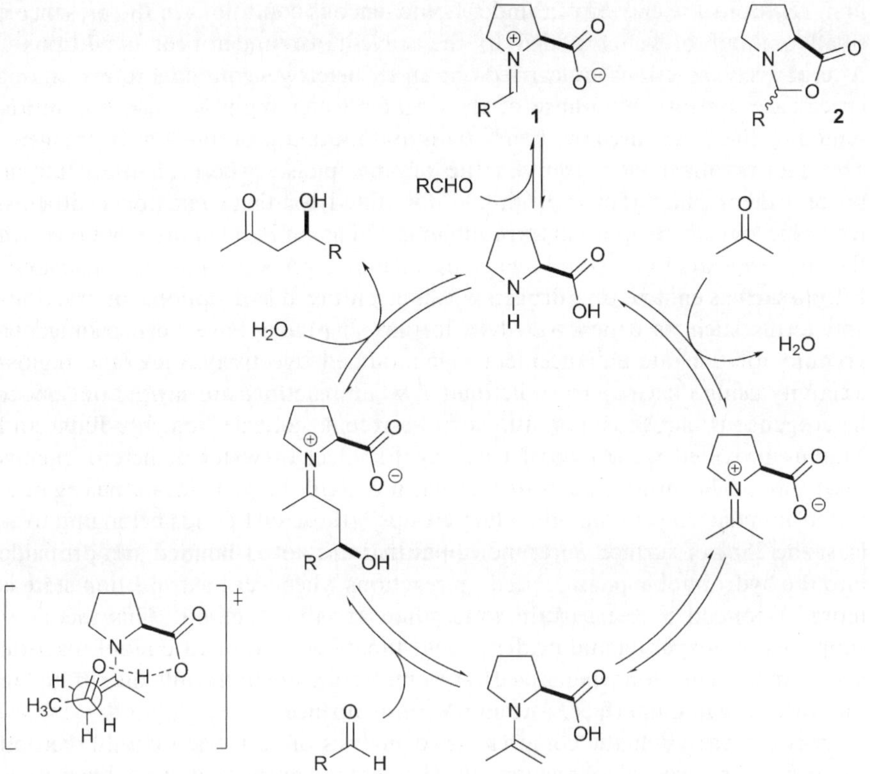

Scheme 1.1

The overall mechanistic picture is, however, more complex than the cycle shown in Scheme 1.1, as shown by the elegant work by Armstrong, Blackmond and co-workers.[36] They demonstrated that the presence of water in trace amounts is necessary, mainly to inhibit the off-cycle formation of oxazolidinones **2** that originate from iminium ions **1** formed by a side reaction of proline with the acceptor aldehyde. Formation of **2** has the consequence of reducing the available amount of proline. Thus, though water has a negative effect on the formation of the enamine, too, its presence ensures an increase in the extant catalyst loading and, consequently, yields increase as well.

Having established the positive effect of the presence of water, let us examine which protocols are to be adopted and which are deleterious in terms of catalytic activity and overall stereocontrol.[37] The first protocols reported in DMSO did not focus on the presence of water but it is likely that the required trace amount was present as contaminant of the solvent used. The procedure to avoid is a homogeneous reaction in water as solvent using water-soluble carbonyl compounds: in this case reactions likely proceed under general base catalysis mechanisms, resulting in very poor conversions and lack of enantiocontrol. Thus, two useful approaches for reactions in water are available: in the first, reactions are carried out under homogeneous conditions in the presence a small amount of water added to the solvent, or under neat conditions.[38] Alternatively, reactions are carried out under heterogeneous conditions. In this latter case, the organic phase of the organic/water biphasic system is represented by the donor used in molar excess and the acceptor, the limiting reagent. The catalyst must be present in the organic phase, where all the catalytic process takes place. Thus, proline is not suited for these reaction protocols, while the more hydrophobic tryptophan is.[39] The biphasic approach has led in the last few years to outstanding achievements in terms of catalytic efficiency.[40] Effects such as enhanced hydrogen bonding, enforced hydrophobic interactions and terms such as concentrated hydrophobic phases have been claimed to account for the rate enhancement and modified selectivity, including regioselectivity using unsymmetrical ketones,[41] when reactions are carried out under heterogeneous aqueous conditions.[40] A recent contribution by Jung and Marcus provided a theoretical basis to the effect of water in heterogeneous reactions.[42] The authors propose that when reactions take place in the organic/water boundary layer, "dangling OH groups" (those OH bonds belonging to at least the 25% of surface water molecules that are not H-bonded and protrude into the hydrophobic phase) speed up reactions whenever the transition state is more H-bonded to the surface water than are the reactants. This theory is supported by experimental evidence that the direct asymmetric aldol reaction occurs in the emulsion where the catalyst molecules are uniformly distributed in the water/organic interface, forming a chiral surface.[43]

Of course, a systematic comprehensive analysis of all literature published on the asymmetric cross aldol reaction goes beyond the purpose of this chapter, thus attention is focused on the best achievements reported for catalysts working under neat conditions in the presence of an aqueous phase (Scheme 1.2). Generally, *anti*-diastereoselectivity is observed, while the opposite *syn*-diastereoselectivity is often exhibited by α-hydroxyketones.

Table 1.1 shows a selection of catalysts developed in 2006–2007, which operate under biphasic aqueous conditions. For previous literature, readers are directed to an excellent review by List and co-workers.[32a] In terms of catalyst performance, since 2006 most of the best achievements have been reported for *O*-modified *trans*-4-hydroxy-L-proline derivatives **3–5**, including the inclusion complex **6** of a proline derivative and a β-cyclodextrin, or catalysts consisting of chiral amines **7** and **8** or (thio)amides **9–14** derived from L-proline. The *O*-protected serine **15a**, threonine **15b** and threonine amide **16** complete the list of catalysts.

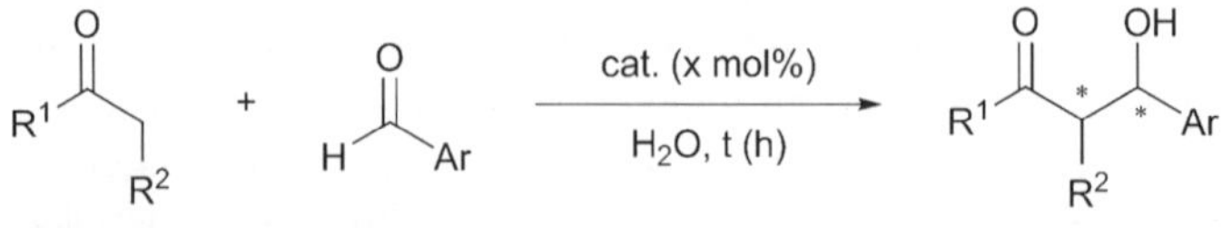

Scheme 1.2

Table 1.1 Aldol reaction organocatalysts working under biphasic aqueous conditions.

Catalyst	Ref.	Catalyst	Ref.
3	44	**4**	45
5	46	**6**	47
7	48	**8**	49
9	50	**10**	51
11	52	**12**	53
13	54	**14a**: R = *i*-Bu **14b**: R = Ph	55
15a: R¹ = H **15b**: R¹ = Me	56	**16**	57

The general strategy adopted was to install hydrophobic substituents on the proline structure, resulting in abatement of the original hydrophilicity of the reference amino acid. Catalysts **3–16** do not dissolve appreciably in water, and form together with the donor/acceptor pair an organic phase. Even though the catalytic cycle takes place in the water-saturated concentrated hydrophobic phase (or in the organic/water boundary layer), the presence of the aqueous phase is essential to ensure the best results. This condition was also verified in the case of a recyclable polystyrene-supported L-proline material used as catalyst in direct asymmetric aldol reactions that take place only in the presence of water.[58] Confirmation that formation of a biphasic system is a necessary condition for good results is given by the fact that water-miscible ketones afford moderate yield and stereoselectivity in water.[51a] An example from Maya *et al.* reports exceptional stereochemical results using acetone as donor and **14** as the catalyst; in this case brine is used as the aqueous phase to abate acetone miscibility with water by salting out.[55] The effect of salts as additives was also examined using thioamide **9**.[50] Catalyst **16**, a threonine amide with the same chiral amine present in **14b**, was applied to *syn*-selective aldol reactions of dihydroxyacetone and aldehydes. In this paper Barbas reported a typical example of how organocatalysis can be managed. Using water-soluble dihydroxyacetone, the best conditions involve a homogeneous solution in a polar organic solvent (Scheme 1.3, top). In contrast, the much more hydrophobic disilylated dihydroxyacetone gives the best results under heterogeneous conditions, using brine as the aqueous phase (Scheme 1.3, bottom).[57] This catalyst also allows the use of aliphatic aldehydes as acceptors.

The rational design of the catalyst may also address, in addition to performance optimization, recyclability issues. For this reason the fluorous pyrrolidine sulfonamide **8** was investigated. The hydrophobicity imparted by the fluorous chain makes **8** an ideal catalyst for heterogeneous aqueous protocols (Scheme 1.4). Both aldehydes and ketones serve as donor units. The catalyst is separated by treating the crude reaction mixture with fluorous silica gel; all the

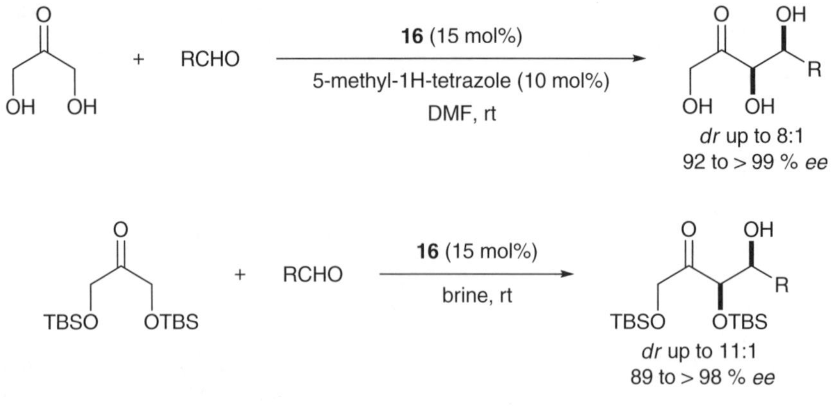

Scheme 1.3

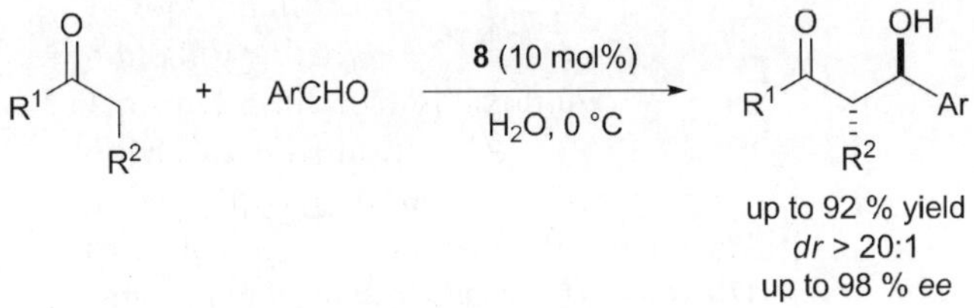

up to 92 % yield
dr > 20:1
up to 98 % *ee*

Scheme 1.4

non-fluorous products are washed out with CH_2Cl_2, then **8** is recovered by washing with THF and recycled.

If maximizing catalyst hydrophobicity is the key to success in the organocatalysed aldol reaction under liquid/liquid biphasic aqueous conditions using catalysts **3–16**, this property does not generally provide practical recycling procedures.

The utilization of tagged catalysts is a far superior strategy,[59] as testified by the dramatic increase in interest in this field. Different tags can be imagined, the most prominent one being a solid phase based on inorganic materials or, alternatively, on polymers. Polymers of choice can be either insoluble in the reaction medium, an approach that resembles the concept of heterogenization of homogeneous catalysts, or soluble therein [*e.g.* poly(ethylene glycol) (PEG)], in which case they are often removed by precipitation when a second solvent is added. Other catalyst tags include ionic groups, such as ionic liquid derived groups, and perfluorinated groups. These strategies exploit the high affinity of these tags to non-conventional reaction media such as ionic liquids (ILs) or perfluorinated solvents that can be poorly miscible with the organic phase.

The tag strategy was followed by us when we studied catalysts **17**. The best conditions devised for **17** involved a heterogeneous mixture of the donor/acceptor pair and **17** in water. The idea was to confer **17** the correct hydrophobocity/hydrophilicity profile that allows it to move from water to the organic phase consisting of the donor/acceptor pair in the reaction stage, then, in the work-up stage, to move back from the organic solvent used to separate the aldol products into water (Scheme 1.5). Such a delicate balance of solubility properties, shared by classic phase-transfer reagents,[60] is much easier to address by a rational design of a task-specific ionic liquid (TSIL).[61] A tag containing a charged group, *e.g.* an imidazolium group, was installed on the 4-OH group of *trans*-4-hydroxy-L-proline. The bis(trifluoromethylsulfonyl)imide (Tf_2N^-) ion conferred to **17** the correct solubility properties to apply a simple catalyst recycling procedure: it involved preliminary removal of the ketone and water under reduced pressure, followed by addition of ~ 0.8 mL of water and extraction with ether.[62] The aqueous phase contained **17**, which could be recharged with the reactants. In particular **17a** was recycled for 5 runs without any loss of catalytic activity and stereochemical performances.[62b]

Ionic liquids (ILs), beside affording new molecular supports for catalytically active species,[62] are entering the field of organocatalysis as solvent by offering (i) a new medium for checking the performance of existing organocatalysts in

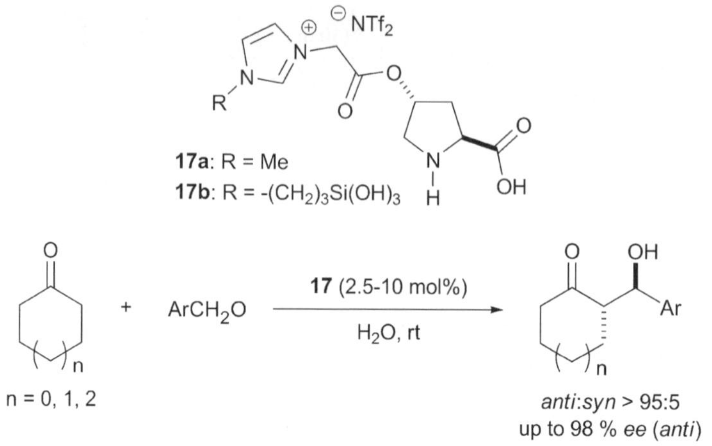

NTf_2

17a: R = Me
17b: R = -(CH$_2$)$_3$Si(OH)$_3$

17 (2.5-10 mol%)

H$_2$O, rt

n = 0, 1, 2

anti:syn > 95:5
up to 98 % ee (anti)

Scheme 1.5

their unique coulombic environment[63] and (ii) new opportunities to develop bi- and polyphasic processes. Focusing on the proline-catalysed direct asymmetric aldol reaction shown in Scheme 1.1, a few examples are reported where proline itself has been used in [bmim][BF$_4$].[64] Proline performance did not enjoy relevant improvements on passing from DMSO to the IL phase.[32g–i] The advantage offered by the IL was a technological improvement, since the easy separation of the catalyst in the IL phase from the aldol products extracted in ether allowed the authors to recycle the catalyst four times with no significant losses of activity and/or selectivity. To optimize catalyst recovery and reuse – a primary goal in catalytic process design from a green chemistry point of view – solid covalently supported ionic liquid phases (SILP) have been proposed as an attractive solution.[65] The surface of a solid support material is modified with a monolayer of covalently attached molecules of ionic liquids. Treatment of this surface with additional ionic liquid furnishes a multiple layer of free ionic liquid on the support that serves as the reaction phase in which a classic homogeneous catalyst is dissolved. This strategy, first developed for continuous flow gas-phase metal-catalysed reactions,[66] was exploited by Gruttadauria to develop a reusable catalytic system where proline, confined on a silica-bound IL layer, is readily separated from the reaction products and recycled.[67] Scheme 1.6 shows the concept of supported ionic liquid catalysis used in the aldol addition of acetone to aldehydes. On a silica surface are covalently bonded molecules carrying an ionic tag. An IL, identified by positive and negative charges, forms a layer on this modified silica surface and proline dissolves into this layer. In this supported phase the catalytic process takes place.

Much stronger, however, is the electrostatic interaction of the ion pairs of an IL with a catalyst possessing an ionic tag installed on its scaffold, *e.g.* **17**. In this coulombic environment either acceleration or inhibition can be observed with respect to the use of molecular solvents. In the absence of a robust theoretical

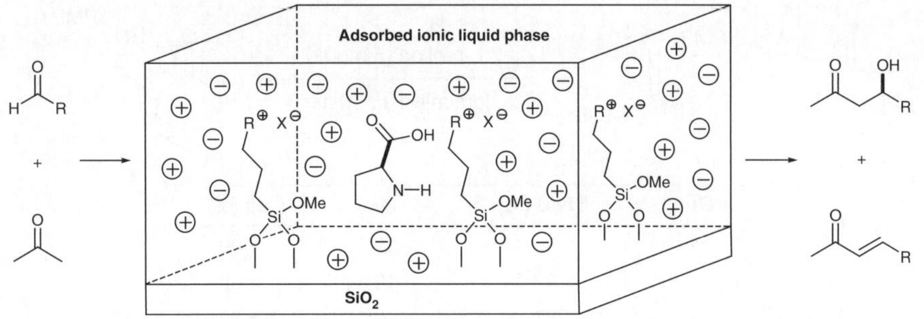

Scheme 1.6

framework that allows us to anticipate the kinetic course of a reaction in ILs, the use of ionic liquids as solvents is still based on a trial and error approach. Thus, the aldol reaction of acetone, cyclohexanone and protected dihydroxyacetone with aromatic aldehydes was studied in [bmim][Tf$_2$N], using **17a,b** as catalysts.[68] Here, using 5 mol.% of **17a** the reaction shown in Scheme 1.1 afforded the *anti*-aldol product in 78% yield, 70:30 *anti/syn* ratio and 94% ee (*anti*), after 24 h at room temperature. Results using the ionic tagging strategy are better in terms of yields and stereocontrol than those originally reported in DMSO as solvent, considering that catalyst loading fell from 30 to 5 mol.%.[32g–i]

1.2.2 Asymmetric Mannich Reactions

Beside the cross aldol reaction, the Mannich reaction, too, has been the object of successful efforts using organocatalysis.[69] The use of small organic molecules such as proline, cyclohexane diamine and *Cinchona* alkaloid-derived catalysts has proven extraordinarily useful for the development of asymmetric Mannich reactions in traditional polar solvents such as DMSO, DMP, DMF, *etc.*[32a,69] However, very few studies have been conducted so far in non-conventional solvents.

Proline-catalysed direct asymmetric Mannich reactions of *N*-PMP protected α-imino ethyl glyoxylate with various aldehydes and ketones in [bmim][BF$_4$] afforded β-amino acid derivatives with excellent yields and enantioselectivities, with the exception of hydroxyacetone.[70] The astonishing enantiocontrol is due to an efficient activation of the azomethine group by hydrogen bonding between the carboxyl group of proline and the imine nitrogen. The process developed allowed a facile product isolation, catalyst recycling (Scheme 1.7), and was characterized by significantly improved reaction rates, *ca.* 4- to 50-fold with respect to common polar organic solvents, likely due to the stabilization of the charged transition state of the rate-limiting process in the ionic liquid environment. As a consequence, catalyst loading could be ranged from 5 to 1 mol.% and the reaction time consistently reduced (0.5–6 h) with respect to the related aldol reactions.

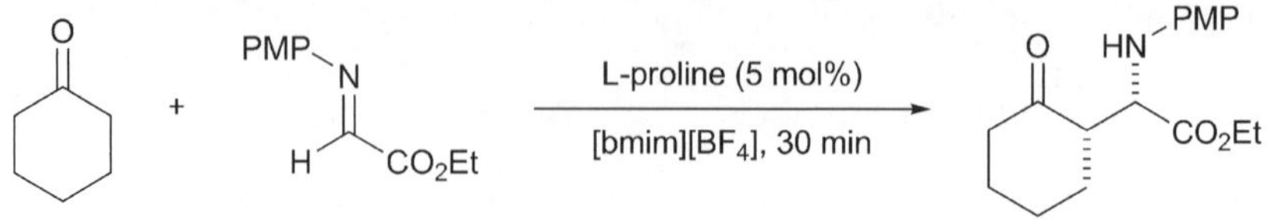

Run	Yield (%)	*dr*	*ee (%)*
1	99	> 19:1	> 99
2	92	> 19:1	> 99
3	87	> 19:1	> 99
4	83	> 19:1	> 99

Scheme 1.7

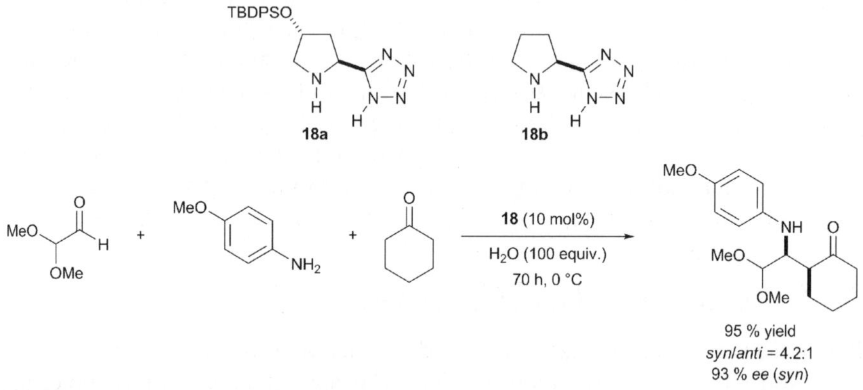

18a

18b

95 % yield
syn/anti = 4.2:1
93 % *ee* (*syn*)

Scheme 1.8

Three-component Mannich reactions involving other imines also worked well in [bmim][BF$_4$],[70] but even more efficient protocols involve solvent-less conditions, in the presence of water; again we deal with aqueous biphasic systems. Catalyst **15a** (Table 1.1) is suitable for such an application,[71] as well as siloxytetrazole **18a**, derived from a combination of the structural features of **3** (Table 1.1) and tetrazole derivative **18b**.[72] Interestingly, while neither **3** nor **18b** were suitable for such transformation, the hybrid catalyst **18a** gave the excellent results shown in Scheme 1.8.

The use of water in Mannich reactions is counterintuitive since it is supposed to shift to the left the imine formation reaction. However, as in the case of aldol reactions catalysed by species **3**–**12** (Table 1.1), the success is due to the biphasic heterogeneous conditions adopted – all three components of the Mannich reaction and the catalyst forming an organic phase where the process takes place, the water/organic interphase providing a local environment rich in "free OH bonds" that can co-activate the substrate.[42]

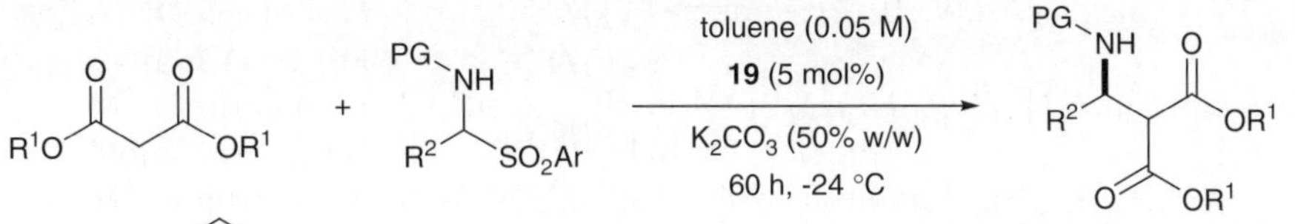

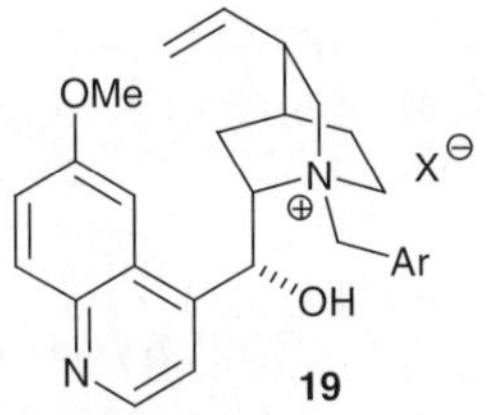

Scheme 1.9

Run	R¹	R²	PG	Yield (%)	ee (%)
1	Me	Ph	Boc	85	90
2	Me	Ph	Cbz	97	94

A related example of an organocatalytic asymmetric Mannich reaction that makes use of an aqueous biphasic system (aq. K_2CO_3/toluene) was reported by Ricci *et al.* Catalysts **19** derive from quinine and act as typical phase-transfer reagents. Active methylene compounds are used as donors, and *N*-Boc and *N*-Cbz protected α-amido sulfones as precursors of *N*-protected imines (Scheme 1.9).[73]

1.2.3 Asymmetric Michael Reactions

Beyond aldol and related reactions based on the intermediate formation of enamines, the next chapter of organocatalysis is founded on the transient formation of iminium ions resulting from the interaction of primary and secondary amines with carbonyl compounds.[74] Iminium catalysis is based on the fact that iminium salts are more electrophilic than the corresponding aldehydes or ketones, and, hence, the reversible formation of the iminium salt activates the carbonyl component toward nucleophilic attack. Both primary and secondary amines can be used, although secondary amines tend to dominate the field. Primary amines always require an external acidic co-catalyst; the use of an acid as co-catalyst is also documented with secondary amines. While the examples previously discussed report enamine chemistry protocols compatible with the aqueous environment, there have been few successes so far in aqueous systems, iminium activation being less compatible with the presence of water. The first study, by Jørgensen and co-workers, reports on the reaction between *tert*-butyl-3-oxo-butyric ester and cinnamaldehyde in distilled H_2O, which was efficiently catalysed by 2-{bis[3,5-bis(trifluoromethyl)phenyl]trimethylsilanylo-xymethyl}pyrrolidine (**20**) (10 mol.%), generating Michael adducts in 71–97% yield and 84–94% ee at room temperature (Scheme 1.10).[75] The authors highlighted that the reaction occurred in the organic phase constituted by the α,β-unsaturated aldehyde, the β-ketoester and the catalyst. This hypothesis was confirmed by the fact that the reaction could be also performed under solvent-free conditions, with 90% yield and 94% ee. In related studies the same authors

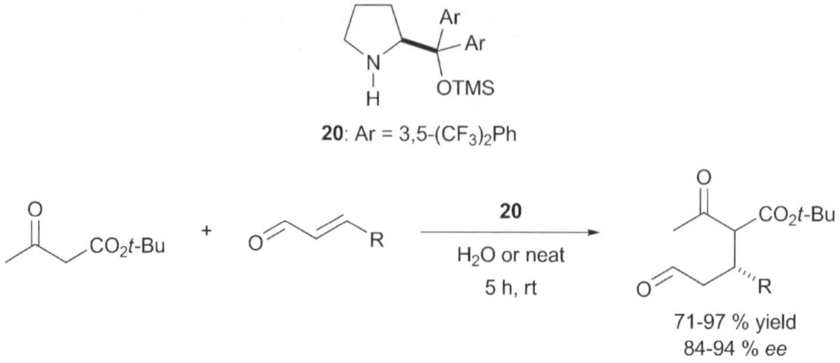

20: Ar = 3,5-(CF$_3$)$_2$Ph

Scheme 1.10

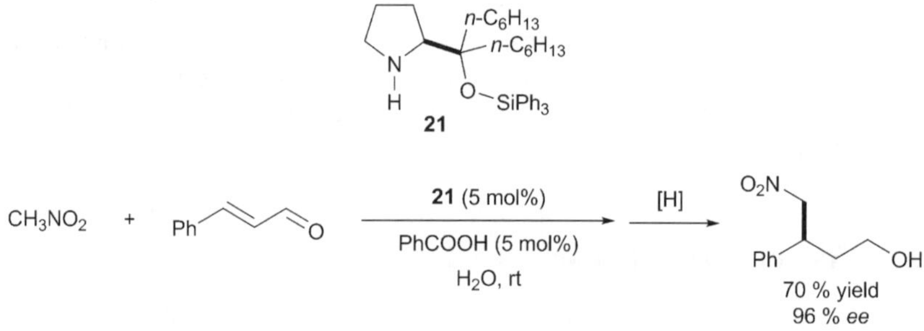

21

Scheme 1.11

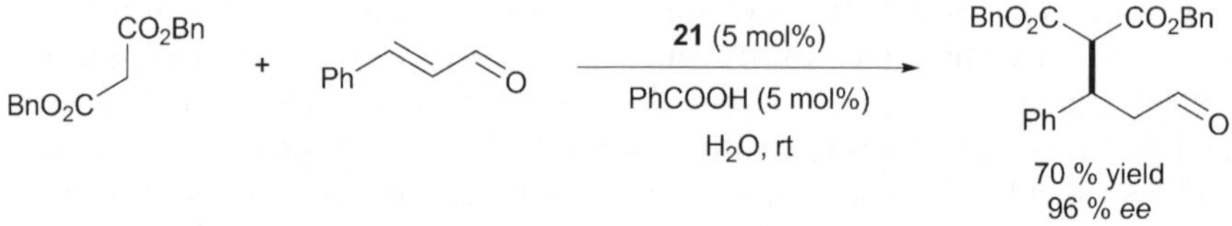

Scheme 1.12

found that the same aqueous biphasic conditions can be not extended, unfortunately, to malonates and malononitriles.[76]

Palomo *et al.* have extended the range of nucleophiles amenable in asymmetric organocatalysed Michael reactions of nitronates and malonates to α,β-unsaturated aldehydes.[77] Inspired by the prolinol motif, they proposed a series of derivatives, among which **21** was the most efficient catalyst for the nitro-Michael reaction of Scheme 1.11 and the Michael reaction of Scheme 1.12. These reactions required the use of benzoic acid as co-catalyst to optimize results.

The Michael addition to nitrostyrenes is, so far, the benchmark reaction where catalyst designers compete in proposing new protocols, including protocols in non-conventional solvents.

Scheme 1.13 shows a general reaction that often includes an additive to the catalyst, generally a Brønsted acid, to speed up the catalytic process.

General comments about the Michael reaction procedures are analogous to those developed in the aldol and Mannich reactions.[78] For example, the *O*-TMS-protected diphenylprolinol compound **20** in cooperation with benzoic acid catalyses the asymmetric Michael addition of aldehydes to nitroalkenes, in a simple, practical and efficient procedure. Benzoic acid promotes the rapid formation of the enamine intermediate and the reaction takes place in the highly concentrated organic phase of the aqueous biphasic system.[79]

Table 1.2 lists other catalysts that work in the presence of an aqueous phase, under heterogeneous conditions. The reactions take place in the concentrated

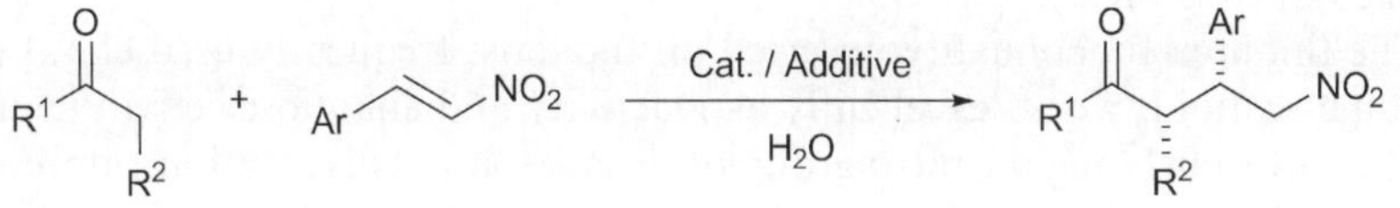

Scheme 1.13

Table 1.2 Organocatalysts for the Michael addition of carbonyl compounds to nitroalkenes, under biphasic aqueous conditions.

Catalyst	Ref.	Catalyst	Ref.
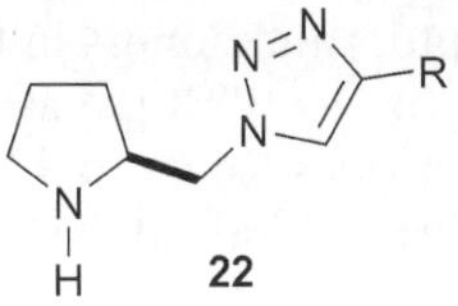 **22** R = COOMe, *n*-C$_5$H$_{11}$, Ph R = , R = 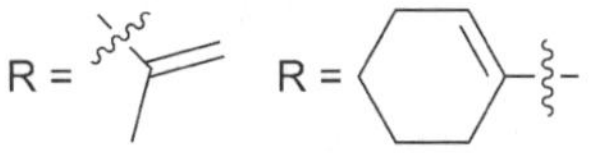	80	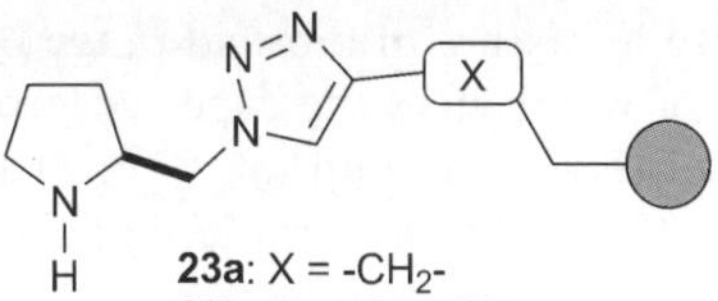 **23a**: X = -CH$_2$- **23b**: X = -C$_6$H$_4$CH$_2$-	81
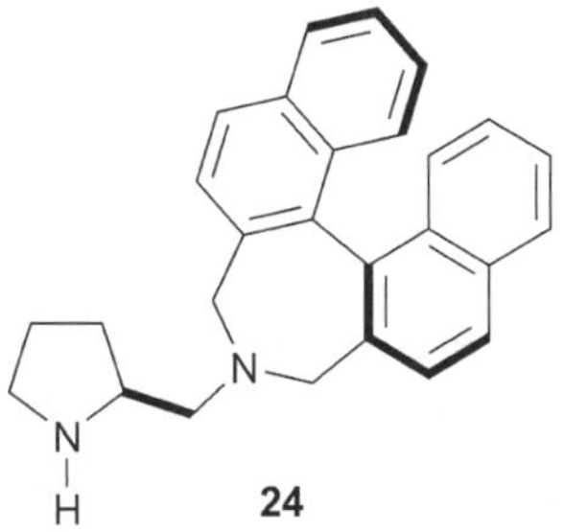 **24**	82	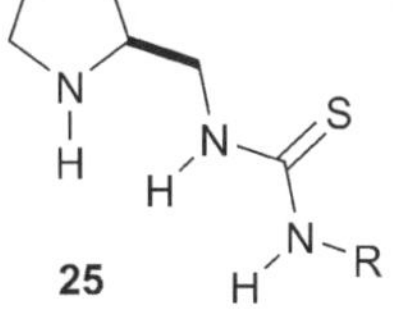 **25** R = 1-Naphthyl, cyclohexyl, 3,5-(CF$_3$)$_2$Ph, *t*-Bu, 4-Tolyl	83

organic phase, consisting of both the Michael donor and acceptor, and the catalyst.

Based on "click chemistry", pyrrolidine catalysts such as **22**, containing a triazole ring, have been shown to efficiently promote a highly diastereoselective and enantioselective Michael addition of ketones to nitroalkenes.[80] The triazole moiety is essential to promote the reaction in water with excellent yields and enantiocontrol. A solid-phase version, **23**, has also been proposed. Here the 1,2,3-triazole ring, constructed through a click 1,3-cycloaddition, also grafts the chiral pyrrolidine monomer onto the polystyrene backbone and provides a structural element capable of conferring the catalyst with high catalytic activity and enantioselectivity, particularly in the case of ketones as donors. Catalytic performance is optimized by using water and DiMePEG as an additive. A basic advantage is the easy recyclability.[81]

With chiral diamine **24**, in the form of a trifluoroacetate salt, the classic aqueous biphasic protocol has been successfully applied to the asymmetric Michael reaction of ketones with both aryl and alkyl nitroolefins. Brine is used as the aqueous phase.[82]

The thiourea functionality, inserted on the most frequently used chiral pyrrolidine scaffold, works excellently as reactivity and enantioselectivity control co-factor by chelating the nitro group of the acceptor. This solution, adopted in **25**, provides a family of robust catalysts that afford high yields (up to 98%) and great stereoselectivities (up to 99 : 1 dr and 99% ee) in direct Michael additions of ketones to various nitroolefins in water.[83]

We now discuss applications of the ionic tagging strategy to the design of catalysts for the direct asymmetric Michael addition of carbonyl compounds to nitroalkenes. Table 1.3 shows a few examples.

All the examples reported consist of chiral pyrrolidines containing a pyridinium **26** or methylimidazolium side chain **27**, **28**. Catalyst **26** when added directly to a mixture of a broad range of Michael acceptors and donors in the absence of water gives the expected products in yields (up to 100%), diastereoselectivities (*syn*/*anti* up to > 99 : 1) and enantioselectivities (ee up to 99%) that are affected by the structure of the counter-ion. These chiral catalyst are easily recovered and reused.[84]

In this context, Luo and co-workers proposed several imidazolium tagged pyrrolidines: **27** and **28a,b** are used in 15 mol.% loading under neat conditions, in the presence of 5% of TFA as co-catalyst.[85] These ionic-tagged catalysts could be recycled easily by exploiting their solubility properties that allow a quantitative precipitation from diethyl ether. The surfactant inspired variant **29a**, on the other hand, was designed to catalyse Michael additions to nitroalkenes in water (20 mol.%) without using any organic solvent or additional additive.[86] The same authors reported the reaction of 4-methylcyclohexanone and nitrostyrenes (Scheme 1.14) catalysed by **29b**, precursor of the previously analyzed **29a**.[87] Reactions are conducted under neat conditions, using 15 mol.% **29b** in conjunction to 5 salicylic acid as additive. A successful desymmetrization reaction resulted *via* an organocatalytic Michael addition

Table 1.3 Ionic-tagged pyrrolidines for the asymmetric Michael reaction under neat conditions or in water

Pyrrolidine	Ref.	Pyrrolidine	Ref.
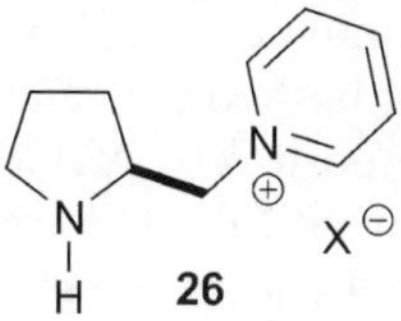 **26** X = BF$_4$, PF$_6$, NTF$_2$	84	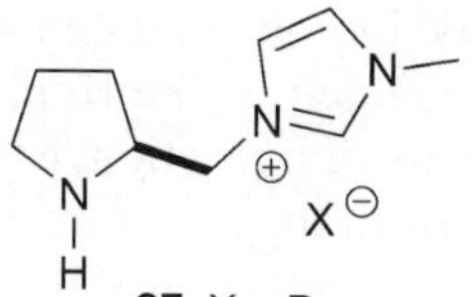 **27**: X = Br	85
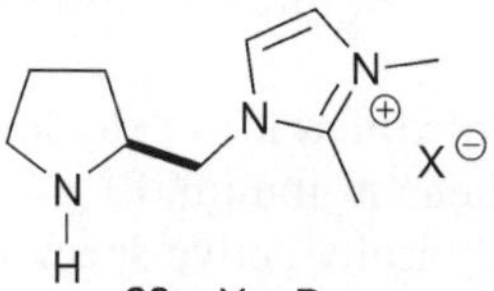 **28a**: X = Br **28b**: X = BF$_4$	85	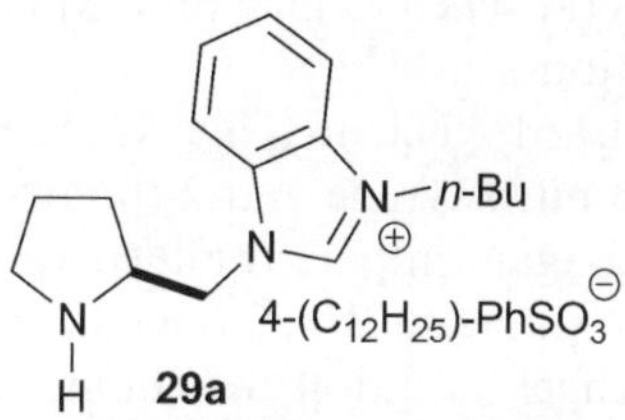 **29a** 4-(C$_{12}$H$_{25}$)-PhSO$_3^{\ominus}$	86
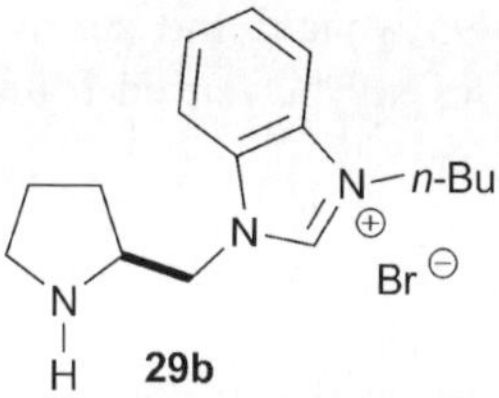 **29b**	87	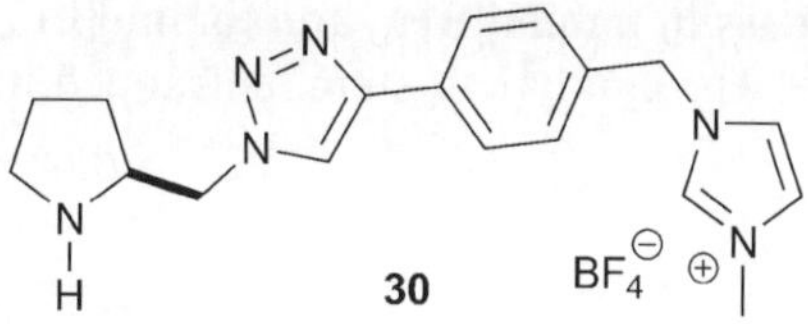 **30** BF$_4^{\ominus}$	88

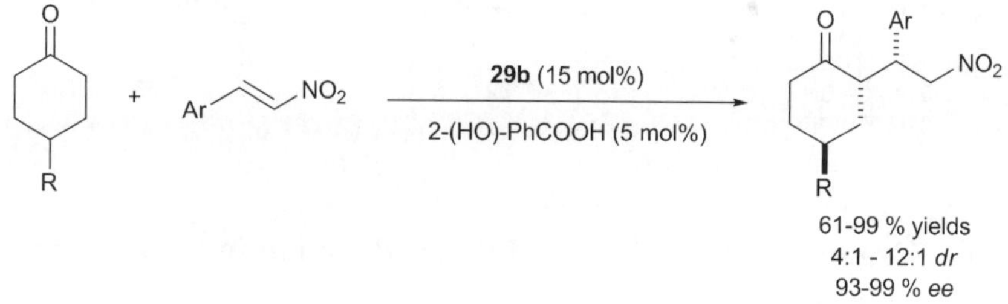

Scheme 1.14

that was characterized by good diastereoselectivity (4 : 1–12 : 1 dr) and enantioselectivities (93–99% ee).

The recyclability and reusability of **29b** was secured by its facile precipitation with ethyl ether, the solid being reused directly in the next run.

The last catalyst of Table 1.3, **30** is easily identified as a hybrid species containing motifs of both **22/23** and **27**. It contains, indeed, the methyltriazolyl ring connected to the chiral pyrrolidine unit, as well as the imidazolium tag, in analogy to **27**. The reaction shown in Scheme 1.12 was efficiently catalysed by 15 mol.% of **30** and 5 mol.% of TFA under neat conditions. Again, catalyst recovery by precipitation with ether allowed four recycles without loss of catalytic performances.[88]

Obvious reaction solvents for ionic-tagged catalysts are ionic liquids, which can establish strong electrostatic interactions not only with the starting catalyst but also with all the ionic intermediates characterizing the catalytic cycle. The coulombic environment created by the IL may strongly affect both kinetic and thermodynamic parameters of a reaction, either negatively or positively. A rule of thumb may be drawn from the literature. Reactions that benefit from ILs are more often those involving transition states with an enhanced charge separation with respect to reactants. Organocatalytic reactions based on the enamine/iminium ion general scheme, including Michael additions, fall into this rough category of reactions.

L-Proline catalysed efficiently the Michael additions of different aldehydes and ketones to β-nitrostyrene and 2-(β-nitrovinyl)thiophene in [bmim][PF$_6$],[89] but ionic tagging again imparted added value to a catalytically active species that works in ILs. Catalysts **31–33** offer an example of very high efficiency in the asymmetric Michael addition reactions of unmodified cyclohexanone to nitroalkenes in [bmim][BF$_4$] and [bmim][PF$_6$] with up to 95% yield and nearly 100% ee. The catalytic system, entrapped in the IL phase, can be reused four times.[90]

31a: X = Br
31b: X = PF$_6$

32

33

1.2.4 Organocatalytic Baylis–Hillman Reaction in Non-conventional Solvents

The Morita–Baylis–Hillman (MBH) reaction is an important 100% atom economic transformation that allows the formation in one step of a flexible allylic alcohol motif.[91] Efforts in this field have been directed recently to the solution of two problems: to enhance the generally sluggish reaction rate and to achieve asymmetric catalytic versions.[92] Scheme 1.15 gives the catalytic cycle of the MBH reaction. The catalyst is a highly nucleophilic tertiary amine, generally DABCO, or a tertiary phosphine, which adds to the α,β-unsaturated electrophile in a 1,4 fashion to deliver an enolate that, in turn, adds to the aldehyde. A critical step is the proton transfer from the enolizable position to the oxygen atom; this process is catalysed by an alcohol that plays the role of a proton shuttle between the two positions. Water has also been reported to strongly speed up the reaction at a well-defined concentration.[93] Moreover, the

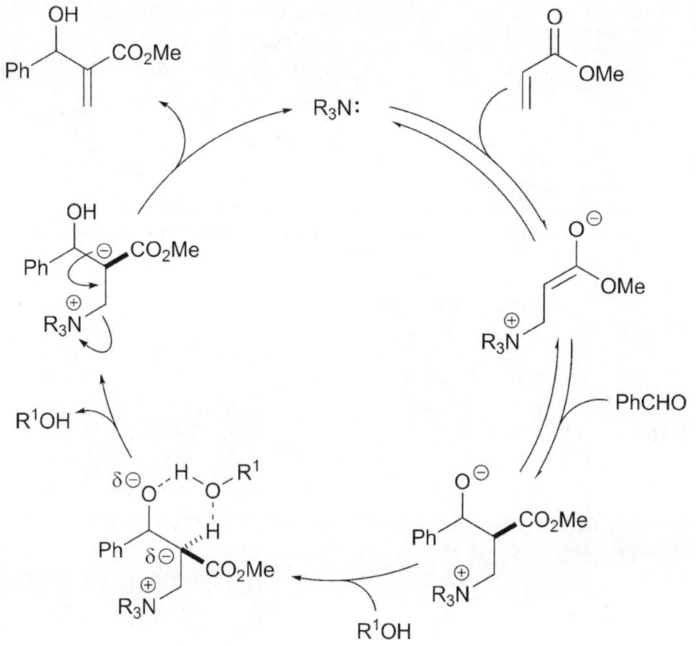

Scheme 1.15

use of surfactants promotes the MBH reaction without the use of organic co-solvents and at rt with beneficial effect on the reaction rates.[94] Once the β-ammonium-substituted enolate is formed, a fast β-elimination takes place, delivering the MBH condensation product.

The formation of zwitterionic intermediates that could probably be stabilized in the IL environment made ILs promising solvents for the MBH reaction. The IL anion plays a significant role in determining reaction rates. For example, in the series [bmim][X] the following reactivity orders were observed: BF_4 > PF_6 > Cl > NTf_2, and [bmim][NTf_2] > [bdmim][PF_6] > CH_3CN. The authors suggest that the C(2)–H hydrogen bond, absent in the case of [bdmim][PF_6], plays an important role in facilitating the new carbon–carbon bond forming reaction, by stabilizing the zwitterionic intermediates formed during the MBH reaction.[95]

When THF is replaced with an IL, in particular [epy][BF_4], in the DABCO-catalysed condensation of aromatic or aliphatic aldehydes with acrylonitrile (Scheme 1.16),[96] the IL-based protocol offered several advantages, *i.e.* milder reaction conditions, shorter reaction times, higher yields, and a simple work-up procedure that allows reuse of the solvent and DABCO dissolved in it. Scheme 1.16 reports a direct comparison between the use of [epy][BF_4] and other solvents or ILs.

A more elaborate molecular design is at the basis of a solvent-less reaction protocol. The ionic-tagging strategy, combined with the installation of a

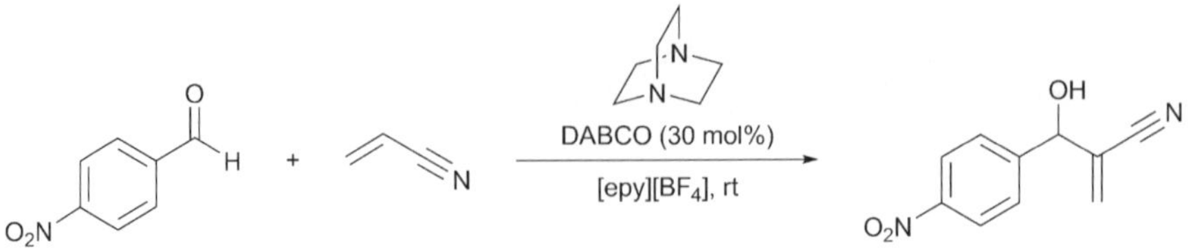

Run	Solvent	t (h)	Yield (%)
1	THF	48	32
2	[bmim][BF$_4$]	12	48
3	[epy][BF$_4$]	2	92

[epy] = ethyl-pyridinium

Scheme 1.16

suitable tertiary amine and of a proton donor, delivered structure **34**.

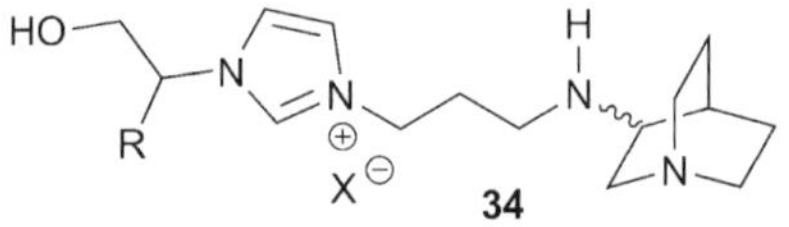

R = H, Me   X = Br, BF$_4$, PF$_6$

34

One arm of the imidazolium scaffold contains the catalytic centre, a bridgehead nitrogen atom that possesses the required nucleophilicity, and the second arm contains a primary alcohol capable of speeding up the critical proton transfer step that leads to the β-ammonium enolate intermediate, the direct precursor of the final Baylis–Hillman product.[97] The reaction of aliphatic and aromatic aldehydes with acrylates is carried out under solvent free conditions at room temperature, affording 80–95% yields of adducts after 8–12 h at rt. Moreover, catalyst **34** can be readily recovered from the reaction mixture and reused at least six times without significant loss of catalytic activity.

In an important achievement Vo-Thanh and co-workers realized that chiral ionic liquids such as **35** can act as chiral inducers for the asymmetric MBH reaction.[98] The *N*-octyl-*N*-methylephedrinium trifluoromethanesulfonate salt **35**, used as solvent in the DABCO-mediated reaction of methyl acrylate and benzaldehyde, produced the corresponding allylic alcohol in 60% yield and 44% ee after 7 days at 30 °C.

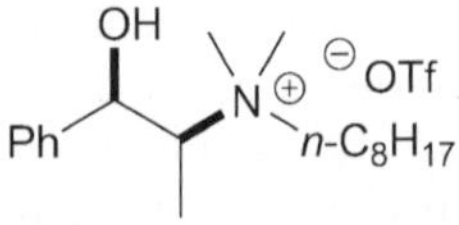

35

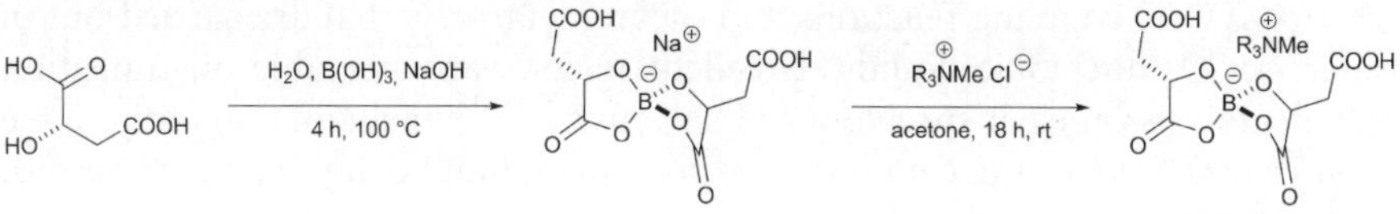

Scheme 1.17

36: R = n-C$_8$H$_{17}$

Scheme 1.18

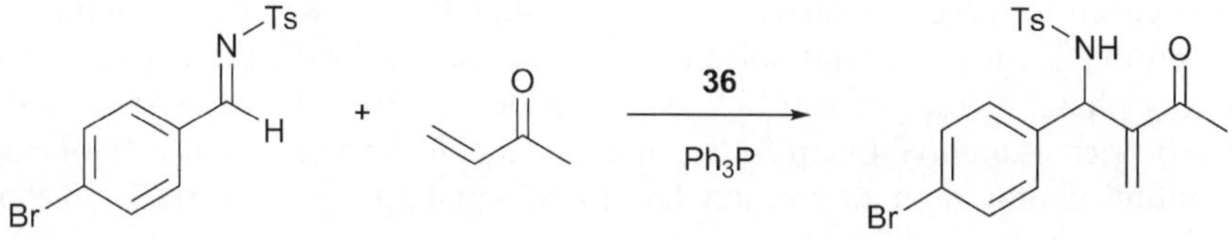

84 % *ee*

In a breakthrough in IL chemistry directed to applications in asymmetric catalysis using chiral reaction media, Leitner and co-workers developed an enantioselective aza-Baylis–Hillman reaction, where enantiocontrol was ensured by the use of IL **36** as solvent.[99] Scheme 1.17 shows the synthesis of the chiral anion. This is the first example in the literature of ees of the order of magnitude of 85% due to the use of a chiral solvent. The imine and the catalyst (10 mol.%) are dissolved in the IL, then methyl vinyl ketone is added and the reaction is simply carried out by stirring at rt for 24 h (Scheme 1.18).

1.3 Transition Metal Catalysed Reactions in Green Solvents

Homogeneous catalysis using transition metal complexes developed as a science in the second half of the 20th century. Homogeneous catalysis expanded by orders of magnitude the range of methodologies for making carbon–carbon bonds, and it has played a fundamental role in asymmetric synthesis, polymer manufacture, in bulk and specialty chemicals, and in pharmaceuticals. As previously outlined, the major drawback of homogeneous catalysis, the costly and difficult catalyst recycling and product separation, is one of the main reasons that industry prefers heterogeneous catalysis. Immobilization of the homogeneous catalyst in a multiphase operation (*e.g.* two-liquid phase approach) offers promising opportunities. The two-liquid phase approach rests on the proper choice of distribution coefficients of the products with immiscible solvents. In this chapter we discuss catalytic reactions in non-conventional solvents that address multiphase strategies. Examples are limited to typical C–C bond forming reactions, such as the hydroformylation and carbonylation reactions, C–C coupling reactions and olefin metathesis that are carried out in water, scCO$_2$ and ionic liquids. Excellent reviews are available on transition metal catalysis on each specific green solvent,[19-23,100] including an entire issue of *Advanced Synthesis & Catalysis*[101] where much more details can be accessed. All the examples quoted, from an operational point of view, can be divided into three classes: (i) the catalyst operates in a green polar medium and the products form a second immiscible phase that can be easily separated; (ii) reactions take place in a heterogeneous biphasic (or triphasic) system, with the catalyst confined in one phase and the products extracted in the second one; (iii) a homogeneous process is carried out conventionally followed by extraction of the products with a second solvent that is immiscible with the catalyst-containing phase.[102] In the ideal biphasic reaction scenario, the catalyst solution may be seen as an investment for a potential technical process or as a "working solution" if only a small amount has to be replaced after a certain time of application.

Catalyst discovery and implementation in transition metal catalysed reactions is even more the combined result of multidisciplinary efforts that join experience on the labile nature of the coordination spheres of complexes with technical efforts to understand the accompanying subtle shifts in their electronic structures by *in situ* spectroscopy, computational competence in providing reliable mechanistic frameworks, up to engineering efforts to develop miniature devices and microreactors for continuous productions capable of working reproducibly for long periods. Given the sheer number of publications that have appeared in the area over the last 5 years we have restricted our analysis to a few representative examples that appeared since 2000 up to early 2008.

Sections 1.3.1–1.3.3 discuss representative hydroformylation processes in water, ILs and scCO$_2$. Unfortunately, strict comparison between the efficiencies of the alternative solutions proposed is not possible due to the number of reaction parameters: syngas pressure, ligand/metal ratios, catalyst loading, olefin/metal ratio, temperature and so on. Moreover, catalysts performances are expressed in different studies in different terms. For clarity, we anticipate that the activity and selectivity of a hydroformylation catalyst can be expressed either in terms of its turnover number (TON) and the ratio of linear to branched aldehyde (n/i) in the product mixture, or in terms of the yield of n-butyraldehyde, which is the combination of the n/i and the TON. In other studies a comparison between TOF values is reported.

1.3.1   Hydroformylation and Carbonylation Reactions in Aqueous Biphasic Systems

The solubility of transition metal catalysts in water is determined by their overall hydrophilic nature, which may arise either as a consequence of the charge of the complex ion as a whole or may be due to the good solubility of the ligands. Examples of ionic organometallics soluble in water are

$Na^+[Re(CO)_5]^-$ and $[(\eta^5\text{-}C_5H_5)Fe(CO)_2\text{-}\{P(C_6H_5)_3\}]^+I^-$. They are precipitated by large counter-ions such as $[P(C_6H_5)_4]^+$, $[B(C_6H_5)_4]^-$ and $[(C_6H_5)_3P{=}N{=}P(C_6H_5)_3]^+$.

The transition metal catalysts themselves are known to be stabilized by tertiary phosphines in their low oxidation states at one or more stages along the catalytic cycle. For this reason the most studied ligands in aqueous phase catalysis are derived from water-insoluble tertiary phosphines by installing on them ionic groups, typically sulfonate, but also sulfate, phosphonate, carboxylate, quaternary ammonium, phosphonium, as well as neutral polyether or polyamide substituents, *etc.*

The hydroformylation process (Scheme 1.19),[103] namely the production of aldehydes from the reaction of an olefin with syngas (produced by the steam reforming of natural gas),[104] begins with the generation of a coordinatively unsaturated metal hydrido carbonyl complex such as $HCo(CO)_3$ and $HRh(CO)(PPh_3)_2$.

The first hydroformylation unit employing an aqueous biphasic system, the "Ruhrchemie/Rhône-Poulenc oxo process" (RCH/RP), went on stream in July 1984 with an initial capacity of 100 000 tons per year. It was based on the use of tris(*m*-sulfonatophenyl)phosphine (**37**, TPPTS), an ideal ligand modifier for the oxoactive $HRh(CO)_4$. Despite a broad array of solubilizing ligands, sulfonated derivatives of ligands containing aryl groups have proven most successful, because of their outstanding solubility in water. Sulfonated phosphines (Table 1.4), readily accessible by direct sulfonation, are well soluble in water

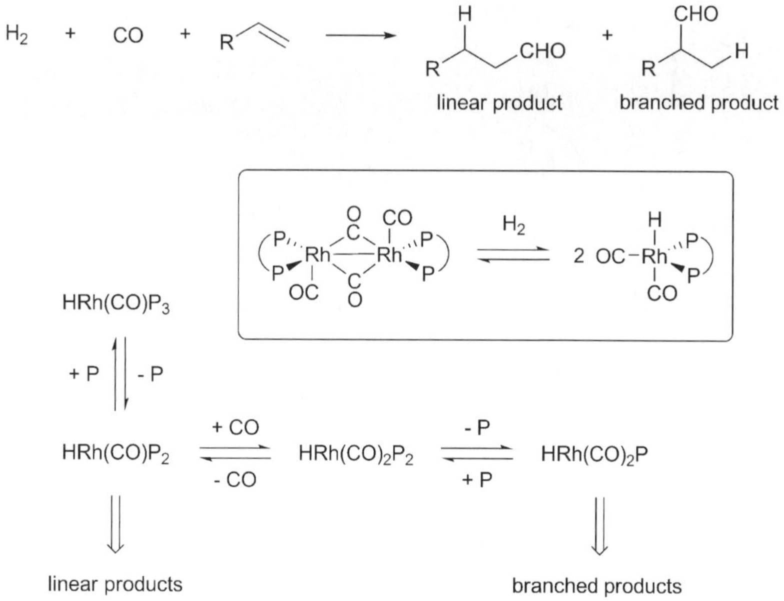

Scheme 1.19

Table 1.4   Examples of functionalized phosphines for biphasic aqueous hydroformylations.

Catalyst	Ref.	Catalyst	Ref.

37 (TPPTS) — Ref. 105

38: R = Ph, Me, n-C$_6$H$_{13}$; n = 1, 2 — Ref. 105

39 (BINAS) — Ref. 106

40 (BISBIS) — Ref. 107

41 — Ref. 108

42 — Ref. 109

43: n = 18, 25 — Ref. 110

44: n = 16, 25 — Ref. 110

45 — Ref. 111

46 — Ref. 112

over a wide pH range, (*e.g.* TPPTS has a solubility in water of $\sim 1.1\,\mathrm{kg\,L^{-1}}$) while in their ionized form they are insoluble in common non-polar organic solvents. Without any expensive preformation steps, three of the four CO ligands can be substituted by the water-soluble TPPTS, which yields the hydrophilic oxo catalyst HRh(CO)[TPPTS]$_3$.[105] Because of the solubility of the

Rh(I) complex in water and its insolubility in butanal, the plant unit is essentially reduced to a continuous stirred tank reactor followed by a phase separator (decanter) and a stripping column. Propene and syngas are initially added to the catalyst containing aq. solution, then, once the reaction has taken place, the reaction products and the catalyst phase pass into the decanter, where the aqueous catalyst solution and the butanal phase are separated.

Apart the use of water as solvent and the excellent atom economy of the process, the ecological benefits of the RCH/RP process with respect to the classic ones can be summarized as follows:

- a higher selectivity toward the desired products,
- lower energy consumption,
- efficient recovery of catalyst (loss factor $= 1 \times 10^{-9}$),
- very low ligand toxicity (LD_{50}, oral $> 5\,g\,kg^{-1}$),
- almost zero environmental emissions.

Phosphines of general structure **38** allow a fine tuning of the hydrophilic *versus* the hydrophobic properties by varying the number of sulfonated rings. The hydrophilicity is of course ranked in the order $n = 0 > n = 1 > n = 2$, from the trisulfonated through to the monosulfonated species. However, molecular design applied to the identification of new water-soluble ligands continuously inspires new target ligands: for example, tris(hydroxymethyl)phosphane, after condensation with various amino acids, provided a family of water-soluble ligands (Scheme 1.20) that were checked for the rhodium-catalysed two-phase hydroformylation of propene.[113] The regioselectivity with these ligands was easily controlled by tuning the pH of the reaction mixture.

A very useful application of BINAS (**39**) was reported in the hydroformylation of internal alkenes in a biphasic water system. At pH 7 and 8 a significant increase in aldehyde yield (72–73%; $TON = 1460$, $TOF = 61\,h^{-1}$) and an excellent regioselectivity (n/i 99:1) are observed. The regioselectivity in favour of the linear aldehyde (up to 99%) is optimized by a careful control of the pH and CO partial pressure.[114] Interestingly, the water-soluble catalyst formed by interaction of Rh(CO)$_2$acac and **39** leads to significantly higher regioselectivity than with similar catalysts soluble in organic solvents. The obtained n/i selectivities exceed all known literature data and the catalyst can be easily reused several times.

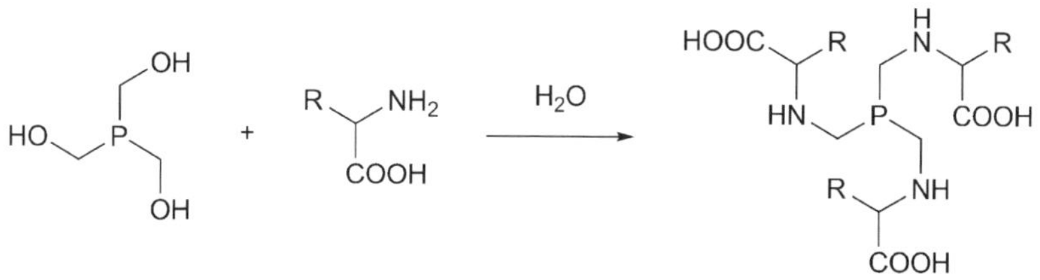

Scheme 1.20

The RCH/RP process cannot be applied to long-chain olefins ($C \geq 6$) which are not soluble enough in water, making their mass transfer to the aqueous catalyst phase the rate-limiting step. The search for more effective catalytic systems for higher olefins remains an active industrial research field. The following strategies have been adopted: (i) design of new ligands with a tunable amphiphilicity profile for aqueous biphasic systems; (ii) use of microreactors, which preserve the advantages of the two-phase technology, and allow us to achieve a process intensification because of their superior intrinsic interface areas between different phases; (iii) extending attention to other non-conventional solvents such as scCO$_2$ and ILs.

For example, 1-octene can be hydroformylated in the interfacial organic/water region with a considerably enhanced rate compared to the classic biphasic hydroformylation. This approach involves the use of both TPPTS and triphenylphosphine. Interaction of triphenylphosphine and the TPPTS-based catalyst takes place at the liquid–liquid interface (Scheme 1.21). A new catalytic species containing two TPPTS and one triphenylphosphine ligand is formed in the liquid/liquid boundary layer, where it can access the reactants present in the organic phase in significantly higher concentrations with respect to the aqueous phase. This new version of the hydroformylation reaction in aqueous biphasic systems resulted in a 10–50-fold increase in overall reaction rate.[115]

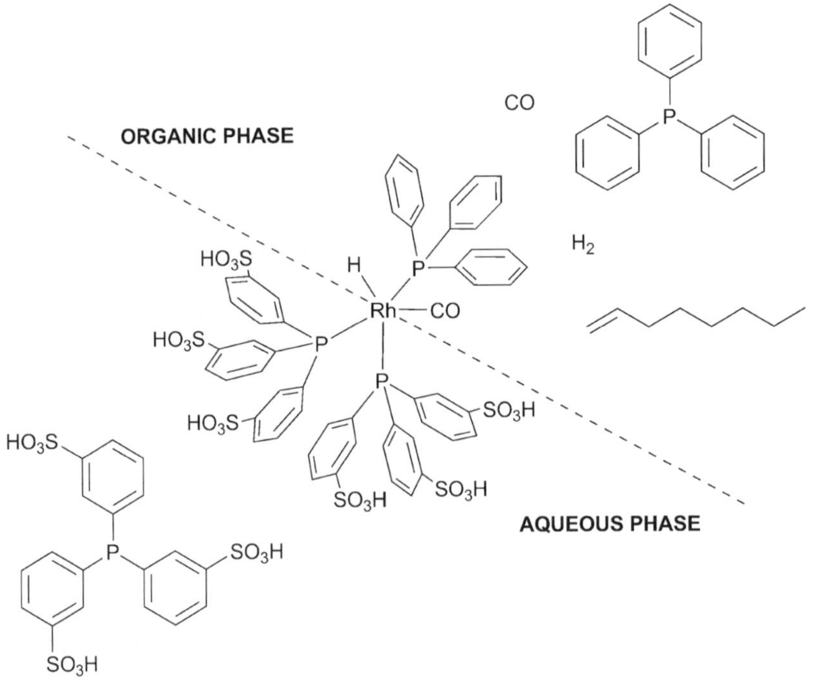

Scheme 1.21

Special chelating ligands, such as **BINAS (39)** and **BISBIS (40)**, were found to be very useful ligands for the two-phase hydroformylation of higher alkenes. Further examples are the non-ionic, water-soluble phosphines **43** and **44**, which act as ligands for the rhodium-catalysed hydroformylation of higher alkenes under biphasic conditions. The rhodium catalyst combined with these ligands gave an average TOF of $182\,h^{-1}$ for 1-hexene. More importantly, recovery and re-use of catalyst is possible because of the inverse temperature-dependent water solubility of the phosphines.[110]

To improve TONs and TOFs, additives are used to overcome the mass transfer problems encountered in biphasic catalysis. Thus, $CoCl_2(TPPTS)_2$ catalyses the hydroformylation of 1-octene and 1-decene in an aqueous biphasic medium in the presence of cetyltrimethylammonium bromide (CTAB), which increases conversion (95%) of 1-octene and 1-decene with a higher aldehyde selectivity (n/i ratio ranges from 2 to 3).[116]

Methylated cyclodextrins promote the hydroformylation of higher olefins, too. Molecular dynamics simulations show that the reaction takes place "right" at the interface and that cyclodextrins act as both surfactants and receptors that favour the meeting of the catalyst and the olefin. The methylated cyclodextrin adopts specific amphiphilic orientations at the interface, with the wide rim pointing towards the water phase. This orientation makes easier the formation of inclusion complexes with the reactant (1-decene), the key reaction intermediate $[Rh(H)CO(TPPTS)_2\text{-decene})]^{6-}$ and the reaction product (undecanal).[117]

The hydroformylation of ω-alkene carboxylic acid methyl esters catalysed by a Rh/TPPTS system (Scheme 1.22) in a biphasic medium does not require additives with low molecular substrates such as methyl 4-pentenoate, whereas methyl esters of higher ω-alkene carboxylic acids such as methyl 13-tetradecenoate require the presence of surfactants as mass-transfer promoters.[118] Surfactants, indeed, decrease the interfacial tension, forming aggregates above the critical micellar concentration that speed up the catalytic process by increasing the interfacial area.

Johnson-Matthey Co. has reported that oleic acid methyl ester or linoleic acid methyl ester can be hydroformylated in micellar media using a water-soluble rhodium complex of monocarboxylated triphenylphosphine **45** as catalyst.[111] As a further example, polyunsaturated linolenic acid methyl ester can be hydroformylated to the triformyl derivative with a selectivity of 55% with a Rh/TPPTS catalytic system in the presence of CTAB (Scheme 1.23).[119]

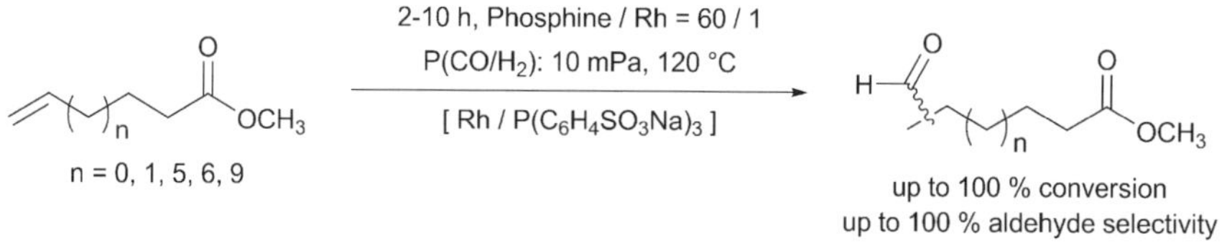

Scheme 1.22

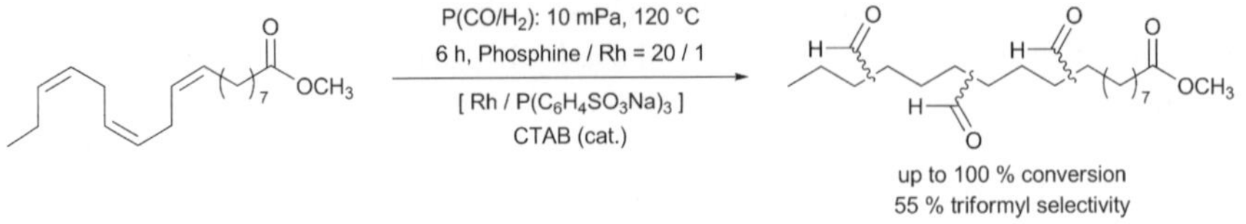

P(CO/H$_2$): 10 mPa, 120 °C

6 h, Phosphine / Rh = 20 / 1

[Rh / P(C$_6$H$_4$SO$_3$Na)$_3$]

CTAB (cat.)

up to 100 % conversion
55 % triformyl selectivity

Scheme 1.23

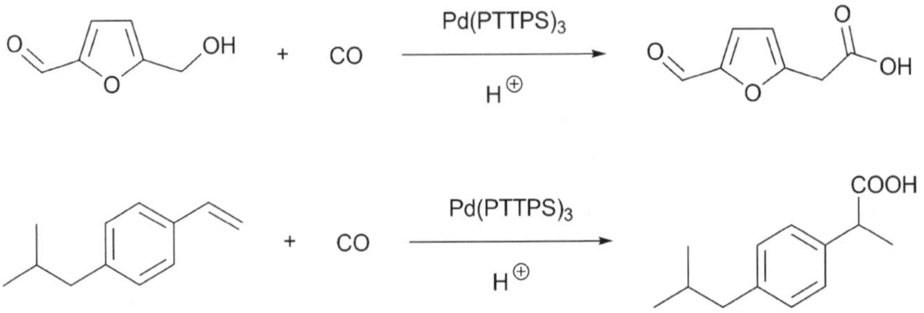

Scheme 1.24

Using RuCl(CO)(TPPTS)(BISBIS) the biphasic aqueous hydroformylation of higher olefins in the presence of the cationic surfactant CTAB ensures a TOF > 700 h^{-1} and regioselectivity > 96% for the linear aldehyde.[120] Piperazinium cationic surfactants were also successfully applied as catalysis promotion agents in the aq. biphasic hydroformylation of higher olefins.[121] The property of surfactant and ligand can be assumed by the same molecule, *e.g.* disulfonated cetyl(diphenyl)phosphine **42**. 1-Dodecene is hydroformylated in water/toluene (3 : 1) under mild conditions [olefin/Ru = 2500, CO/H$_2$ = 1, P(CO + H$_2$) = 15 bar, **42**/Ru = 10] with TOF = 188 h^{-1}.[109] Another approach to micellar catalysis is based on the use of non-ionic surfactants. The water-soluble ligand **46** in combination with RhCl$_3 \cdot$ 3H$_2$O catalyse the hydroformylation of 1-decene at 120 °C, affording a 99% yield of aldehyde (olefin/Ru = 1000, CO/H$_2$ = 1, P(CO + H$_2$) = 50 bar, **46**/Ru = 4). Using an olefin/Ru of 26 600, the reaction affords an 80% yield with a TOF = 4218 h^{-1}.[112] The catalyst could be efficiently recycled 20 times by exploiting the concept of "thermoregulated phase-transfer catalysis". Upon increasing the temperature the catalyst is first transferred to the organic phase where the hydroformylation process takes place then, at a lower temperature, the catalyst moves back to the aqueous phase, allowing an easy separation of the product.

If cobalt, rhodium and ruthenium complexes are the most frequently used in hydroformylation reactions, most carbonylation reactions employ palladium catalysts. The active water-soluble complex Pd(TPPTS)$_3$ is easily prepared by reducing *in situ* PdCl$_2$/TPPTS with CO in water at room temperature. The carbonylation of alcohols[122] and olefins[123] (Scheme 1.24) requires the presence of a Brønsted acid as co-catalyst, *e.g.* HCl, which is assumed to produce an alkyl halide capable of oxidatively adding to Pd(0) to afford an intermediate alkyl-Pd(II) species.

1.3.2 Hydroformylation and Carbonylation Reactions in Ionic Liquids

As mentioned above, the poor solubility of higher olefins in water still hampers their hydroformylation in aqueous media. Since 2003, ILs have entered the hydroformylation field as an alternative to the aqueous phase thanks to the higher solubilities of long-chain olefins in these non-conventional solvents.[124] Of course, as a prerequisite, the transition metal catalyst has to dissolve in the IL, too. Moreover, an efficient confinement in the IL phase has to occur to prevent catalyst leaching under the conditions of intense mixing characteristic of continuous liquid–liquid biphasic operations. Important advances were made in ligand design, addressing in particular the ionic tagging strategy. A vast number of publications have shown improved potential of ILs with regard to product separation, catalyst recovery, improved catalyst stability and selectivity.[125]

Since the pioneering results by Chauvin *et al.* in 1995,[126] the story of the struggle against problems such as catalyst leaching, deactivation, *etc.*, working in ILs, has been excellently outlined in a review by Haumann and Riisager.[124]

The use of triphenylphosphine as ligand led to acceptable rates in ILs, but with high rhodium leaching into the organic phase. Recourse to sulfonated phosphines such as monosulfonated triphenylphosphine retained the catalyst in the ionic liquid phase but decreased its activity significantly. This drawback was surmounted by the use of **47** (Table 1.5), which was derived from a simple cation metathesis reaction between TPPTS (**37**) and 1-butyl-2,3-dimethylimidazolium chloride [bdmim][Cl] in acetonitrile.[127]

Hydroaminomethylation reactions (Scheme 1.25) were also successfully performed in [pmim][BF$_4$] using a rhodium/sulfoxantphos system by reacting piperidine with different n-alkenes. The resulting amines were obtained in more than 95% yield with TOF of up to 16 000 h^{-1}, along with high regioselectivity for the linear amines, with *n/i* ratios up to 78, and quantitative catalyst recovery.[128] Analogous hydroaminomethylation reactions were also reported to be catalysed by Rh(CO)$_2$acac/BISBIS in [bmim][*p*-CH$_3$-C$_6$H$_4$-SO$_3$] with high conversions and *n/i* ratios ranging from 5 to 36.[129]

The known cobaltocenium complexes **48** and **49** were proposed as ligands for rhodium (Table 1.5). They conferred a lower activity to the catalyst, but Rh leaching was less than 0.2% for both ionic ligands, and recycling experiments showed unchanged activity and selectivity for several runs.

The concept of ionic tagging in aryl phosphine design is made explicit by the structures of ligands collected in Table 1.5.

Pyridinium and guanidinium-based ligands **50** and **51**, respectively, have been applied to the hydroformylation of 1-hexene in [bmim][BF$_4$] and

Table 1.5 Ligands for hydroformylation reactions in ILs.

Ligand	Ref.	Ligand	Ref.
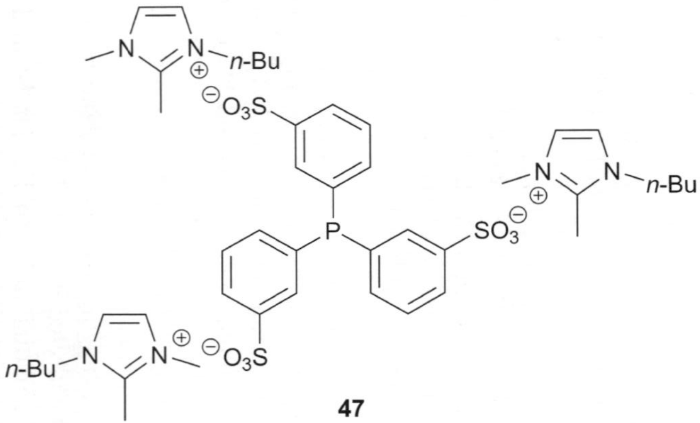 **47**	127	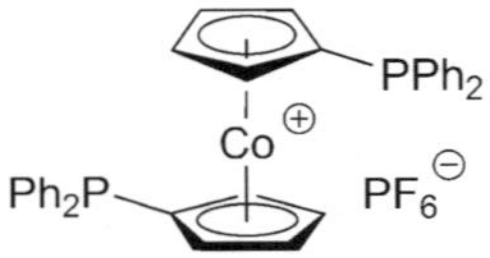 **48**	130
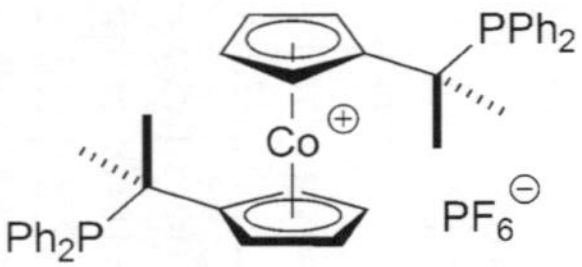 **49**	130	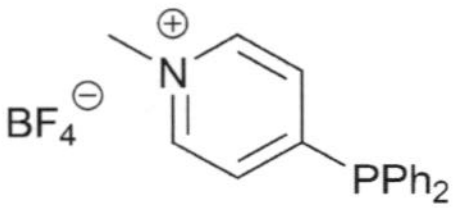 **50**	131

51a: X = BF₄
51b: X = PF₆

51

52

53

54

R =

55

56

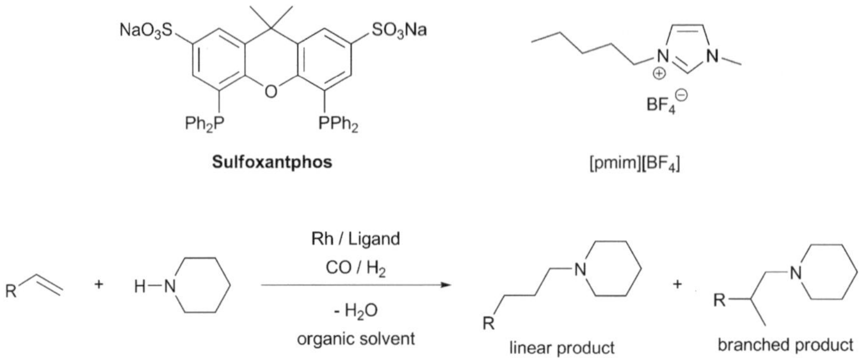

Sulfoxantphos

[pmim][BF$_4$]

Scheme 1.25

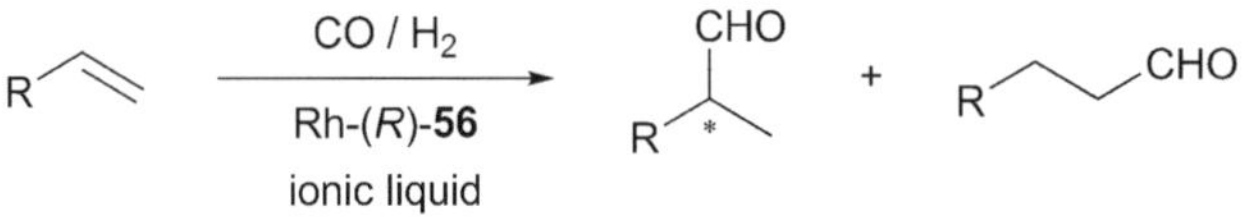

Scheme 1.26

[bmim][PF$_6$]. Their use resulted in moderate selectivity, affording heptanal in 72–80% yield with TOFs ranging between 180 and 240 h^{-1}. Minor rhodium leaching was observed when using the guanidinium-derived ligand **51** in [bmim][BF$_4$].[131]

Ligands **52** and **53** were tested in the biphasic hydroformylation of 1-octene in [bmim][PF$_6$]. With **52** a higher activity (TOF of 552 h^{-1}) was balanced by a lower selectivity (52% nonanal), whereas in the case of **53** the lower TOF (51 h^{-1}) was compensated by an improved selectivity (74%).[132]

Compared to TPPTS, which allowed good immobilization of rhodium but provided much lower rates, the bidentate xanthene-based ligand **54** showed an excellent selectivity towards nonanal of about 95% during all recycling experiments, with no Rh leaching detected.[133] In analogy, ligand **55** gave excellent results in the rhodium-catalysed hydroformylation of 1-octene in [bmim][PF$_6$], with no Rh and P leaching detected throughout seven recycling experiments.[134]

Much more recently, enantiocontrol issues have been addressed in hydroformylation reactions. The first study on stereoselective biphasic hydroformylation in IL appeared in 2007, using the chiral sulfonated ligand **56** in [bmim][BF$_4$] for the enantioselective hydroformylation of vinyl acetate and styrene (Scheme 1.26). The hydroformylation of vinyl acetate resulted in the predominant formation of 2-acetoxypropanal with ee $\geq$ 50%. In the hydrophilic IL [bmim][BF$_4$], the **56** derived catalyst showed 79% conversion with high selectivity for the branched product, while the ee was moderate at around 22%.[135]

The last contribution quoted in this section is a phosphine-free hydro-formylation process based on a liquid triphasic system consisting of isooctane, water and trioctylmethylammonium chloride (TOMAC). The hydroformyla-tion of model olefins required neat $RhCl_3$ only as catalyst precursor. In the triphasic system, the catalyst is confined in the TOMAC phase, likely in the form of an ion pair. Products are obtained in excellent yields ($>90\%$ at 80 °C) and high regioselectivity ($>98\%$) in favour of the branched aldehyde in the case of styrene, while the *exo* isomer was obtained in $>90\%$ selectivity in the case of norbornene. The products were easily removed and the catalyst was recycled several times, with no leaching of rhodium into the organic phase.[136]

Carbonylation of olefins and aryl halides to the corresponding esters or amides find in ILs an ideal media for the palladium-catalysed process.

(*E*/*Z*)-Isomeric vinyl bromides were stereoselectively carbonylated to the corresponding (*E*)-α,β-unsaturated carboxylic acids in [bmim][PF$_6$], while vinyl dibromides underwent hydroxycarbonylation to give monoacids (Scheme 1.27). The catalyst-containing IL phase could be recycled five times in the second reaction of Scheme 1.27.[137]

A group of imidazolium-based ILs with different substituents has been tested as solvent in the hydroethoxycarbonylation of olefins catalysed by PdCl$_2$(PPh$_3$)$_2$. Beside the nature of the IL, the presence of 1,1′-bis(diphenyl-phosphino)ferrocene (dppf) strongly affected the regioselectivity. For example, using [acetonyl-mim][PF$_6$] as the solvent the branched ester was obtained only, while in the presence of dppf the regiochemistry was completely inverted in favour of the linear ester (Scheme 1.28).[138]

A drawback was partial metal leaching in the product separation process, which prevents recycling with identical performance of the catalyst-containing IL phase.

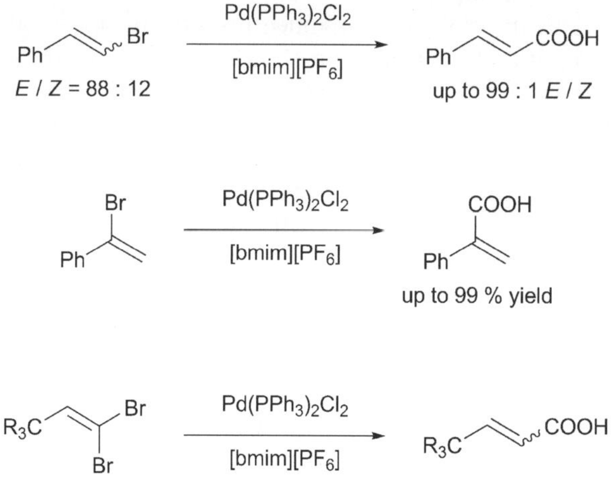

Scheme 1.27

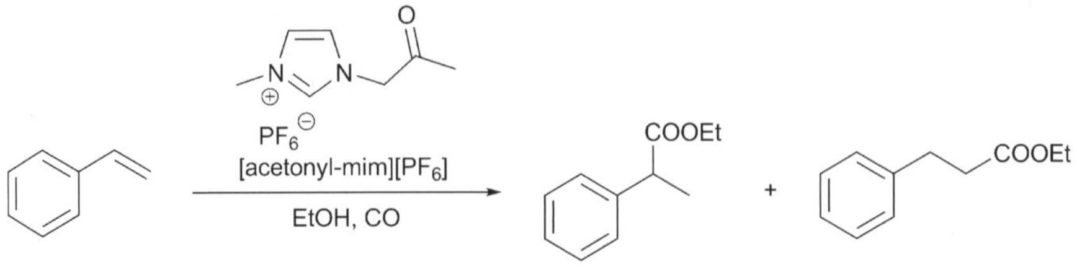

Scheme 1.28

An efficient carbonylation of aniline to the corresponding methyl carbamate was reported to be catalysed by $Pd(phen)Cl_2$ in [bmim][BF_4]. The reaction was carried out by introducing successively, in an autoclave, 1 mg Pd complex catalyst, 1 mL ionic liquid, 1 mL amine, 10 mL methanol and 5 MPa of a mixture of gases (CO 4.5 MPa and O_2 0.5 MPa), followed by heating at 175 °C for 1 h. The reaction took place homogeneously thanks to the good solubility of $Pd(phen)Cl_2$ in the IL (0.11–0.12 g per 100 mL). The *N*-phenyl methyl carbamate was eventually precipitated by adding water to the resulting mixture and filtered off (conversion = 99%, selectivity = 98%, TOF = 4540 h^{-1}). The catalyst-containing IL solution could be reused with slight loss of catalytic activity.[139]

1.3.3 Hydroformylation Reactions in scCO$_2$ and in Ionic Liquids/scCO$_2$ Systems

Supercritical carbon dioxide (scCO$_2$) has been used as the reaction medium for hydroformylation reactions. The most attractive feature of SCFs is that its properties (density, polarity, viscosity, diffusivity, *etc.*) can be varied over a wide range by small changes in P/T values.

Catalysts soluble in scCO$_2$ are examined first. In the late 1990s Leitner and co-workers reported that the hexafluoroacetylacetonate ligand (hfacac) confers its metal complexes a good CO$_2$-philicity – a necessary requisite to promote an efficient catalysis in scCO$_2$. Thus, the rhodium complex [(cod)Rh(hfacac)] exhibited improved activity for hydroformylation in scCO$_2$ compared with that observed in toluene solution; however, the linear/branched selectivity was poor. Terminal and internal double bonds with alkyl and aryl substituents were hydroformylated in scCO$_2$ almost quantitatively at 40 °C at substrate/catalyst ratios of ~800:1.[140] Triaryl phosphines are the most important class of ligands for hydroformylation catalysts; however, most of them, including triphenylphosphine, are not soluble in scCO$_2$ and cannot be used as ligands for the hydroformylation reaction. Since the installation on a molecule of perfluoroalkyl tags confers it with good solubility in scCO$_2$, ligands **59** and **60** (Scheme 1.29) were proposed as modifiers of [(cod)Rh(hfacac)]. With catalysts containing the CO$_2$-philic phosphine (**59a,b**) and phosphite (**60**) ligands, hydroformylation reactions compared favourably with analogous reactions in

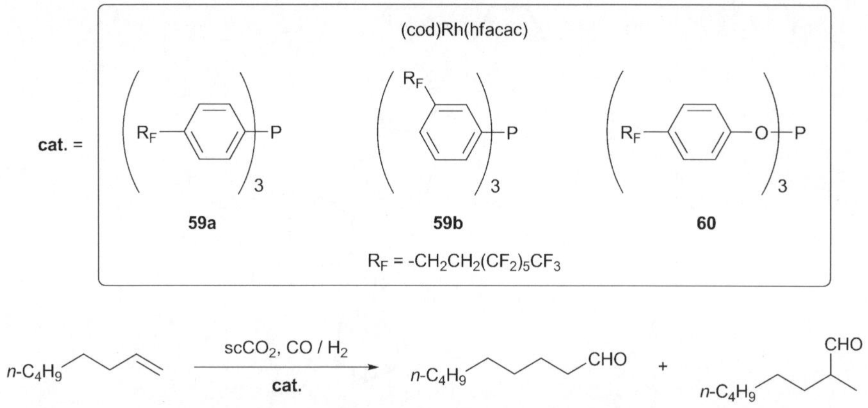

Scheme 1.29

other reaction media, linear selectivity improved up to a maximum n/i ratio of 5.6 : 1 and an efficient separation and recycling was possible, with rhodium leaching in products being limited to $<1\,ppm$.[140]

Ligand design led to the development of the CO_2-philic chiral ligand **61** (Scheme 1.30). Rhodium-catalysed asymmetric hydroformylation of styrene in the presence of **61** could be performed for eight successive runs – the system being still active after a total turnover number of more than 12 000 catalytic cycles per rhodium centre. Both the catalytic reaction and the extraction step made use of $scCO_2$, allowing quantitative recovery of the product free of solvent, with rhodium content in the products ranging from 0.36 to 1.94 ppm.[141] The branched aldehyde was obtained in up to 93.6% ee and with more than 1:9 inverse n/i regioselectivity.

A simple way to overcome the insolubility of TPPTS in $scCO_2$ is based on the use of supercritical mixtures containing water. For example, the hydroformylation of propene has also been studied in supercritical CO_2/H_2O and in supercritical propene/H_2O mixtures using $Rh(acac)(CO)_2$ and TPPTS as the catalytic system.[142] Visual observation of the reaction in both systems revealed that the reaction occurred under homogeneous conditions since a single phase was present at supercritical T and P. The work-up was very simple, indeed when the pressure and temperature were lowered to the ambient value a biphasic system was formed. Products and the catalyst were easily separated, the products being in the organic phase and the rhodium catalyst in the aqueous phase. Rhodium leaching in the organic phase was $\sim 1.0 \times 10^{-6}\,mg\,mL^{-1}$. Table 1.6 compares results of the hydroformylation of propene in $scCO_2/H_2O$, in sc propene/H_2O and under classic aqueous biphasic conditions (see Section 1.3.1).

$ScCO_2$ has also found an interesting application not as the solvent for a hydroformylation reaction but as a switch to halt the reaction, induce a phase separation and separate products with negligible metal leaching. Briefly, the hydroformylation of 1-octene was conducted neat in the presence of a catalyst

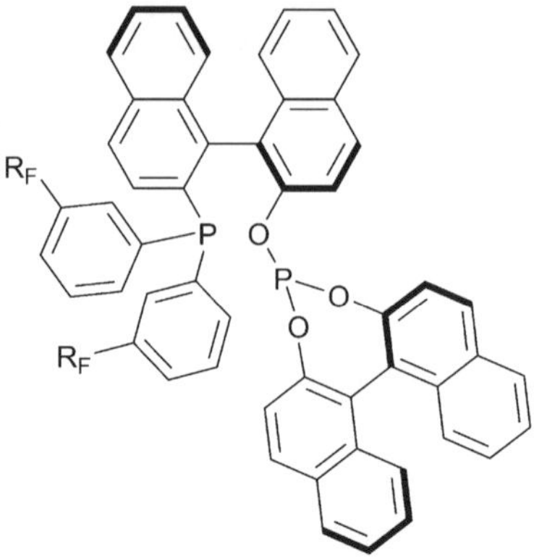

61: (*R*,*S*)-3-H^{2}F^6-BINAPHOS

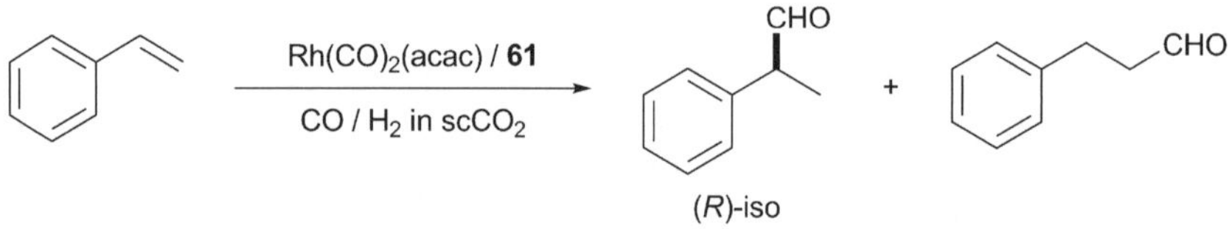

> 90 % conversion in eight consecutive runs; total TON > 12000

Scheme 1.30

Table 1.6 Comparison of catalytic properties for biphasic and supercritical reaction systems

Reaction system	*TON [g-aldehyde (g-Rh h)$^{-1}$]*	*n/i Ratio*
Biphasic^a	76.3	3.2
scCO$_2$/H$_2$O^b	190.1	4.3
sc propene /H$_2$O^c	601.4	8.4

^{a}[Rh] = 30 ppm, P/Rh = 50, propylene = 2 g, H$_2$O = 5 mL, cyclohexane = 5 mL, CO/H$_2$ = 1:1 (3.5–4 MPa), reaction pressure = 4.0–6.0 MPa, T = 100 °C, t = 8 h.
^{b}[Rh] = 15 ppm, P/Rh = 20, CO$_2$ = 13 g, propylene = 2 g, H$_2$O = 0.4 mL, ethanol = 0.4 mL, CO/H$_2$ = 1:1 (3.5–4 MPa), reaction pressure = 12.0–14.0 MPa, T = 55 °C, t = 6 h.
^{c}[Rh] = 15P ppm, /Rh = 20, propylene = 13 g, H$_2$O = 0.4 mL, ethanol = 0.4 mL, CO/H$_2$ = 1:1 (3.5–4 MPa), reaction pressure = 6.0–10.0 MPa, T = 117 °C, t = 4 h.

formed *in situ* from [Rh(acac)(CO)$_2$] (P/Rh = 5:1) and the scCO$_2$-insoluble ligand MeO-PEG750PPh$_2$.[143] After 2 h at T = 70 °C and p(H$_2$/CO) = 50 bar, CO$_2$ was introduced. The reaction stopped completely when the CO$_2$ density reached 0.57 g mL^{-1}, a yellow-orange solid formed, the catalyst, and the aldehydes (n/i = 2.5:1) were isolated quantitatively in solvent-free form with low levels of metal contamination (*ca.* 5 ppm) by extraction with scCO$_2$. The strength of this procedure was that the reaction/separation sequence was

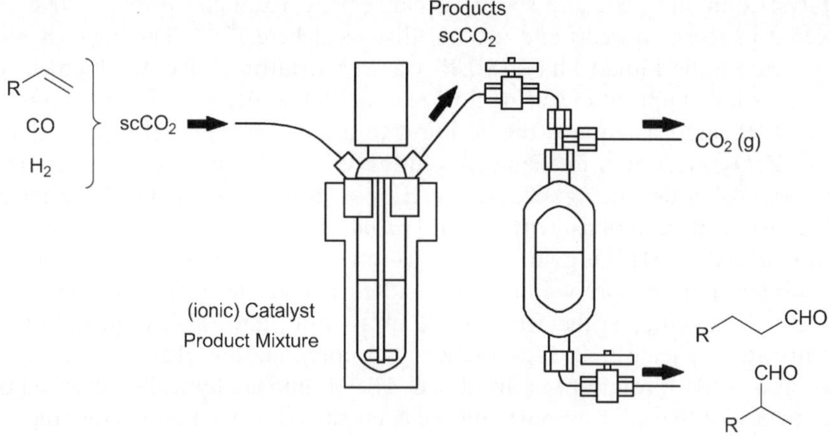

Scheme 1.31

repeated successfully six times with no significant changes in conversion or selectivity (conversion > 99%, $n/i = 2.3{:}1{-}2.5{:}1$).

A technologically highly attractive approach to the design of new hydroformylation processes exploits CO_2-insoluble catalysts under continuous flow conditions or using microreactor technology.[144] A hydroformylation process of 1-octene was developed in a continuous flow system using scCO$_2$ as the transport vector for substrates and products under relatively mild conditions for this kind of reaction, namely at 100 °C and 125 bar, with TOFs up to 240 h^{-1} (Scheme 1.31).[145] The catalytic system was prepared by mixing [Rh(acac)(CO)$_2$] and [Rmim][Ph$_2$P-p-C$_6$H$_4$SO$_3$]. The R group in the imidazolium ion was optimized for a correct balance of solubility in 1-octene (the highest solubility, the fastest reaction) and in scCO$_2$ (the lowest solubility, the least catalyst leaching into product). The pentyl group was the best compromise, and the process operated at 100 °C, 140 bar with the following typical flow rates: CO$_2$ (0.65 nL min^{-1}), syngas (1:1; 3.72 mmol min^{-1}), 1-octene (0.2 mL min^{-1} = 1.27 mmol min^{-1}). Observed rates were on the order of 160–240 catalyst turnovers per hour with low rhodium leaching over a 12 h period at a total pressure of 125–140 bar.

The authors pointed out that this process is potentially emissionless since the scCO$_2$ can, in principle, be recycled.[145]

1.3.4 Hydroformylation and Carbonylation Reactions Promoted by SILP Catalysts

To apply continuous flow technologies, the choice of solid catalysts is highly recommended not only for the easier separation process involved but also for the use of fixed bed reactors.[146] In this section attention is focused only on the most recent examples of solid-phase assisted catalysis using ionic liquids as the catalyst-containing bed and scCO$_2$ as carrier gas. Examples prior to 2006 are covered in recent reviews and are not discussed here.[146,147] The concept of a Supported Ionic Liquid Phase (SILP) was very fruitful to the development of new gas-phase applications using flow reactor technologies.[148] The preparation of a SILP catalyst requires the addition to a support material (SiO$_2$, Al$_2$O$_3$, TiO$_2$, ZrO$_2$, *etc.*) of a methanolic solution of a catalyst precursor, *e.g.* [Rh (acac)(CO)$_2$], a ligand, *e.g.* **62**, and an IL, *e.g.* [bmim][n-C$_8$H$_{17}$OSO$_3$] (Scheme 1.32). After methanol is removed *in vacuo*, the resulting solid material is characterized. A SILP is defined by (i) its ionic liquid loading α, defined as the IL volume/support pore volume ratio and correlated to the film thickness, (ii) the metal content, *i.e.* the Rh/support mass ratio, and (iii) the ligand/metal molar ratio. Scheme 1.32 gives a schematic representation of the surface cross section of a SILP catalyst in a fixed bed. The IL film is physically adsorbed on the surface of the solid support and the catalyst is dissolved in this microlayer. Since the film has the size of the diffusion layer, all metal complexes are involved in the catalytic reaction, which takes place under homogeneous

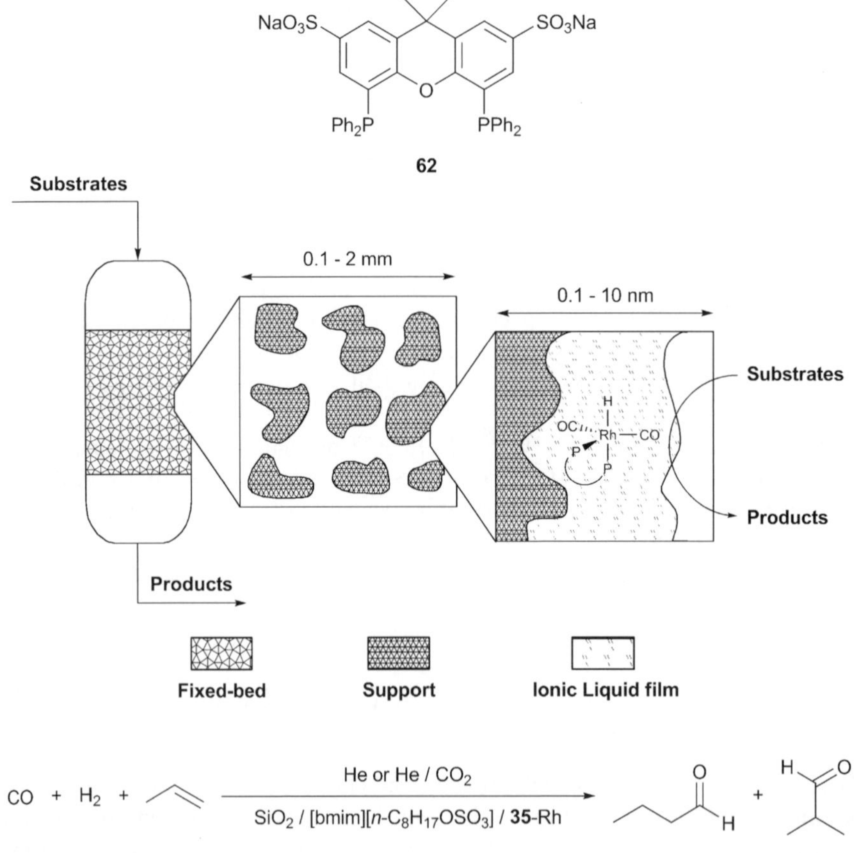

Scheme 1.32

conditions. When SILP particles are used as the fixed bed of a flow reactor, reagents enter the IL film, they react under homogeneous conditions and products, eventually, are desorbed into the carrier gas stream.

Reactions are run for a time that depends on catalyst half-life. Stabilities exceeding 700 h time-on-stream have been recorded.[149] Wasserscheid and co-workers optimized the hydroformylation of propene catalysed by a silica-supported phosphane **62**-Rh complex in the IL [bmim][n-C$_8$H$_{17}$OSO$_3$]. Using helium or He/CO$_2$ mixtures as carrier gas, TOF values ranged from 16 to 46 h^{-1} under different reaction conditions (reagent partial pressures, support pre-treatment, *etc.*), while selectivity in favour of the linear aldehyde was constantly around 94–95%.

1-Butene has been also successfully hydroformylated under these conditions using the SILP catalyst, which exhibited a higher activity and selectivity with respect to propene.[150]

Among various conditions examined in a continuous fixed bed reactor, when 0.1 wt% of Rh is used, the TOF is 17 and 324 h^{-1} at 80 and 120 °C, respectively, while the % of linear aldehyde decreases from 99.9% to 97.7%. From the determination of the full rate law, a first order with respect to Rh is found, meaning that mass transport from the gas to the IL phase is not limiting the reaction rate; moreover, the overall kinetics picture is consistent with a homogeneous Rh-catalysed reaction. This observation confirms the potentiality of SILP-catalysis, which combines the kinetic advantages of homogeneous conditions (the reaction takes place in the IL liquid film confined on the support surface with high specific area) with the practicality of solid heterogeneous catalysis. Indeed recyclability and the reduced amount of IL needed with respect to liquid–liquid processes make the SILP strategy competitive from an economic and an environmental point of view.

Higher alkenes, too, can be hydroformylated over a fixed bed SILP catalyst. In this case the carrier gas is scCO$_2$, which is soluble in ILs but does not dissolve ionic compounds.[151] For example, 1-octene, CO and H$_2$ are mixed in scCO$_2$ and flowed through a tubular reactor. The reactor was packed with the catalytic system generated from [Rh(acac)(CO)$_2$] and [propylmim][Ph$_2$P-p-C$_6$H$_4$SO$_3$] dissolved in [omim][Tf$_2$N] which, in turn, is supported on silica gel.

Under optimized conditions, rates corresponding to 800 molsubst (mol catalyst^{-1}) h^{-1} are observed, with $n/i \sim 3$. The catalyst remains stable over a 40 h reaction time with less than 0.5 ppm of rhodium leaching.[151]

A fixed-bed SILP process has the major advantage over existing batch technology of demanding smaller amounts of expensive metal catalyst and ionic liquids. This approach proved useful also for conducting continuous, fixed-bed gas-phase methanol carbonylation at industrially relevant reaction conditions.[152] The SILP rhodium iodide complex catalyst used was [bmim][Rh(CO)$_2$I$_2$]-[bmim][I]-SiO$_2$, prepared by one-step impregnation of a silica support with a methanolic solution of the [bmim][I] and the dimer [Rh(CO)$_2$I]$_2$. The reactant gas stream consisting of CO and vaporized methanol/methyl iodide feed (3:1 wt%) passed through the SILP catalyst bed at 180 °C. At $P = 20$ bar and after 2 h, 99% of conversion was obtained with a

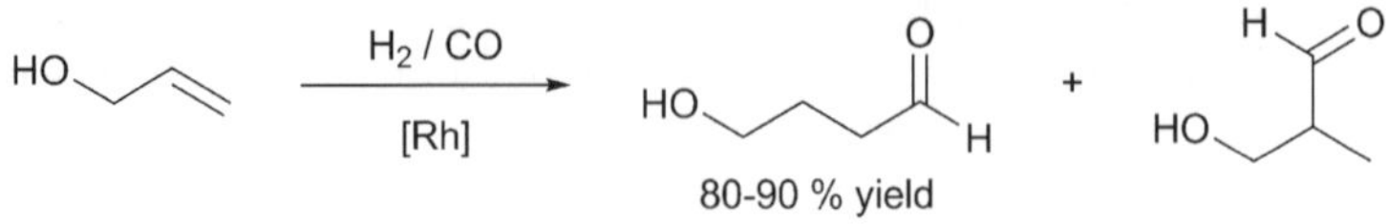

$$\text{HO}\diagdown\diagup\diagup \quad \xrightarrow[\text{[Rh]}]{\text{H}_2\,/\,\text{CO}} \quad \text{HO}\diagdown\diagup\diagdown\diagup\!\!=\!\!O_H \quad + \quad \overset{\text{H}}{\diagup}\!\!=\!\!O ,\ \text{HO}$$

80-90 % yield

Scheme 1.33

production rate of $21\,\mathrm{mol\,L^{-1}\,h^{-1}}$ and a product selectivity for methyl acetate of 75%.

SILP catalysts are versatile materials that can be also used under standard batch conditions. Thus, allyl alcohol was hydroformylated with a SILP Rh/PPh$_3$ system, affording 80–90% of linear aldehydes with n/i up to 31 (Scheme 1.33).[153]

The SILP catalyst was prepared simply by adding to a CH$_2$Cl$_2$ solution of HRhCO(PPh$_3$)$_3$ and PPh$_3$, in a 1:7 ratio, [bmim][PF$_6$] and then silica previously calcined at 450 °C for 24 h, and finally removing the solvent under vacuum. The catalyst was suspended in water used as the reaction solvent and allyl alcohol; the mixture was charged into a 100 mL stainless steel high-pressure autoclave and finally the reactor was flushed with a CO/H$_2$ (1:1) mixture at 25 °C, then heated to 80 °C and pressurized up to 4 MPa of CO/H$_2$ (1:1) for 5 h. After depressurization, the catalyst was simply recovered by filtration. Yields of linear aldehyde reached 90% with a n/i ratio in the 10–30 range.

Another SILP catalyst used under batch conditions employed mesoporous MCM-41 as the solid support. The catalyst was derived from [Rh(CO)$_2$(acac)] and TPPTS (1:5 mol ratio) in the desired IL. The excellent catalytic performance of this SILP catalyst in the hydroformylation of C$_6$–C$_{12}$ linear alkenes (TOF up to 500 h^{-1}) was determined by the large surface area and uniform mesopore structure of MCM-41 and was almost independent of the type of IL used: [bmim][BF$_4$], [bmim][PF$_6$] and 1,1,3,3-tetramethylguanidinium lactate.[154]

1.4   Olefin Metathesis Reactions

The metathesis reactions of olefins,[155] enynes and alkynes,[156] has been the process, together with palladium-catalysed cross-coupling reactions, that has had the most profound impact on the formation of carbon–carbon bonds and the art of total synthesis in the last quarter of the 20th century.[157]

Scheme 1.34 summarizes the most characteristic metathesis reactions. The following abbreviations are used: CM (cross metathesis), RCM (ring-closing metathesis), ROM (ring-opening metathesis), ROMP (ring-opening metathesis polymerization) and ADMET (acyclic diene metathesis polymerization).

The olefin metathesis story started with the advent of well-defined catalysts, most notably the Grubbs ruthenium catalysts.[158] These are ruthenium carbenes that undergo a [2 + 2] cycloaddition to olefins to produce a metallacyclobutane intermediate. Cycloreversion of the metallacyclobutane in the opposite sense leads to olefin metathesis (Scheme 1.35).

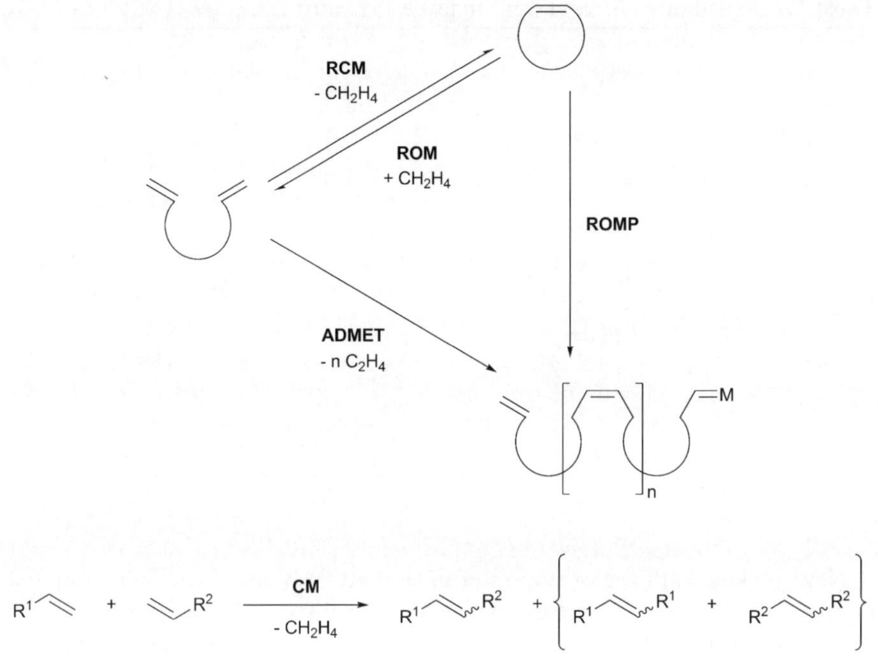

Scheme 1.34

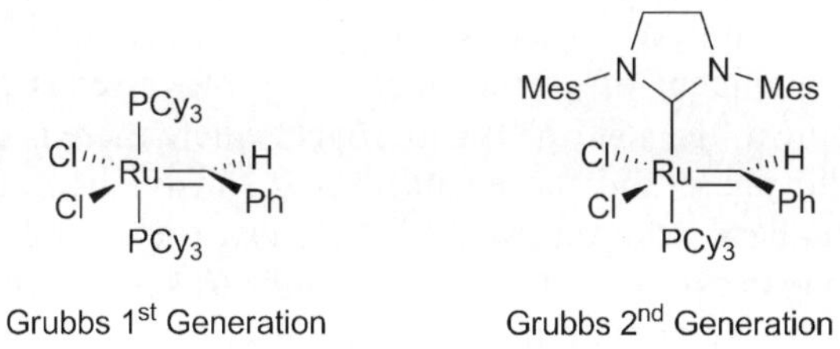

Grubbs 1st Generation

Grubbs 2nd Generation

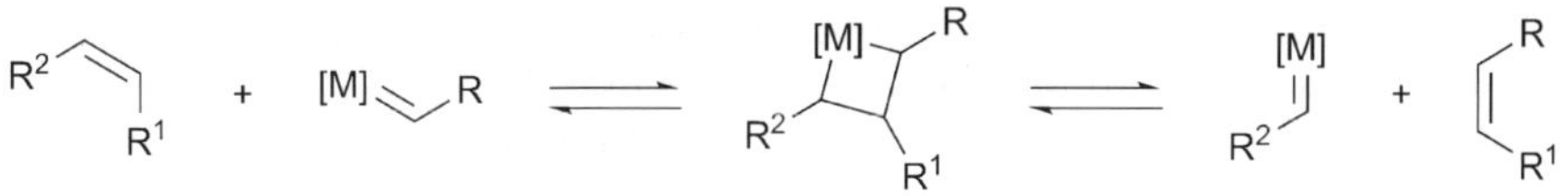

Scheme 1.35

Some of these catalysts have been demonstrated to be able to promote ROMP and RCM in polar solvents such as water and methanol.[159]

Further ruthenium catalysts such as those developed by Grubbs (**63**,[160] **64**,[161] and **65**,[162] Cy = cyclohexyl), Nolan (**66**),[163] Hoveyda (**67**,[164] **68**,[165] Mes = 2,4,6-trimethylphenyl), Blechert (**69**),[166] and Grela (**70**)[167] are collected in Table 1.7.

Table 1.7 Structures of common metathesis catalysts.

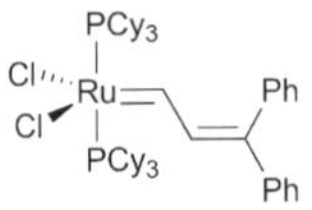

<table>
<tr><td>63</td><td>64</td><td>65</td><td>66</td></tr>
<tr><td>67</td><td>68</td><td>69</td><td>70</td></tr>
</table>

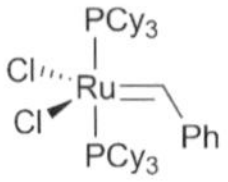

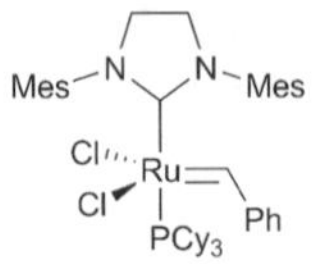

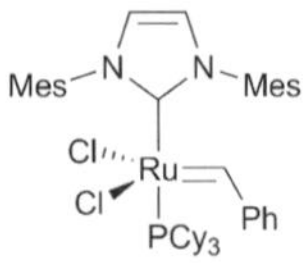

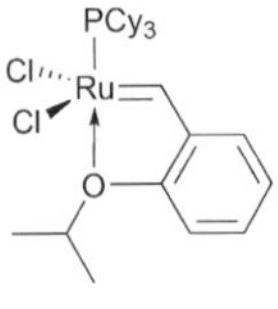

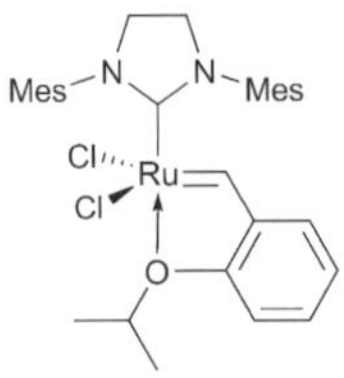

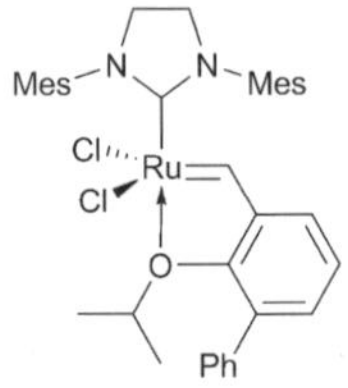

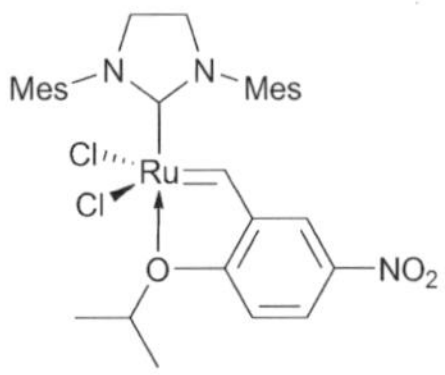

Nevertheless, with the exception of uses in ROMP processes, only a limited number of industrial processes use olefin metathesis. This is mainly due to difficulties associated with removing ruthenium from the final products and recycling the catalyst. To tackle these problems, there is tremendous activity in this area, dealing with supported or tagged versions of homogeneous catalysts.[168]

The second drawback of metathesis reactions is that they usually require large amounts of catalyst, typically 20 mol.% ruthenium.[169] For all these reasons, the development of greener metathesis processes is of great importance. In this context, the solid-phase approach, where there is a covalent bond between the solid phase and the ligand, has been investigated and the most important results have been reviewed.[170] Actually, the immobilization technique of homogeneous catalysts *via* covalent bonds was not particularly efficient because of catalyst degradation. More promising is the strategy of ionic attachment: the catalyst is tagged with an ionic group that interacts electrostatically with the ionic solid phase.[171] Such an approach to immobilization allows for easy separation and reuse of the catalyst. An example is represented by **71** where the 2-isopropoxy-benzylidene ligand of **67** is further modified by an additional amino group, which was immobilized by treatment with sulfonated polystyrene, forming the corresponding ammonium salt by acid/base reaction (Scheme 1.36).[172]

Supercritical fluids such as $scCO_2$ have also been used in a limited number of studies. For example, it was reported that **63** and **64** catalyse the ROMP of norbornene and cyclooctene in $scCO_2$, as well as the RCM of various dienes (*e.g.* Scheme 1.37).[173]

In the next two sections we limit our analysis to a few recent examples of the use of water or ionic liquids in metathesis reactions using new technical solutions or ionically tagged catalysts.

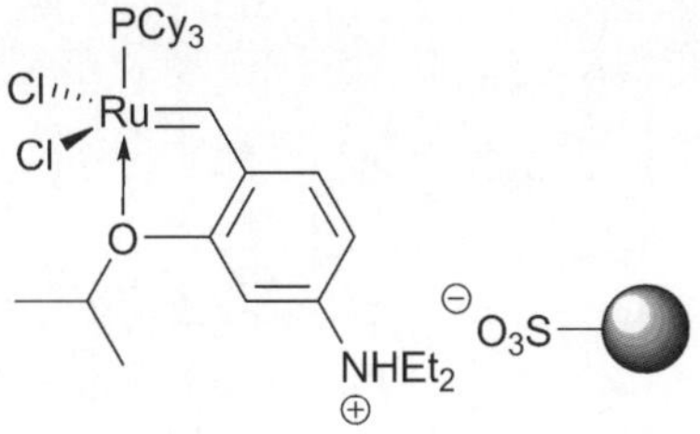
PCy3
Cl
Ru
Cl
O
O3S
NHEt2

71a: polymer prepared by precipitation polymerisation
71b: Dowex 50Wx2

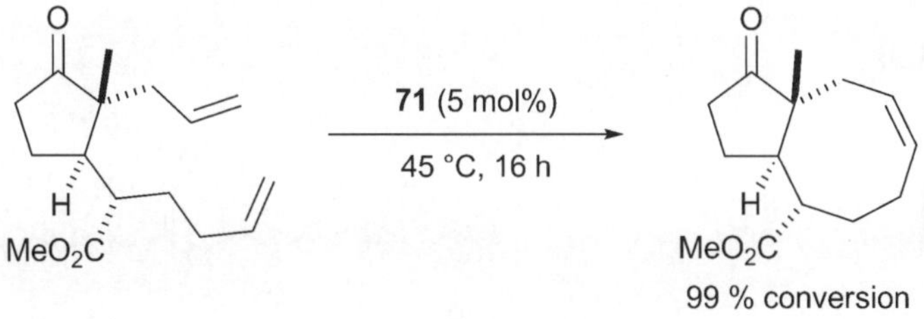
O
H
MeO2C
71 (5 mol%)
45 °C, 16 h
99 % conversion

Scheme 1.36

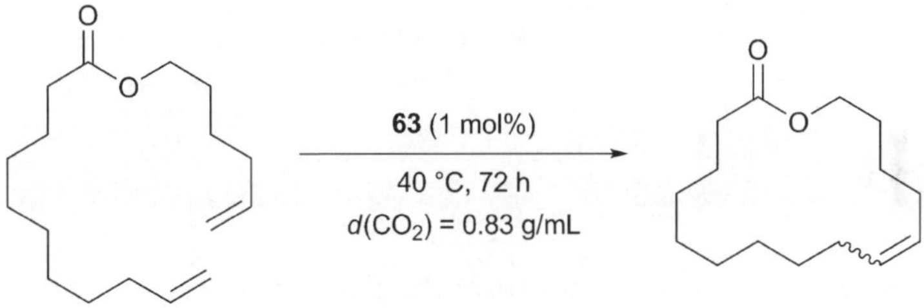
O
O
63 (1 mol%)
40 °C, 72 h
d(CO2) = 0.83 g/mL

Scheme 1.37

1.4.1 Olefin Metathesis in Aqueous Media

We start with studies that aim to overcome solubility problems when a typical metathesis reaction is transferred into water. The first way to introduce water is to use a solvent pair capable of rendering the reaction mixture homogeneous. For example, after an optimization study, the Hoveyda catalyst **68** was found to promote the RCM of various dienes in aqueous DME and acetone solutions. Under these homogeneous conditions, substrates with various substituents underwent RCM to form five-, six- and seven-membered cyclic products.[174]

The example given in Scheme 1.38 shows the excellent results obtained after relatively short reaction times and low catalyst loading (3 mol.%).

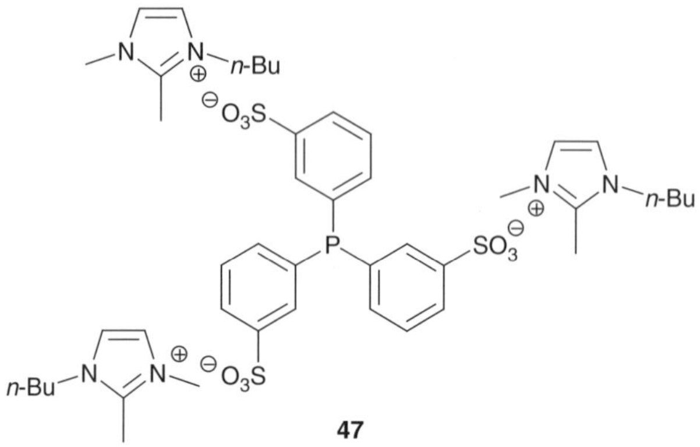

47

Scheme 1.38

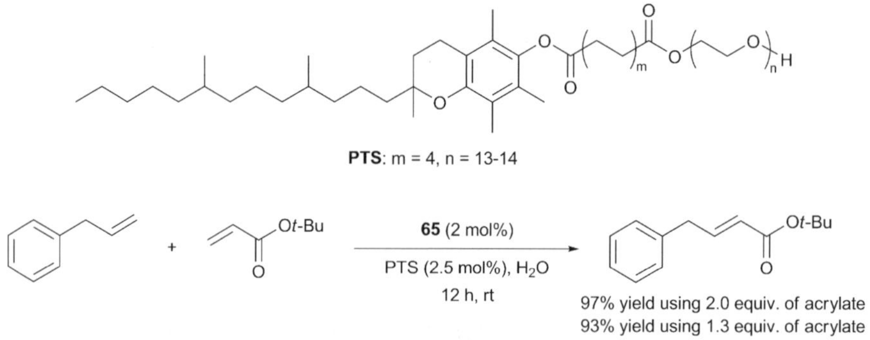

PTS: m = 4, n = 13-14

65 (2 mol%)

PTS (2.5 mol%), H_2O

12 h, rt

97% yield using 2.0 equiv. of acrylate
93% yield using 1.3 equiv. of acrylate

Scheme 1.39

The foregoing solution did not solve the problem of using a VOC since an organic co-solvent is necessary. In any case, the use of pure water presents serious limitations, *e.g.* the limited aqueous solubility of most neutral organic substrates and the use of water-insoluble classic ruthenium-based catalysts. Thus, non-ionic surfactants have been tested as additives, and from a screening of several amphiphiles PEG-600/(*R*)-tocopherol-based diester of sebacic acid (PTS, MW $\sim$ 1200), which in water forms on average 22 nm micelles above its critical micelle concentration (0.28 mg g^{-1}), looked very promising. Adding a preformed solution of only 2.5% (by weight) PTS in water to a mixture of allylbenzene, *tert*-butyl acrylate (2 equiv.) and **65** (2 mol.%) with continuous stirring overnight at rt, the desired CM enoate product was obtained in 97% isolated yield (Scheme 1.39).[175]

Reactions can be followed visually since the starting rose-colored colloidal dispersion of 2 mol.% **65** in 2.5% PTS/water, in the presence of reactants, darkens while the CM product forms over a few hours. Most notably, reactions are carried out at rt in the open air, while most CM reactions are done at higher

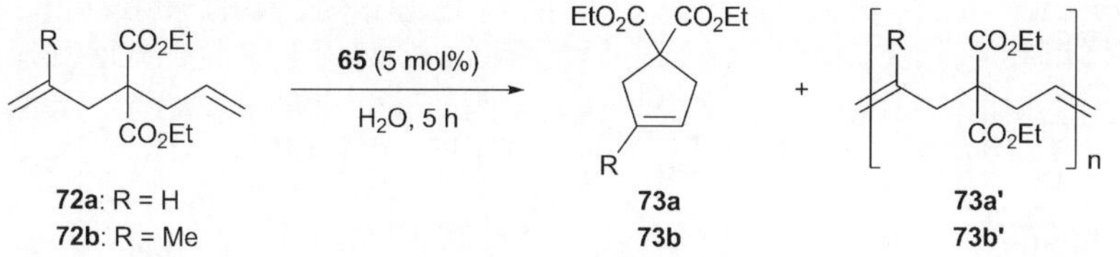

72a: R = H
72b: R = Me

73a
73b

73a'
73b'

R	Conditions	Yield (%)		R	Conditions	Yield (%)	
		73a	**73a'**			**73b**	**73b'**
H	(((/ H_2O, 40 °C	99	0	Me	(((/ H_2O, 40 °C	99	0
H	neat, 25 °C	81	18	Me	neat, 25 °C	93	6

Scheme 1.40

temperatures. The E/Z selectivity is comparable to that typically observed in organic media and functional groups such as allylic silanes, unprotected alcohols, amino acid derivatives and epoxides are tolerated.

However, what about optimizing a heterogeneous reaction avoiding co-solvents and surfactants? One answer to this question is to form a stable emulsion with the aid of ultrasonication.[176] A user-friendly protocol for effecting olefin metathesis in water at rt has been reported in the absence of any co-solvents and surfactants. In this protocol, ultrasonication of a water-insoluble substrate (or substrates) floating on water affords an emulsion, in which a smooth catalytic metathesis took place after addition of water-insoluble commercially available Grubbs catalyst **65**. Ring closing (RCM), enyne and cross metathesis (CM) of various substrates proceeded in water using acoustic emulsification. Scheme 1.40 shows a typical experiment carried out using dienes **72a** and **72b** (which are insoluble in water) in water and under neat conditions. Interestingly, reactions in water did not produce any polymerization products, which conversely were formed in appreciable amounts under neat conditions.

A more creative approach to catalyst design when water is the targeted solvent for a catalytic process is the tag strategy. A few representative examples of charged tags for metathesis catalysts working in water and ionic liquids (next section) will be discussed. As usual, the specific interactions between the tag and the other phase can either allow for easy removal of the catalyst during workup or represent an immobilization strategy.

Let us examine first the principal sites of tagging ruthenium complexes (Scheme 1.41): they include covalent ligands **A** [phosphines, N-heterocyclic carbenes (NHCs) or pyridine], anionic ligands **B**, or the carbene ligand **C** (either in the aromatic moiety or at the alkoxy ligand).

The most recent literature is organized on the basis of the selected tagging site.

In most tagging approaches to catalyst design the ionic group is installed on the carbene ligands, *i.e.* the benzylidene fragment. Table 1.8 collects a few examples.

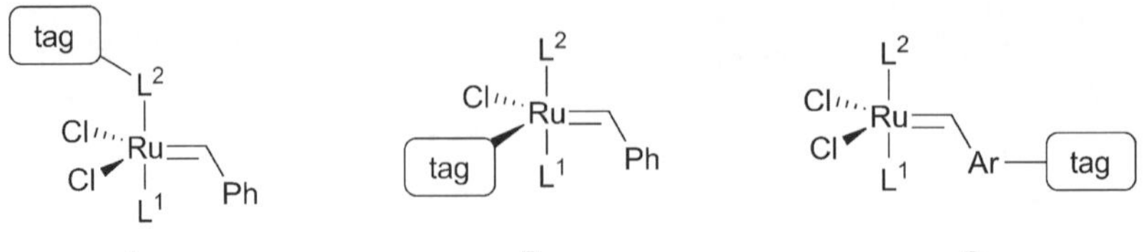

Scheme 1.41

tag	second phase
$-SO_3^{\ominus}$	cationic exchange resin
$-NMe_3^{\oplus}$	anionic exchange resin
$-(CF_2)_nCF_3$	perfluorinated solvent
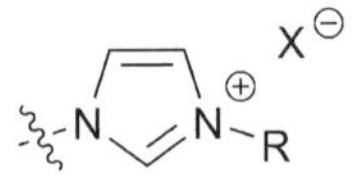 (imidazolium) $X^{\ominus}$	ionic liquid

Table 1.8 Structures of some carbene-tagged ligands.

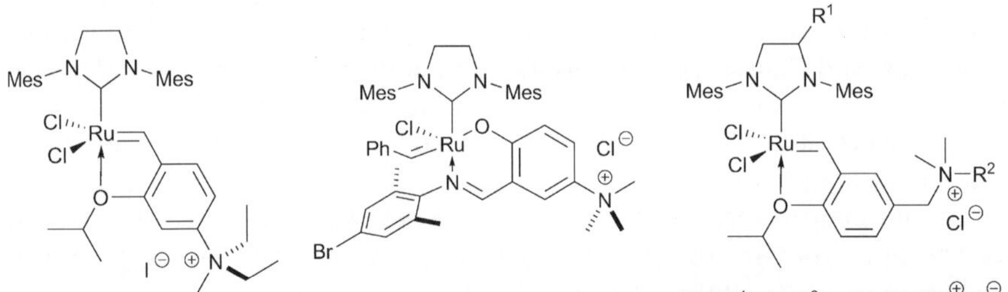

74

75

76a: R^1 = H, R^2 = $CH_2CH_2NH_3^{\oplus}$ $Cl^{\ominus}$

76b: R^1 = $CH_2NH_3^{\oplus}$ $Cl^{\ominus}$, R^2 = CH_3

When the ammonium group is connected directly to the benzylidene ring, *e.g.* in **74**[177] and **75**,[178] the electron density on the chelating oxygen atom of the 2-propyloxy group decreases and weakens the O–Ru coordination, facilitating faster initiation of the catalytic cycle. The other effect of the ammonium ions is to enhance the solubility of the catalyst in polar media. Therefore, **74–76** can find applications for metathesis reactions both in traditional solvents as well as in aqueous media. For example, **74** can be used in $MeOD/D_2O$ (2:5 v/v) or in neat water, where it forms quasi-emulsions. In water, reactions are conducted in air under the following conditions: catalyst (1–5 mol.%), H_2O, substrate conc $= 0.2$ M, $258\,°$C. Several CM and RCM reactions using **74** have been reported to give 40–99% yields after 5 min to 8 h reaction times. Catalyst **75**, a ruthenium complex bearing both a NHC and a salicylaldimine ligand, is effective for diene and enyne RCM in protic media. The enhanced stability engendered by the salicylaldimine ligand allows trimethylammonium-functionalized complex **75** to achieve clean RCM of *N,N*-diallylamine hydrochloride in CD_3OD/D_2O (2:1 v/v) with the highest conversion yet reported in an aqueous environment (Scheme 1.42).

The last two catalysts, **76a,b**, have been recently developed by Grubbs.[179] The difference between them is that **76b** is much more soluble in water than is **76a**. Moreover, the decomposition half-life of **76b** in water is over a week at ambient temperature under inert conditions. Both catalysts **76a** and **76b** are highly competent ROMP catalysts, rapidly transforming *endo*-norbornene monomer into a polymeric product (Scheme 1.43).

Both catalysts also mediate RCM reactions in aqueous media (Scheme 1.44), while neither catalyst is sufficiently stable for the practical aqueous CM reaction.

The NCH carbene ligand has also been exploited as tagging site. Two examples, **77** and **78**, are quoted here, where the tag is a hydrophilic non-ionic PEG chain (Table 1.9).[180]

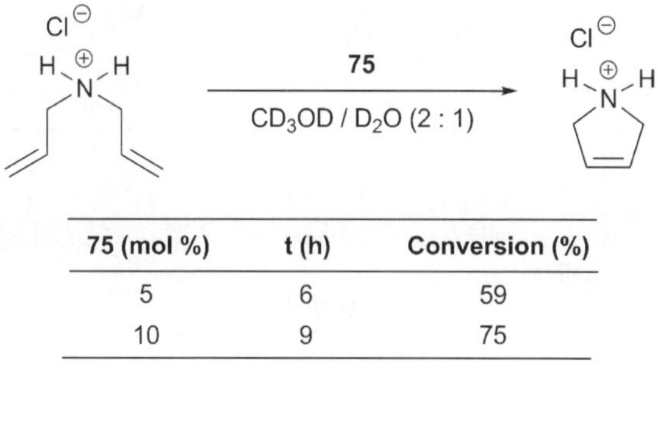

75 (mol %)	t (h)	Conversion (%)
5	6	59
10	9	75

Scheme 1.42

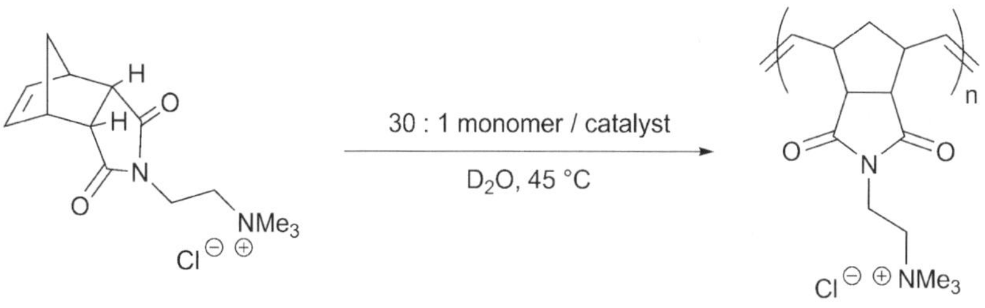

Scheme 1.43

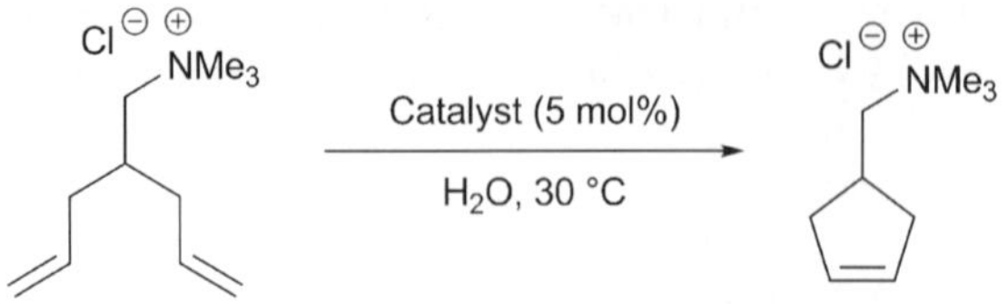

Catalyst	t (h)	Conversion (%)
76a	12	> 95
76b	24	> 95

Scheme 1.44

Table 1.9 N-Heterocyclic carbene (NHC)-tagged ligands.

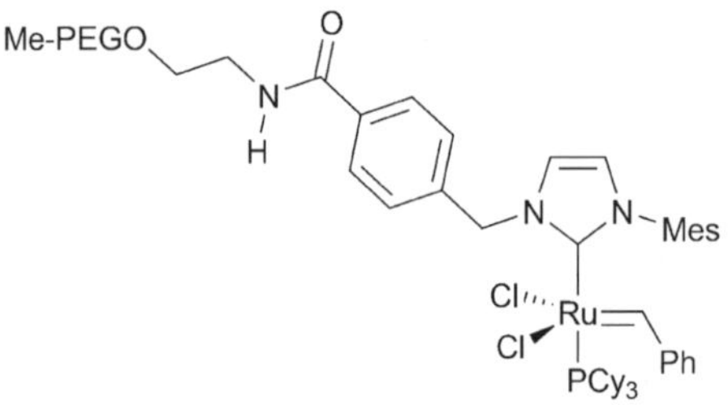

77

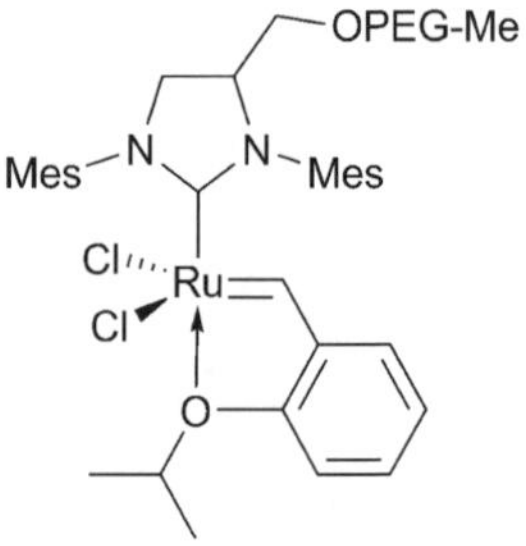

78

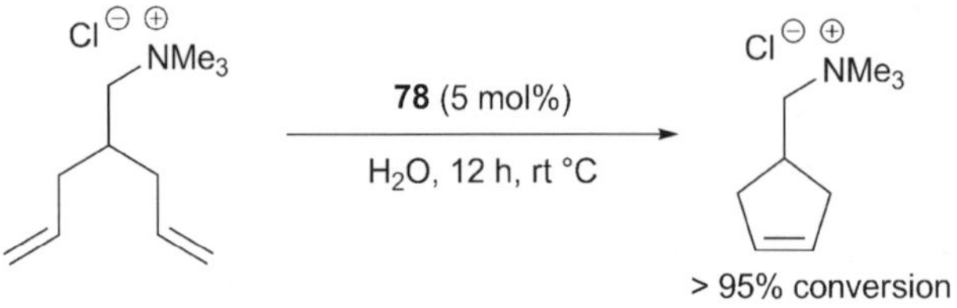

Scheme 1.45

While catalyst **77** containing a tagged NHC-based ligand showed good activity in aqueous ring-opening metathesis polymerization (ROMP) reactions,[181] catalyst **78** showed unprecedented RCM activity with water-soluble dienes in water, yielding the corresponding five- and six-membered rings in good to excellent yields. An example in reported in Scheme 1.45.

Catalyst **78** was also the first water-soluble catalyst active in cross-metathesis in water, as witnessed by the homodimerization reactions reported in Scheme 1.46.

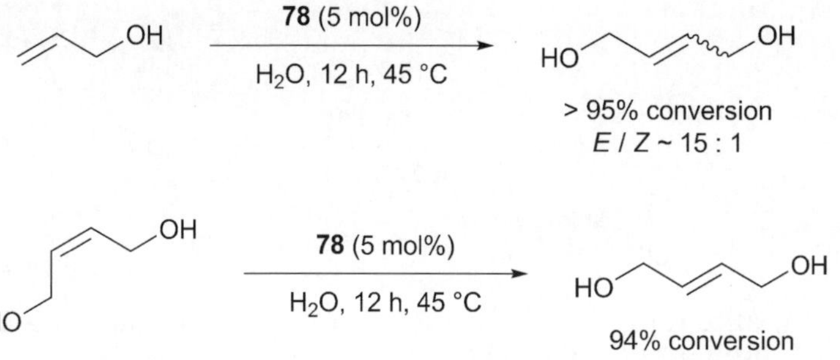

> 95% conversion
$E / Z \sim 15 : 1$

94% conversion

Scheme 1.46

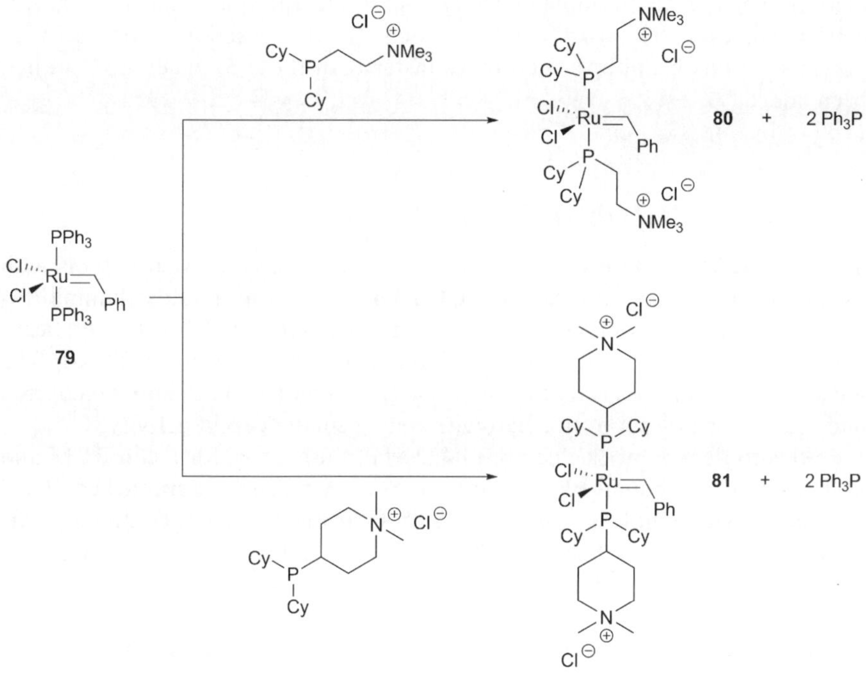

80 + 2 Ph$_3$P

79

81 + 2 Ph$_3$P

Scheme 1.47

The last tagging site available is represented by phosphines. Indeed, Grubbs' group proposed catalysts **80** and **81** generated from Grubbs I catalysts **79** by ligand exchange with the corresponding ammonium tagged phosphines (Scheme 1.47).[182]

Alkylidenes **80** and **81** are soluble in protic, high-dielectric solvents such as water, methanol and water/THF mixtures, and are completely insoluble in other common organic solvents, including acetone, THF and benzene. It was

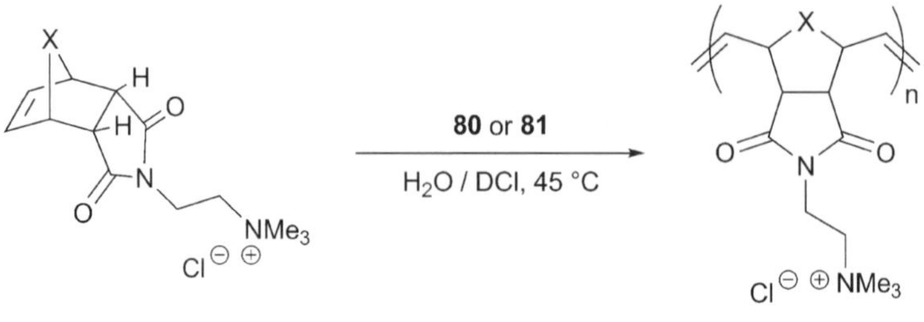

82: X = CH$_2$
83: X = O

> 95% conversion

Scheme 1.48

found that monomers **82** and **83** (Scheme 1.48) could be quantitatively polymerized when small amounts of DCl (from 0.3 to 1.0 equiv. relative to alkylidene complexes **80** or **81**) were added to the reaction mixture, with polymerizations being up to ten times faster than those to which no acid had been added.

1.4.2 Olefin Metathesis in Ionic Liquids

The first studies examined refer to the transfer of classic CM protocols with Grubbs I and II generation catalysts **64** and **65** in two ionic liquids, [bmim][PF$_6$] and [bmim][BF$_4$]. Compared to the same reactions in CH$_2$Cl$_2$ significant enhancements in the reactivity, yield and reaction rate were achieved. The advantage of using ILs is that the catalyst/IL system could be simply recovered and reused for at least four cycles with only a small drop in activity.[183]

In this context Schrock's catalyst **84** has been used, too, for olefin RCM and CM reactions in [bmim][PF$_6$] as solvent. Several reactions approached 100% conversion with 5 mol.% of catalyst at 75 °C, using molecules containing various functional groups (Scheme 1.49). A new method to extract small amounts of organic products from IL by carrying out a Soxhlet extraction using PDMS (polydimethyl siloxane) thimbles has also been reported.[184]

The assistance of microwave heating has been also proposed to accelerate RCM reactions using classic ruthenium-based catalysts. The reaction can be rapidly conducted in either ionic liquids, such as [bmim][BF$_4$], or in a microwave transparent solvent such as dichloromethane.[185]

However, ionic tagging is a winning strategy in ILs, too.

An efficient RCM reaction protocol has been developed under biphasic conditions using catalyst **85** in a 25/75 (v/v) [bmim][PF$_6$]/toluene mixture (Scheme 1.50).[186] The reaction proceeded at rt and a high level of recyclability (up to eight cycles) was obtained with excellent conversions after 3 h. A strength of this procedure was the very low residual ruthenium levels (1.2 –22 ppm) in the toluene extract, the average value over eight cycles (7.3 ppm) being approximately an order of magnitude lower than the best previously reported methods.

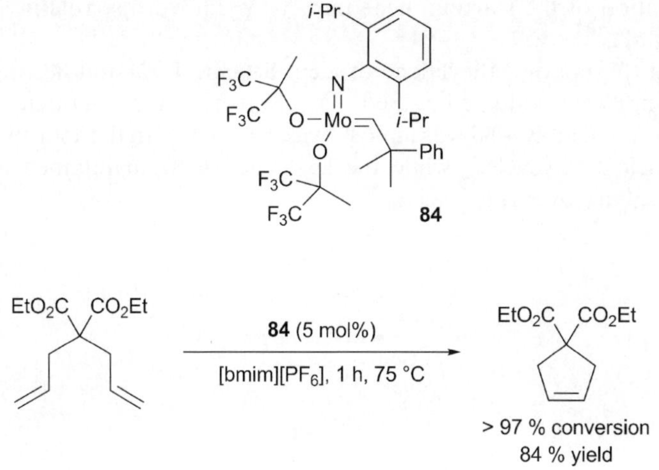

84

> 97 % conversion
84 % yield

Scheme 1.49

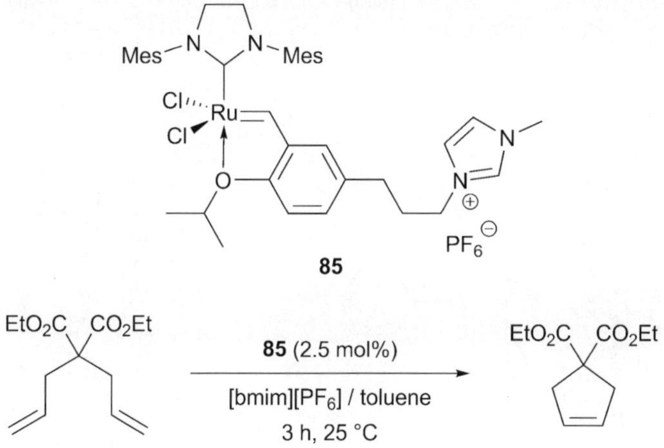

85

85 (2.5 mol%)

[bmim][PF$_6$] / toluene
3 h, 25 °C

Cycle	Conversion (%)	Cycle	Conversion (%)
1	> 98	5	> 98
2	> 98	6	> 98
3	> 98	7	95
4	> 98	8	95

Scheme 1.50

An alternative solution to ionic tagging to confine a catalyst into an IL phase is that of using a cationic metathesis complex, *e.g.* ruthenium allenylidene salt (**86**). Treatment of diallyltosylamide at 80 °C for 5 h dissolved in [bmim][OTf] in the presence of 2.5 mol.% of pre-catalyst **86** led to the transformation of the starting diene into *N*-tosyldihydropyrrole in 97% yield (Scheme 1.51).[187]

Catalyst **87** (tag on salicylidene oxygen; Scheme 1.52) and **88** (tag on salicylidene ring) were evaluated in the RCM of diethyl diallylmalonate at rt using 5 mol.% in 1 h. Catalyst **88** was almost twice as reactive in the 1st run but could not be efficiently recycled, while the less reactive **87** maintained an almost constant activity over four cycles.[188]

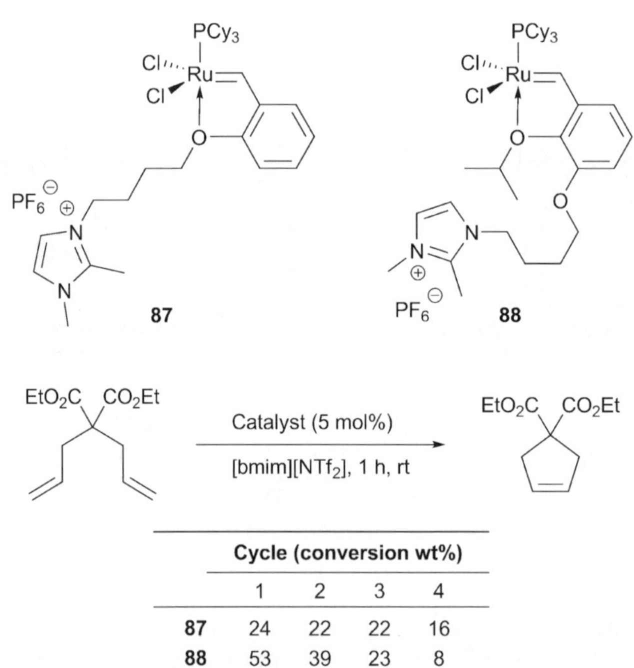

Scheme 1.51

	Cycle (conversion wt%)			
	1	2	3	4
87	24	22	22	16
88	53	39	23	8

Scheme 1.52

Recyclable imidazolium-tagged ruthenium catalysts **89** and **90** have been developed to perform olefin RCM and CM in ILs [bmim][PF$_6$] and [bmim][Tf$_2$N]. A high level of recyclability combined with a high reactivity were obtained in the RCM of various di- or tri-substituted and/or oxygen-containing dienes (Scheme 1.53). Extremely low residual ruthenium levels were detected in the RCM products (average of 7.3 ppm per run).[189]

Similar results have been reported for the closely related complex **91**, which was an efficient catalyst for the RCM of di-, tri- and tetrasubstituted diene and enyne substrates. The reactions were performed in a homogeneous mixture of [bmim][PF$_6$]/CH$_2$Cl$_2$ (1:9 v/v) as the solvent and in the presence of 1 mol.% of catalyst **91** at 45 °C (Scheme 1.54). Catalyst **91** combined the advantages of high reactivity and a high level of recyclability and reusability with only a very slight loss of activity.[190]

1.5 Palladium-catalysed Cross Coupling Reactions in Non-conventional Solvents

Palladium-catalysed cross-coupling reactions of organometals represent the most important and efficient carbon–carbon bond construction methodology

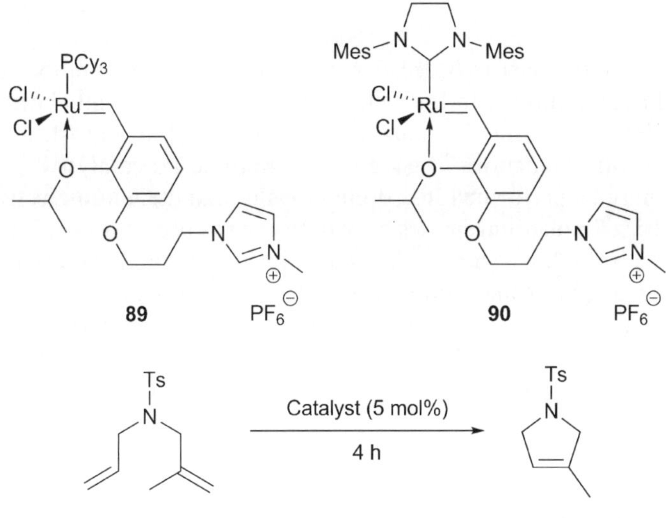

Catalyst	Solvent	T (°C)	Cycle (conversion %)						
			1	2	3	4	5	6	7
89	[bmim][PF$_6$]	60	92	92	73	45	-	-	-
90	[bmim][PF$_6$]	60	> 98	97	91	68	16	-	-
90	[bmim][PF$_6$]	40	79	-	-	-	-	-	-
90	[bmim][NTf$_2$]	40	> 98	> 98	91	89	74	50	-

Scheme 1.53

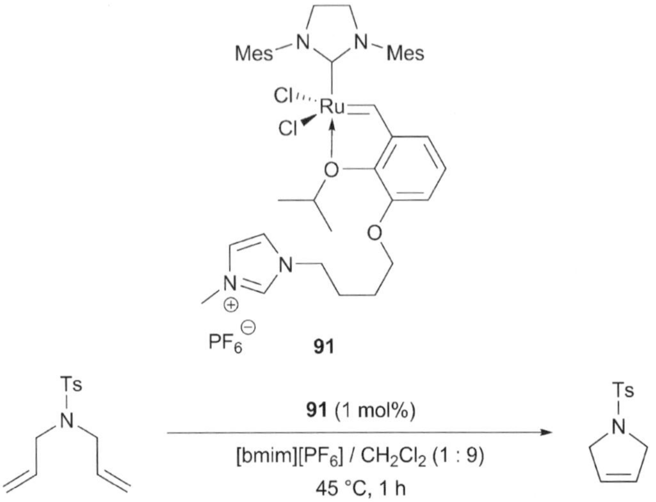

Cycle	1	2	3	4	5	6	7	8	9	10
Conversion %	> 98	> 98	98	98	96	94	92	92	92	90

Scheme 1.54

developed over the past two decades, as evidenced by the number of named reactions in connection with this field.[191] Further developments born from the Mizoroki–Heck,[192] Stille,[193] Suzuki–Miyaura,[194] Tsuji–Trost,[195] Sonogashira[196] and other reactions[197] gave rise to multiple Pd-catalysed transformations that may be performed in a domino fashion and enantioselectively.[198] A huge number of palladium species have been used as pre-catalysts. They include (i) solid not soluble species,[199] (ii) solubilized palladium nanoparticles, (iii) metalated dendrimers and complexes with well-defined nanosized ligands,[200] (iv) metal–ligand complexes (preassembled or produced *in situ*), and (v) simple palladium salts like $Pd(OAc)_2$. Nearly every form of palladium has been demonstrated to work as a pre-catalyst in the Heck and other Pd-catalysed cross coupling reactions.[201]

In the past decade protocols for palladium-catalysed cross coupling reactions have been successfully developed in alternative media that make catalyst recovery easier and minimize environmental impact. A comprehensive review covers the most important contributions in this field.[100e]

The palladium-catalysed arylation of olefins with aryl halides, the Heck reaction, is usually performed in polar solvents such as acetonitrile or dimethyl sulfoxide, in combination with a base and a Pd(II) pre-catalyst that may or may not be associated with a phosphorus ligand. Given that quaternary ammonium or phosphonium salts are known to increase reaction rates, ILs emerged as promising solvents for this reaction.[25,202] In the case of imidazolium-based ionic liquids, the solvent, beside providing an unusual coulombic environment, can also play a catalytically active role, *e.g.* by delivering imidazolylidene ligands.[203] In this context, formation of Pd-carbene complexes from the interaction of [bmim][Br] with $Pd(OAc)_2$ has been reported in the literature to be responsible for the enhanced efficiency of the Heck reaction reported in Scheme 1.55.[204]

Beneficial effects were exerted by $[bmim][BF_4]$ on the rate of the Suzuki reaction. When the Suzuki reaction shown in Scheme 1.56 was carried out in a toluene/water/EtOH (4:2:1 v/v ratio) mixture using 3 mol. of $Pd(PPh_3)_4$ the corresponding 4-methylbiphenyl was obtained in 84% after 6 h, equating to a TOF of $5\,h^{-1}$. When the same reaction was carried out in $[bmim][BF_4]$ with a lower catalyst loading (1.2 mol.%), a 92% yield was obtained in 3 h. In particular, after 10 min 68% of the product was formed, corresponding to an initial TOF of $170\,h^{-1}$.[205]

Even better results in terms of catalyst recyclability have been obtained in the low viscosity $[bmim][NTf_2]$, which plays a dual role as that of the reaction medium and precursor of palladium carbene catalyst (Scheme 1.57).[206]

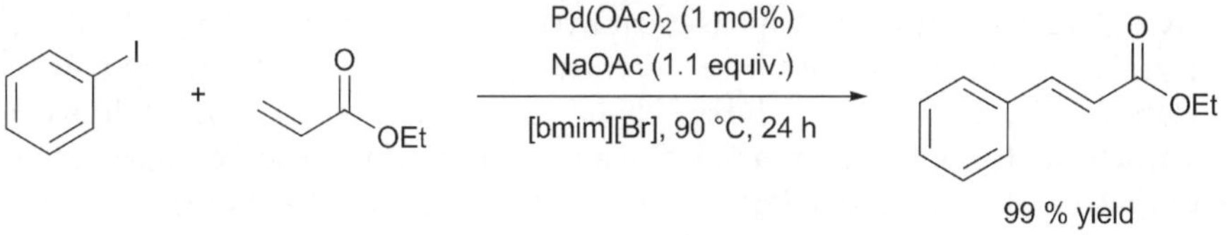

Scheme 1.55

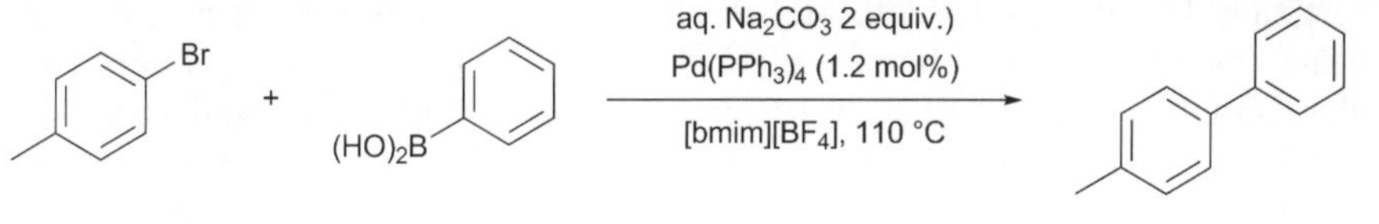

Scheme 1.56

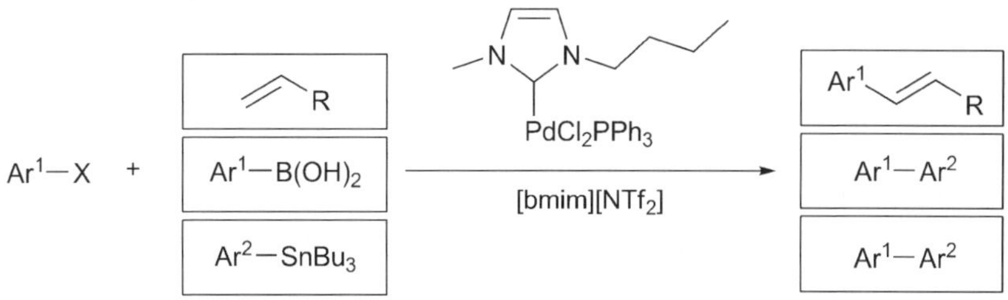

Scheme 1.57

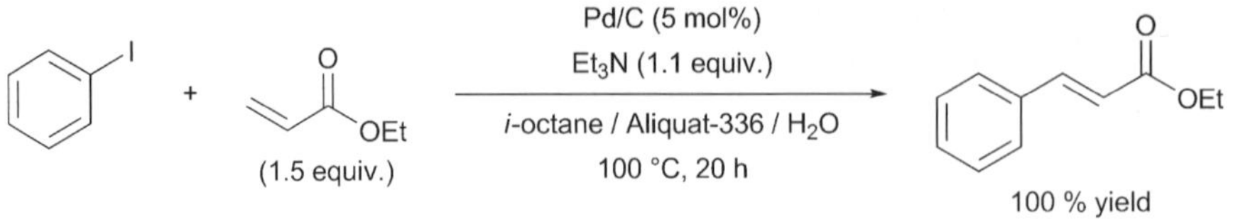

Scheme 1.58

Further evidence of the acceleration enjoyed by the Heck reaction in an ionic phase (*N*-methyl-*N*,*N*,*N*,-trioctylammonium chloride or Aliquat-336) has been reported by Tundo and co-workers in a triphasic catalytic system.[207] The arylation of electron-poor olefins was catalysed by palladium supported on charcoal (Pd/C) in the heterogeneous *i*-octane/Aliquat-336/water system (Scheme 1.58). The use of phosphines was not necessary. Aliquat-336 trapped the solid-supported catalyst, ensuring an efficient mass transfer between the bulk phases, which resulted in an increase in reaction rate of an order of magnitude, compared to the reaction in the absence of the ionic liquid.

A different example of triphasic catalysis for the Heck, Stille and Suzuki reactions relied on a three-phase microemulsion/sol–gel transport system.[208] Gelation of an *n*-octyl(triethoxy)silane, tetramethoxysilane and Pd(OAc)$_2$ mixture in a H$_2$O/CH$_2$Cl$_2$ system led to a hydrophobicitized sol–gel matrix that entrapped a phosphine-free Pd(II) precatalyst. The immobilized precatalyst was added to a preformed microemulsion obtained by mixing the hydrophobic components of a cross coupling reaction with water, sodium dodecyl sulfate and a co-surfactant, typically *n*-propanol or butanol. This immobilized palladium catalyst was leach proof and easily recyclable.

Examples of task-specific ILs designed for Pd-catalysed cross coupling reactions are reported here. 1-(2-Hydroxyethyl)-3-methylimidazolium tetrafluoroborate, [hemim][BF$_4$], has been reported as an efficient and recyclable reaction medium for the Heck reaction. The olefination of iodoarenes and bromoarenes with olefins generated the corresponding products in good to excellent yields under phosphine-free reaction conditions. After separation of the product, fresh starting materials were charged into the recovered Pd-containing ionic liquid phase. The reactions proceeded quantitatively for six cycles, without significant loss of catalytic activity.[209] Scheme 1.59 shows the effect of both the cation and the anion on the chemical yield.

Dyson and co-workers have prepared a family of nitrile-functionalized pyridinium[210] and imidazolium salts[211] and investigated their reactions with PdCl$_2$.[212] When nitrile-functionalized imidazolium chlorides, *e.g.* **92a**, are used, a tetrachloropalladate dianion salt **93** results; in contrast, when the counter-ion is PF$_6^-$, BF$_4^-$ or Tf$_2$N$^-$, as in **92b–d**, the reaction of nitrile-functionalized imidazolium salts and PdCl$_2$ in a 2:1 ratio gives complexes **94** in which two nitrile groups are coordinated to a metal centre (Scheme 1.60).

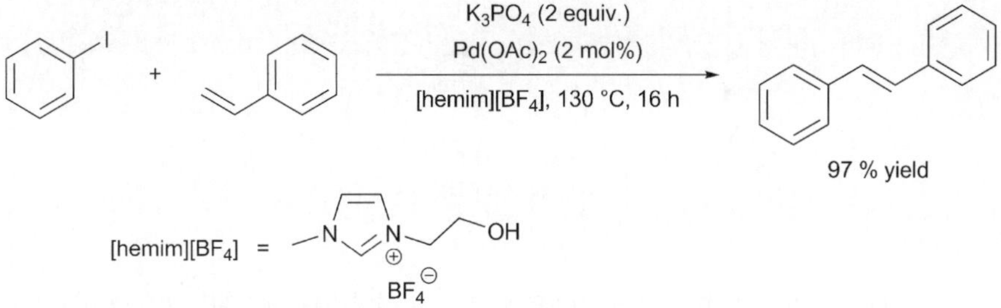

Scheme 1.59

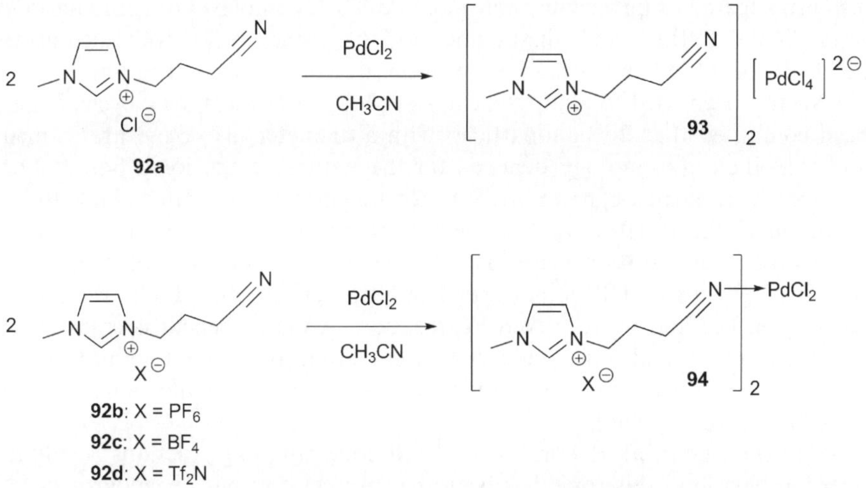

Scheme 1.60

Several nitrile functionalized ILs Pd-complexes, including **93** and **94**, were found to be active as catalysts for carbon–carbon bond forming reactions such as Suzuki and Stille cross-coupling and the Heck reaction. For example, **93** ensured a virtually quantitative conversion of phenyl iodide and phenyl boronic acid to biphenyl in [bmim][BF$_4$] or [bmim][PF$_6$] and aqueous sodium carbonate after heating at 110 °C for 12 h. Comparing the catalytic performance of PdCl$_2$ in [bmim][BF$_4$] and **92c**, the catalyst stability is increased by the latter, which prevented formation of the palladium black observed in the former system. Moreover, using a nitrile-functionalized IL, metal leaching during the extractive work-up is greatly reduced, falling from ∼100 ppm using PdCl$_2$/[bmim][BF$_4$] to ∼5 ppm using **92c**.

The same authors reported the superior activity of PdCl$_2$ dissolved in **95**, in both the Heck and the Stille reaction.[212]

95a: X = BF$_4$
95b: X = Tf$_2$N

The Heck reaction between iodobenzene and ethyl acrylate proceeds efficiently using PdCl$_2$ dissolved in **95b**, with cholinium acetate as the base and ammonium formate as reducing agent. *Trans*-ethyl cinnamate is produced in 60% after heating at 80 °C for 1 h.

The Stille coupling of (tributylvinyl)stannane with iodobenzene has been also investigated using various preformed Pd-TSIL complexes or combinations of PdCl$_2$ and TSILs. The highest conversion in styrene (96%) is obtained using PdCl$_2$ dissolved in **95a**. In this last reaction, transmission electron microscopy (TEM) has been used to study the catalyst phase after a catalytic cycle. It has been confirmed that Pd nanoparticles with a diameter of $\sim$5 nm are formed and that their morphology depends on the nature of the ionic liquid. For example, nanoparticles grown in TSIL **92c** are perfectly separated, while those produced in [bmim][BF$_4$] are aggregated, forming nanoclusters of $\sim$30 nm. These observation are considered as evidence of the stabilizing effect exerted by the nitrile groups in TSILs; in particular nitrile groups coordinate the nanoparticle surface and prevent their aggregation. A major benefit is enjoyed by catalyst recovery and reuse; indeed the ionic liquid phase containing Pd-particles can be recycled up to ten times, maintaining almost the same activity ($\sim$90% yield of styrene).[213,214]

What the true catalyst is in Pd-catalysed cross coupling reactions is still an open problem and this topic has been the subject of recent review articles.[215] Metal nanoparticles (Met · NPs) prepared in and stabilized by ILs are emerging as catalysts for reactions under multiphase conditions.[214,216]

In ILs, indeed, conditions capable of stabilizing palladium nanoparticles (Pd · NPs) are easily achieved.[217] Scheme 1.61A depicts a shell of ions formed around the NP, which stabilizes it by an "electrosteric" (electrostatic plus steric) effect. In Scheme 1.61B stabilization is ensured by a task-specific ionic liquid *via* surface-ligation by suitable groups, *e.g.* nitriles.[210] The effect of the shell composed of either covalently bound ligands or associated ion provides solubility and stability. The thicker the ligand shell, the higher is the interparticle spacing and the lower is the tendency to agglomeration and cluster growth.

For example, Pd · NPs stabilized by [bmim][PF$_6$] and generated from [PdCl$_2$(cod)] under hydrogen catalysed an efficient Suzuki cross-coupling reaction, with no palladium leaching and up to ten cycles.[218] Surprisingly, the Pd catalyst "improves" upon lowering of the Pd loading. This observation was rationalized in terms of an equilibrium between Pd · NPs serving as a catalyst reservoir and small monomeric or dimeric catalytically active palladium

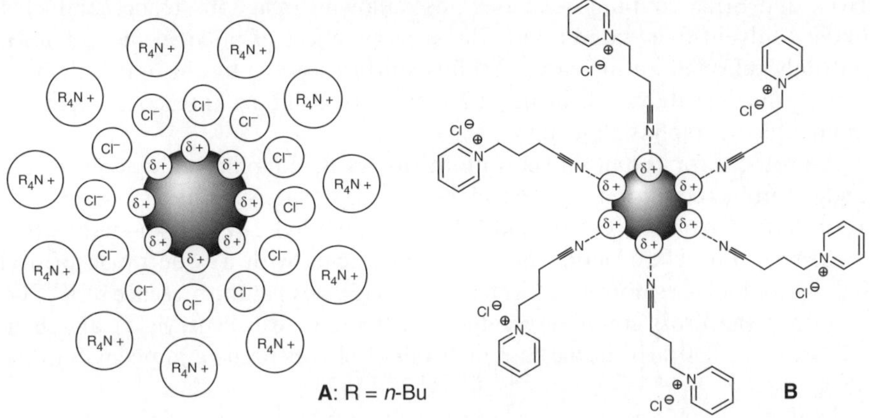

Scheme 1.61

Scheme 1.62

entities, which, when the Pd concentration is too high, agglomerate into inactive Pd black.[219]

As a rule of thumb, the larger the surface area, the smaller the Pd NP, and the greater its catalytic activity. Polymers such as polyvinylpyrrilidone (PVP), poly(vinyl alcohol) (PVA) and copolymers stabilize Pd · NPs, allowing us to approach an optimum balance between stability and catalytic activity. Ionic polymers have been also used. Dyson, for example, reduced a solution of $PdCl_2$ in a homogeneous mixture of poly[1-methyl-3-(4-vinylbenzyl)imidazolium] [NTf₂] (**96**, Scheme 1.62) and TSIL **92d** (see Scheme 1.60) with $NaBH_4$. A black solution results, consisting of Pd NPs with an average size of 5.0 ± 0.2 nm, as established by TEM.[220]

The solution was inert to moisture and could be stored for 2 years with no trace of degradation. These Pd · NPs were excellent pre-catalysts for Suzuki,

Heck and Stille coupling reactions, also allowing reactions to be conducted under "solvent-free" conditions. The synergic effect of polymer **96** and TSIL not only reflected an increased stability during storage but also an enhanced activity during catalysis. Scheme 1.62 shows a typical Suzuki reaction occurring in an aqueous biphasic medium.

A protocol for room temperature Suzuki cross-coupling in aqueous media under aerobic conditions was developed using a water-soluble, air-stable catalyst derived from $Pd(OAc)_2$ and 1,1,3,3-tetramethyl-2-N-butylguanidine (**97**) (Scheme 1.63). The reaction of arylboronic acids with a wide range of aryl halides took place smoothly in water/ethanol (3:2 v/v) in the presence of K_2CO_3 to afford the cross-coupling products in good to excellent yields and high TONs, up to $850\,00\,h^1$ in the case of 1-iodo-4-nitrobenzene and phenylboronic acid.[221]

In the previous sections we discussed the advantages of using hydrophilic catalysts in water or in aqueous-biphasic systems, which provide the opportunity to easily recover the catalyst soluble in water from the organic product streams dissolved in the hydrophobic phase, and potentially recycle it. The feasibility of cross-coupling reactions catalysed by hydrophilic palladium catalysts has been reviewed recently.[222] The first experiments used TPPTS or catalysts derived from TPPTS (**37**, Section 1.3.1) and other structurally similar hydrophilic phosphines, which were effectively applied to cross coupling protocols involving aryl iodides and activated aryl bromides.[223] The reviews by Shaughnessy focus on hydrophilic ligands developed up to 2005 that are

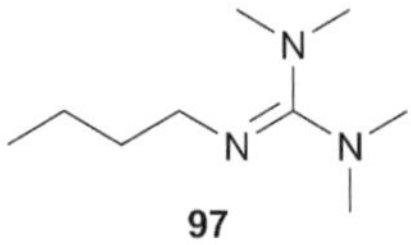

97

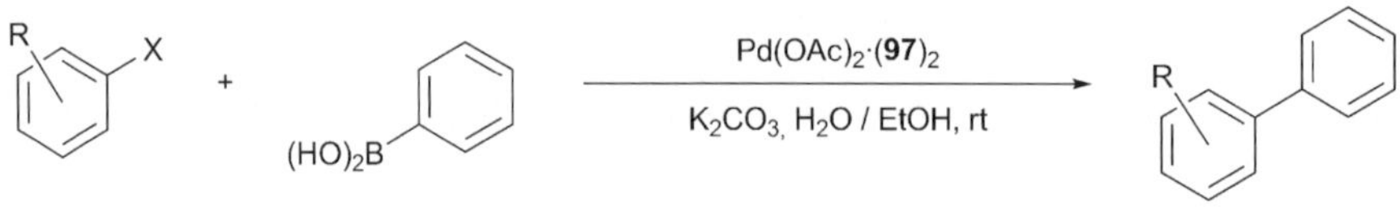

R	X	Pd (mol%)	t (h)	yield (%)
H	I	0.001	20	99
MeO	I	0.001	30	72
NO$_2$	I	0.0001	20	85
H	Br	0.001	30	35
H	Br	0.1	2	88
MeO	Br	1.0	1	98

Scheme 1.63

effective in coupling reactions of aryl bromides and chlorides under mild conditions in aqueous solvents.[222]

The first examples of water-soluble alkylphosphines that were capable of promoting Suzuki couplings of aryl bromides in aqueous solvents at room-temperature were monodentate phosphines **98** and **99** (Scheme 1.64). Catalysts derived from them showed high activity towards aryl bromides, with TONs up to 10 000 at rt, which increased up to 734 000 at 80 °C, while the more sterically demanding ligand **99** gave catalysts with a superior activity towards aryl chlorides.[224]

Phosphine **98** combined with palladacycles **100** and **101** gave hydrophilic active species for the Suzuki coupling of aryl bromides (Scheme 1.65). Deactivated aryl bromides, *e.g.* 4-bromoanisole at 80 °C after 4 h with catalyst loadings as low as 0.02 mol.% Pd, gave high yields of coupled biphenyls using the **100/98** system. The catalyst derived from **101/98** was efficiently reused for a total of 12 reaction cycles (80 °C, 1 h) with an average yield of 98%. This degree of recyclability is one of the highest in the literature for a Suzuki coupling of an aryl bromide using a homogeneous, hydrophilic catalyst system. The role of the palladacycle precursor in determining the catalyst activity and stability is essential.[225]

A different solution to the use of hydrophilic ligands is possible for the Suzuki reaction of aryl bromides in water. The complex Pd(dppf)Cl$_2$ (dppf = 1,1'-bis(diphenylphosphino)ferrocene, **102**) was indeed an extremely active catalyst, simply using an aqueous solution of K$_2$CO$_3$ as the reaction medium. Excellent yields and TONs up to 870 000 were recorded for the reaction of 1-bromo-4-nitrobenzene and phenylboronic acid at 110 °C for 48 h (Scheme 1.66).[226] The choice of the ferrocenyl ligand in Pd(dppf)Cl$_2$, a weakly electron-

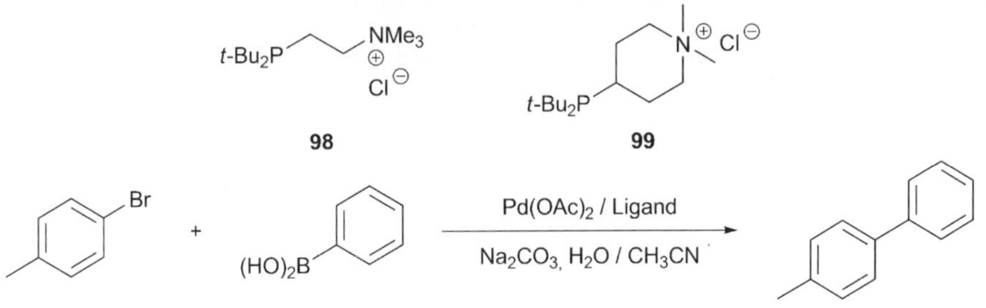

Ligand	Pd (mol%)	T (°C)	t (h)	yield (%)	TON
99	0.0098	23	24	> 99	10100
98	0.00103	80	4	> 99	95600
99	0.00101	80	4	> 99	97800
99	0.0000995	80	4	73	734000

Scheme 1.64

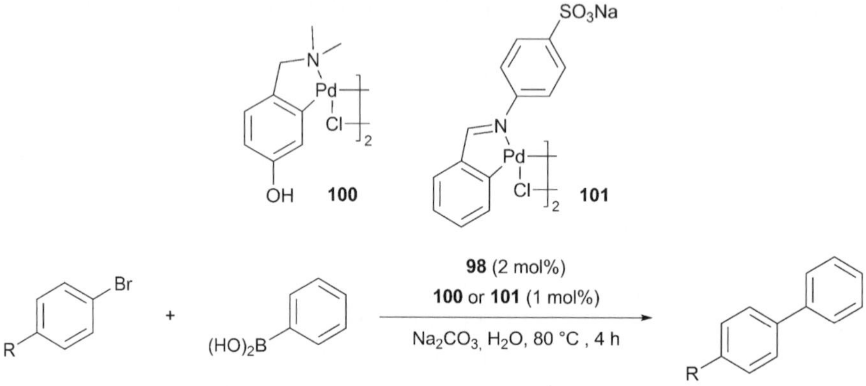

SO3Na
N
Pd
Cl
2
OH
100
N
Pd
Cl
2
101
Br
R
+
(HO)2B
98 (2 mol%)
100 or 101 (1 mol%)
Na2CO3, H2O, 80 °C , 4 h
R

Palladacycle	R	yield (%)
100	OCH_3	95
101	OCH_3	93
100	OH	90
101	OH	95
101	CH_3	90

Scheme 1.65

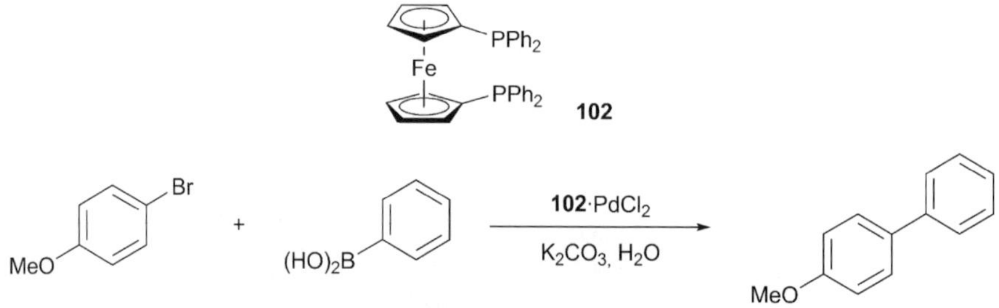

PPh2
Fe
PPh2
102
Br
MeO
+
(HO)2B
102·PdCl2
K2CO3, H2O
MeO

102 (mol%)	T (°C)	t (h)	conversion / yield (%)
0.1	60	0.5	88 / 85
0.1	80	0.5	100 / 98
0.01	110	3	98 / 96
0.001	110	48	82 / 78

Scheme 1.66

donating bidentate triarylphosphine with a rigid and blocked conformation, imparted activity and stability to the catalyst.

Recovery and reuse of **102** was also possible by using a 20% aq. solution of PEG-2000 as the reaction media for the cross-coupling of 2-bromotoluene with phenylboronic acid. After stirring for 1 h, 2-methylbiphenyl was extracted from

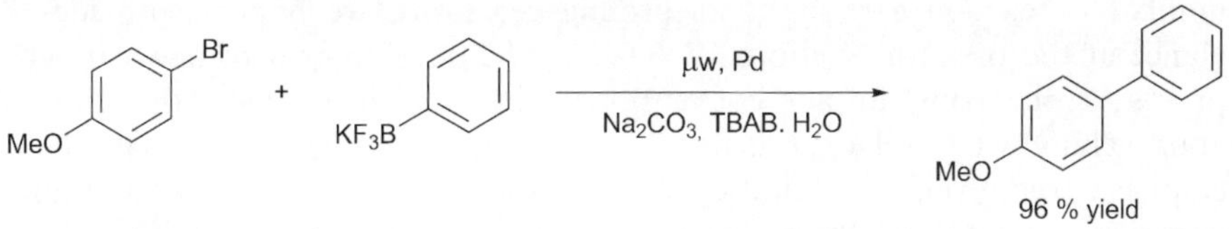

Scheme 1.67

the reaction mixture with *n*-pentane, allowing the recycling of the palladium catalyst three times without any significant loss of catalytic activity.

However, efforts were also directed to the development of phosphine-free catalysts. For example, the ligand-free $Pd(OAc)_2$ promoted Suzuki reaction in water was reported using microwave heating. In this way, a low palladium loading (0.4 mol.%) was required and the cross coupling proceeded quickly (5–10 min reaction time), using boronic acids and aryl iodides, bromides and chlorides.[227]

Potassium aryltrifluoroborates proved to be useful candidates for microwave-promoted Suzuki reactions in aqueous environments (Scheme 1.67).[228]

Reactions were run in a sealed tube, using 1 mmol aryl halide, 1.0 mmol potassium phenyltrifluoroborate, 2.5 ppm Pd, 3.7 mmol of Na_2CO_3, in ethanol/water (2 mL, 1:1 v/v). Applying a 300 W microwave irradiation, the temperature was ramped to 150 °C and held at this value for 5 min, at which point the conversions were complete.

Finally, a Suzuki–Miyaura cross-coupling of *p*-tolylboronic acid and iodobenzene to form 4-phenyltoluene using $scCO_2$ as solvent is reported. The reaction is carried out in continuous-flow mode using a packed high-pressure column containing Pd(II) EnCatTM, which consists of $Pd(OAc)_2$ encapsulated in a porous polyurea matrix. Experiments were carried out at 80–100 °C over pressure ranges up to 250 bar, using a $scCO_2$ stream enriched with additional methanol necessary to solubilize the base Bu_4NOMe. A 81% conversion was obtained under optimized conditions.[229]

1.6 Concluding Remarks

At the beginning of the 21st century organic synthesis is a mature science, capable of assembling molecular structures of exceptional complexity using the tools now available.[230] Paraphrasing Noyori, the chemical community is facing now a main challenge, that of pursuing "practical elegance", that means conjugating efficiency to sustainability.[231] Thus, the synthesis of a molecule is no longer the unique goal, but the way we produce it must be environmentally compatible, or, as Noyori writes, "must fall within the boundary conditions of our planet: it must be safe, environmentally benign, and reasonable in terms of cost, resources, and energy. Without attention to what is now called green chemistry, chemical manufacturing will be unsustainable in this century." Even though, for obvious reasons, we have dealt in this chapter with a limited number of carbon–carbon bond-forming reactions, we hope it provides a glance at the breadth of efforts directed to the development of new catalytic processes occurring in non-conventional reaction media with the aim of approaching what Gladysz defined as the "ideal catalyst".[9] In particular, emphasis was given to a selection of processes that show how the molecular design of new catalytically active species is driven by the quest for efficiency and recyclability. Biphasic catalysis seems to be a powerful answer to this quest with its challenging rational design aimed simultaneously at optimizing the activity/selectivity performance and the best partitioning profile between different phases.[19] Water, scCO$_2$ and ionic liquids afford green opportunities to implement catalysis. In particular, it is our firm belief that the potential in catalysis of ionic liquids, thanks to the easy tuning of their chemical and physical properties, is still unexplored and will provide a powerful tool to fulfil the demand of meeting "practical elegance".

Glossary

acac	acetylacetonate
ADMET	acyclic diene metathesis polymerization
API	active pharmaceutical ingredients
[bdmim]	1-butyl-2,3-dimethylimidazolium
[bmim]	1-butyl-3-methylimidazolium
Boc	*t*-butoxycarbonyl
Cbz	carbobenzyloxy
CM	cross metathesis
Cod	1,5-cyclooctadiene
CTAB	cetyltrimethylammonium bromide
DABCO	1,4-diazabicyclo[2.2.2]octane
DiMePEG	dimethyl poly(ethylene glycol)
DME	1.2-dimethoxyethane
DMSO	dimethyl sulfoxide
DMP	1,3-dimethyl-2-pyrrolidone
DMF	*N*,*N*-dimethylformamide
Dppf	1,1′-bis(diphenylphosphino)ferrocene
[epy]	1-ethylpyridinium
[hemim]	1-(2-hydroxyethyl)-3-methylimidazolium
hfacac	hexafluoroacetylacetonate
IL	ionic liquid
Mes	2,4,6-trimethylphenyl
NP	nanoparticle
NHC	*N*-heterocyclic carbenes
[omin]	1-octyl-3-methylimidazolium
PDMS	polydimethyl siloxane
PEG	poly(ethylene glycol)
[pentylmim]	1-pentyl-3-methylimidazolium

phen	1,10-phenanthroline
PMP	*p*-methoxyphenyl
[propylmim]	1-propyl-3-methylimidazolium
PVA	poly(vinyl alcohol)
PVP	polyvinylpyrrolidone
RCM	ring-closing metathesis
ROM	ring-opening metathesis
ROMP	ring-opening metathesis polymerization
scCO$_2$	supercritical CO$_2$
SILP	supported ionic liquid phase
TEM	transmission electron microscopy
TFA	trifluoroacetic acid
Tf$_2$N	bis(trifluoromethylsulfonyl)imide
TfO	trifluoromethanesulfonate
TMS	trimethylsilyl
TOF	turnover frequency
TOMAC	trioctylmethylammonium chloride
TON	turnover number
TPPTS	tris(*m*-sulfonatophenyl)phosphine
TS	transition state
TSIL	task-specific ionic liquid
VOC	volatile organic compound

References

1. B. M. Trost, *Science*, 1991, **254**, 1471.
2. R. A. Sheldon, *Green Chem.*, 2007, **9**, 1273.
3. (a) J. Augé, *Green Chem.*, 2008, **10**, 225; (b) J. Andraos, *Org. Process Res. Dev.*, 2005, **9**, 149; (c) D. J. C. Constable, A. D. Curzons and V. L. Cunningham, *Green Chem.*, 2002, **4**, 521; (d) D. J. C. Constable, A. D. Curzons, L. M. Freitas dos Santos, G. R. Green, R. E. Hannah, J. D. Hayler, J. Kitteringham, M. A. McGuire, J. E. Richardson, P. Smith, R. L. Webb and M. Yu, *Green Chem.*, 2001, **3**, 7; (e) A. D. Curzons, D. J. C. Constable, D. N. Mortimer and V. L. Cunningham, *Green Chem.*, 2001, **3**, 1; (f) T. Hudlicky, D. A. Frey, L. Koroniak, C. D. Claeboe and L. E. Brammer, *Green Chem.*, 1999, **1**, 57.
4. http://www.epa.gov/greenchemistry/pubs/pgcc/presgcc.html.
5. M. Mours, *ChemSusChem*, 2008, **1**, 59.
6. http://www.ccrhq.org/vision/.
7. P. T. Anastas, J. C. Warner, *Green Chemistry: Theory and Practice*, Oxford University Press, New York, 1998. In 2008 the 24 Principles of Green Engineering and Green Chemistry have been revised: see S. Tang, R. Bourne, R. Smith and M. Poliakoff, *Green Chem.*, 2008, **10**, 268.
8. K. B. Sharpless, *Adv. Synth. Catal.*, 2007, **349**, 2533.
9. J. A. Gladysz, *Pure Appl. Chem.*, 2001, **73**, 1319.

10. N. E. Leadbeater and M. Marco, *Chem. Rev.*, 2002, **102**, 3217.

11. P. T. Anastas and M. M. Kirchhoff, *Acc. Chem. Res.*, 2002, **35**, 686.

12. R. K. Grasselli, *Catalysis Today*, 2005, **99**, 23.

13. H. C. Kolb and K. B. Sharpless, *Drug Discovery Today*, 2003, **8**, 1128.

14. R. K. Grasselli, *Topics Catal.*, 2002, **21**, 79.

15. C. Coperet, M. Chabanas, R. P. Saint-Arroman and J-M. Basset, *Angew. Chem. Int. Ed.*, 2003, **42**, 156.

16. J-M. Basset, C. Coperet, D. Soulivong, M. Taoufik and J. Thivolle-Cazat, *Angew. Chem. Int. Ed.*, 2006, **45**, 6082.

17. (a) *Multiphase Homogeneous Catalysis*, eds. B. Cornils, W. A. Herrmann, I. T. Horvath, W. Leitner, S. Mecking, H. Olivier-Bourbigou, D. Vogt, Wiley-VCH, Weinheim, 2005; (b) *Recovery and Recycling in Homogeneous Catalysis*, eds. D. J. Cole-Hamilton, R. P. Tooze, Kluwer, Dordrecht, 2005; (c) M. Benaglia, A. Puglisi and F. Cozzi, *Chem. Rev.*, 2003, **103**, 3401; (d) F. Cozzi, *Adv. Synth. Catal.*, 2006, **348**, 1367.

18. P. Tundo and A. Perosa, *Chem. Soc. Rev.*, 2007, **36**, 532.

19. For an industrial application see the Shell higher olefin process (SHOP): W. Keim, *Green Chem.*, 2003, 5, 105.

20. R. Šebesta, I. Kmentová and Š. Toma, *Green Chem.*, 2008, **10**, 484.

21. B. Cornils and E. G. Kuntz, *J. Organomet. Chem.*, 1995, **502**, 177.

22. (a) C. M. Gordon and W. Leitner, in: *Catalyst Separation, Recovery and Recycling.*, eds. D. Cole-Hamilton, R. Tooze, Springer, Netherlands, 2006, pp. 215-236; (b) W. Leitner, *Acc. Chem. Res.*, 2002, **35**, 746.

23. (a) P. E. Savage, J. Dunn and J. Yu, *Comb. Sci. Techn.*, 2006, **178**, 443; (b) P. E. Savage, *Chem. Rev.*, 1999, **99**, 603.

24. (a) C-J. Li, *Chem. Rev.*, 2005, **105**, 3095; (b) R. Poli, *Chem Eur. J.*, 2004, **10**, 332; (c) U. M. Lindström, *Chem. Rev.*, 2002, **102**, 2751.

25. (a) H. Weingärtner, *Angew. Chem. Int. Ed.*, 2008, **47**, 654; (b) X. Chen, X. Li, A. Hu and F. Wang, *Tetrahedron: Asymm.*, 2008, **19**, 1; (c) V. I. Parvulescu and C. Hardacre, *Chem Rev.*, 2007, **107**, 2615; (d) S. Chowdhury, R. S. Mohan and J. L. Scott, *Tetrahedron*, 2007, **63**, 2363; (e) R. Giernoth, *Top. Curr. Chem.*, 2007, **276**, 1; (f) Z. C. Zhang, *Adv. Catal.*, 2006, **49**, 153; (g) T. Welton, *Coordin. Chem Rev.*, 2004, **248**, 2459.

26. (a) P. J. Walsh, H. Li and C. A. de Parrodi, *Chem. Rev.*, 2007, **107**, 2503; (b) K. Tanaka and F. Toda, *Chem Rev.*, 2000, **100**, 1025.

27. E. Garcia-Verdugo, J. Fraga-Dubreuil, P. A. Hamley, W. B. Thomas, K. Whiston and M. Poliakoff, *Green Chem.*, 2005, **7**, 294.

28. D. C. Rideout and R. Breslow, *J. Am. Chem. Soc.*, 1980, **102**, 7816.

29. A. Lubineau, J. Augé and Y. Queneau, *Synthesis*, 1994, 741.

30. (a) H. L. Ngo, A. G. Hu and W. B. Lin, *Chem. Commun.*, 2003, 1912; (b) S-G. Lee, Y. J. Zhang, J. Y. Piao, H. Yoon, C. E. Song, J. H. Choi and J. Hong, *Chem. Commun.*, 2003, 2624.

31. (a) A. Dondoni and A. Massi, *Angew. Chem. Int. Ed.*, 2008, **47**, 4638; (b) C. F. Barbas III, *Angew. Chem. Int. Ed.*, 2008, **47**, 42.

32. (a) S. Mukherjee, J. W. Yang, S. Hoffmann and B. List, *Chem. Rev.*, 2007, **107**, 5471; (b) D. Enders and C. Grondal, *Angew. Chem. Int. Ed.*, 2005, **44**, 1210; (c) J. Casas, M. Engqvist, I. Ibrahem, B. Kaynak and A. Córdova, *Angew. Chem. Int. Ed.*, 2005, **44**, 1343; (d) W. Notz, F. Tanaka and C. F. Barbas III, *Acc. Chem. Res.*, 2004, **37**, 580; (e) A. B. Northrup and D. W. C. MacMillan, *J. Am. Chem. Soc.*, 2002, **124**, 6798; (f) M. Movassaghi and E. N. Jacobsen, *Science*, 2002, **298**, 1904; (g) K. Sakthivel, W. Notz, T. Bui and C. F. Barbas III, *J. Am. Chem. Soc.*, 2001, **123**, 5260; (h) W. Notz and B. List, *J. Am. Chem. Soc.*, 2000, **122**, 7386; (i) B. List, R. A. Lerner and C. F. Barbas III, *J. Am. Chem. Soc.*, 2000, **122**, 2395.

33. (a) A. Córdova, W. Zou, P. Dziedzic, I. Ibrahem, E. Reyes and Y. Xu, *Chem. Eur. J.*, 2006, **12**, 5383; (b) A. Bassan, W. Zou, E. Reyes, F. Himo and A. Córdova, *Angew. Chem. Int. Ed.*, 2005, **44**, 7028; (c) A. Córdova, W. Zou, I. Ibrahem, E. Reyes, M. Engqvist and W-W. Liao, *Chem Commun.*, 2005, 3586.

34. (a) B. Rodríguez, A. Bruckmann and C. Bolm, *Chem. Eur. J.*, 2007, **13**, 4710; (b) B. Rodríguez, T. Rantanen and C. Bolm, *Angew. Chem. Int. Ed.*, 2006, **45**, 6924.

35. C. Allemann, R. Gordillo, F. R. Clemente, P. Ha-Yeon Cheong and K. N. Houk, *Acc. Chem. Res.*, 2004, **37**, 558.

36. N. Zotova, A. Franzke, A. Armstrong and D. G. Blackmond, *J. Am. Chem. Soc.*, 2007, **129**, 15100.

37. (a) A. Córdova, W. Notz and C. F. Barbas III, *Chem Commun.*, 2002, 3024; (b) T. J. Dickerson and K. D. Janda, *J. Am. Chem. Soc.*, 2002, **124**, 3220; (c) D. B. Ramachary, N. S. Chowdari and C. F. Barbas III, *Tetrahedron Lett.*, 2002, **43**, 6743.

38. Y. Hayashi, S. Aratake, T. Itoh, T. Okano, T. Sumiya and M. Shoji, *Chem. Commun.*, 2007, 957.

39. Z. Jiang, Z. Liang, X. Wua and Y. Lu, *Chem. Commun.*, 2006, 2801.

40. For a discussion on acceleration effects on the Diels–Alder cycloadditions and Claisen rearrangements of hydrophobic compounds in dilute aqueous solutions, see: S. Narayan, J. Muldoon, M. G. Finn, V. V. Fokin, H. C. Kolb and K. B. Sharpless, *Angew. Chem. Int. Ed.*, 2005, **44**, 3275.

41. X-H. Chen, S-W. Luo, Z. Tang, L-F. Cun, A-Q. Mi, Y-Z. Jiang and L-Z. Gong, *Chem. Eur. J.*, 2007, **13**, 689.

42. Y. Jung and R. A. Marcus, *J. Am. Chem. Soc.*, 2007, **129**, 5492; See also (a) Y. R. Shen and V. Ostroverkhov, *Chem. Rev.*, 2006, **106**, 1140; (b) J. E. Klijn and J. B. F. N. Engeberts, *Nature*, 2005, **435**, 746.

43. L. Zhong, Q. Gao, J. Gao, J. Xiao and C. Li, *J. Catal.*, 2007, **250**, 360. For a general review on microemulsions, see: K. Holmberg, *Eur. J. Org. Chem.*, 2007, 731.

44. (a) S. Aratake, T. Itoh, T. Okano, N. Nagae, T. Sumiya, M. Shoji and Y. Hayashi, *Chem. Eur. J.*, 2007, **13**, 10246; (b) Y. Hayashi, T. Sumiya, J.

Takahashi, H. Gotoh, T. Urushima and M. Shoji, *Angew. Chem. Int. Ed.*, 2006, **45**, 958.

45. Y. Hayashi, S. Aratake, T. Okano, J. Takahashi, T. Sumiya and M. Shoji, *Angew. Chem. Int. Ed.*, 2006, **45**, 5527.

46. J. Huang, X. Zhang and D. W. Armstrong, *Angew. Chem. Int. Ed.*, 2007, **46**, 9073.

47. K. Liu, D. Häussinger and W-D. Woggon, *Synlett*, 2007, 2298.

48. N. Mase, Y. Nakai, N. Ohara, H. Yoda, K. Takabe, F. Tanaka and C. F. Barbas III, *J. Am. Chem. Soc.*, 2006, **128**, 734.

49. L. Zu, H. Xie, H. Li, J. Wang and W. Wang, *Org. Lett.*, 2008, **10**, 1211.

50. D. Gryko and W. J. Saletra, *Org. Biomol. Chem.*, 2007, **5**, 2148.

51. (a) G. Guillena, M. D. C. Hita and C. Nájera, *Tetrahedron: Asymm.*, 2006, **17**, 1493; (b) S. Guizzetti, M. Benaglia, L. Raimondi and G. Celentano, *Org. Lett.*, 2007, **9**, 1247.

52. (a) C. Wang, Y. Jiang, X-X. Zhang, Y. Huang, B-G. Li and G-L. Zhang, *Tetrahedron Lett.*, 2007, **48**, 4281; (b) A. Russo, G. Botta and A. Lattanzi, *Tetrahedron*, 2007, **63**, 11886.

53. Y. Wu, Y. Zhang, M. Yu, G. Zhao and S. Wang, *Org. Lett.*, 2006, **8**, 4417.

54. Y.-Q. Fu, Z.-C. Li, L.-N. Ding, J.-C. Tao, S.-H. Zhang and M.-S. Tang, *Tetrahedron: Asymm.*, 2006, **17**, 3351. See also: S. Sathapornvajana and T. Vilaivan, *Tetrahedron*, 2007, **63**, 10253.

55. V. Maya, M. Raj and V. K. Singh, *Org. Lett.*, 2007, **9**, 2593.

56. (a) X. Wu, Z. Jiang, H-M. Shen and Y. Lu, *Adv. Synth. Catal.*, 2007, **349**, 812; (b) Y-C. Teo, *Tetrahedron: Asymm.*, 2007, **18**, 1155; (c) S. S. V. Ramasastry, K. Albertshofer, N. Utsumi, F. Tanaka and C. F. Barbas III, *Angew. Chem. Int. Ed.*, 2007, **46**, 5572.

57. S. S. V. Ramasastry, K. Albertshofer, N. Utsumi and C. F. Barbas III, *Org. Lett.*, 2008, **10**, 1621.

58. M. Gruttadauria, F. Giacalone, A. Mossuto Marculescu, P. Lo Meo, S. Riela and R. Noto, *Eur. J. Org. Chem.*, 2007, 4688. For a similar approach see: D. Font, C. Jimeno and M. A. Pericàs, *Org. Lett.*, 2006, **8**, 4653.

59. (a) R. Šebesta, I. Kmentová and Š. Toma, *Green Chem.*, 2008, **10**, 484; (b) A. G. M. Barrett, B. T. Hopkins and J. Köbberling, *Chem. Rev.*, 2002, **102**, 3301; (c) J-I. Yoshida and K. Itami, *Chem. Rev.*, 2002, **102**, 3693; (d) S. Bhattacharyya, *Curr. Opin. Drug Discovery Dev.*, 2004, **7**, 752.

60. T. Ooi and K. Maruoka, *Angew. Chem. Int. Ed.*, 2007, **46**, 4222.

61. (a) *Ionic Liquids in Synthesis.* eds. P. Wasserscheid, T. Welton, 2008, Wiley-VCH, Weinheim; (b) D. Zhao and P. J. Dyson, *Chem. Eur. J.*, 2006, **12**, 2122; (c) S. -G. Lee, *Chem. Commun.*, 2006, 1049.

62. W. Miao and T. H. Chan, *Acc. Chem. Res.*, 2006, **39**, 897.

63. V. I. Pârvulescu and C. Hardacre, *Chem. Rev.*, 2007, **107**, 2615.

64. (a) P. Kotrusz, I. Kmentová, B. Gotov, Š. Toma and E. Solcániová, *Chem. Commun.*, 2002, 2510; (b) T-P. Loh, L-C. Feng, H-Y. Yang and J-Y. Yiang, *Tetrahedron Lett.*, 2002, **43**, 8741.

65. C. P. Mehnert, *Chem. Eur. J.*, 2005, **11**, 50.

66. A. Riisager, R. Fehrmann, M. Haumann and P. Wasserscheid, *Eur. J. Inorg. Chem.*, 2006, 695.

67. M. Gruttadauria, S. Riela, C. Aprile, P. Lo Meo, F. D'Anna and R. Noto, *Adv. Synth. Catal.*, 2006, **348**, 82.

68. M. Lombardo, F. Pasi, S. Easwar and C. Trombini, *Adv. Synth. Catal.*, 2007, **349**, 2061.

69. A. Ting and S. E. Schaus, *Eur. J. Org. Chem.*, 2007, 5797.

70. N. S. Chowdari, D. B. Ramachary and C. F. Barbas III, *Synlett*, 2003, 1906.

71. Y-C. Teo, J-J. Lau and M-C. Wu, *Tetrahedron: Asymm.*, 2008, **19**, 186.

72. Y. Hayashi, T. Urushima, S. Aratake, T. Okano and K. Obi, *Org. Lett.*, 2008, **10**, 21.

73. O. Marianacci, G. Micheletti, L. Bernardi, F. Fini, M. Fochi, D. Pettersen, V. Sgarzani and A. Ricci, *Chem. Eur. J.*, 2007, **13**, 8338.

74. (a) A. Erkkilä, I. Majander and P. M. Pihko, *Chem. Rev.*, 2007, **107**, 5416; (b) K. A. Ahrendt, C. J. Borths and D. W. C. MacMillan, *J. Am. Chem. Soc.*, 2000, **122**, 9874.

75. A. Carlone, M. Marigo, C. North, A. Landa and K. A. Jørgensen, *Chem. Commun.*, 2006, 4928.

76. A. Carlone, S. Cabrera, M. Marigo and K. A. Jørgensen, *Angew. Chem. Int. Ed.*, 2007, **46**, 1101.

77. C. Palomo, A. Landa, A. Mielgo, M. Oiarbide, A. Puente and S. Vera, *Angew. Chem. Int. Ed.*, 2007, **46**, 8431.

78. S. Sulzer-Mossé and A. Alexakis, *Chem. Commun.*, 2007, 3123.

79. S. Zhu, S. Yu and D. Ma, *Angew. Chem. Int. Ed.*, 2008, **47**, 545.

80. Z-Y. Yan, Y-N. Niu, H-L. Wei, L-Y. Wu, Y-B. Zhao and Y-M. Liang, *Tetrahedron: Asymm.*, 2006, **17**, 3288.

81. E. Alza, X. C. Cambeiro, C. Jimeno and M. A. Pericàs, *Org. Lett.*, 2007, **9**, 3717.

82. V. Maya and V. K. Singh, *Org. Lett.*, 2007, **9**, 1117.

83. Y-J. Cao, Y-Y. Lai, X. Wang, Y-J. Li and W-J. Xiao, *Tetrahedron Lett.*, 2007, **48**, 21.

84. B. Ni, Q. Zhang and A. D. Headley, *Tetrahedron Lett.*, 2008, **49**, 1249.

85. S. Luo, X. Mi, L. Zhang, S. Liu, H. Xu and J-P. Cheng, *Angew. Chem. Int. Ed.*, 2006, **45**, 3093.

86. S. Luo, X. Mi, S. Liu, H. Xu and J-P. Cheng, *Chem. Commun.*, 2006, 3687.

87. S. Luo, L. Zhang, X. Mi, Y. Qiao and J-P. Cheng, *J. Org. Chem.*, 2007, **72**, 9350.

88. L-Y. Wu, Z-Y. Yan, Y-X. Xie, Y-N. Niu and Y-M. Liang, *Tetrahedron: Asymm.*, 2007, **18**, 2086.

89. P. Kotrusz, Š. Toma, H-G. Schmalz and A. Adler, *Eur. J. Org. Chem.*, 2004, 1577.

90. D-Q. Xu, B-T. Wang, S-P. Luo, H-D. Yue, L-P. Wang and Z-Y. Xu, *Tetrahedron: Asymm.*, 2007, **18**, 1788.

91. D. Basavaiah, A. J. Rao and T. Satyanarayana, *Chem. Rev.*, 2003, **103**, 811.

92. G. Masson, C. Housseman and J. Zhu, *Angew. Chem. Int. Ed.*, 2007, **46**, 4614.

93. H. J. Davies, A. M. Ruda and N. C. O. Tomkinson, *Tetrahedron Lett.*, 2007, **48**, 1461.

94. A. Porzelle, C. M. Williams, B. D. Schwartz and I. R. Gentle, *Synlett*, 2005, 2923.

95. C. A. M. Afonso, L. C. Branco, N. R. Candeias, P. M. P. Gois, N. M. T. Lourenc, N. M. M. Mateus and J. N. Rosa, *Chem. Commun.*, 2007, 2669.

96. S-H. Zhaoa, H-R. Zhang, L-H. Feng and Z-B. Chen, *J. Mol. Catal. A: Chem.*, 2006, **258**, 251.

97. X. Mi, S. Luo, H. Xu, L. Zhanga and J-P. Ghenga, *Tetrahedron*, 2006, **62**, 2537.

98. B. Pégot, G. Vo-Thanh, D. Gori and A. Loupy, *Tetrahedron Lett.*, 2004, **45**, 6425.

99. R. Gausepohl, P. Buskens, J. Kleinen, A. Bruckmann, C. W. Lehmann, J. Klankermayer and W. Leitner, *Angew. Chem. Int. Ed.*, 2006, **45**, 3689.

100. (a) J. P. Hallett, P. Pollet, C. L. Liotta and C. A. Eckert, *Acc. Chem. Res.*, 2008, **41**, 458; (b) S. Keskin, D. Kayrak-Talay, U. Akman and O. Hortacsu, *J. Supercrit Fluids*, 2007, **43**, 150; (c) P. J. Walsh, H. Li and C. Anaya de Parrodi, *Chem. Rev.*, 2007, **107**, 2503; (d) S. Liu and J. Xiao, *J. Mol. Catal. A*, 2007, **270**, 1; (e) C. K. Z. Andrade and L. M. Alves, *Curr. Org. Chem.*, 2005, **9**, 195; (f) F. Alonso, I. P. Beletskaya and M. Yus, *Tetrahedron*, 2005, **61**, 11771; (g) T. Welton, *Coordin. Chem. Rev.*, 2004, **248**, 2459; (h) P. G. Jessop, F. Jo'o and C. C. Tai, *Coord. Chem. Rev.*, 2004, **248**, 2425; (i) D. Prajapati and M. Gohain, *Tetrahedron*, 2004, **60**, 85; (j) Q-H. Fan, Y-M. Li and A. S. C. Chan, *Chem. Rev.*, 2002, **102**, 3385.

101. R. T. Baker, S. Kobayashi and W. Leitner, *Adv. Synth. Catal.*, 2006, **348**, 1337.

102. W. Keim, *Green Chem.*, 2003, **5**, 105.

103. (a) *Modern Carbonylation Methods*, L. Kollár, Ed., 2008 Wiley-VCH, Weinheim; (b) D. Evans, J. A. Osborn and G. Wilkinson, *J. Chem. Soc. A*, 1968, **33**, 3133.

104. T. V. Choudhary and V. R. Choudhary, *Angew. Chem. Int. Ed.*, 2008, **47**, 1828.

105. W. A. Herrmann, J. A. Kulpe, W. Konkol and H. Bahrmann, *J. Organomet. Chem.*, 1990, **389**, 85.

106. H. Bahrmann, H. Bach, C. D. Frohning, H-J. Kleiner, P. Lappe, D. Peters, D. Regnat and W. A. Herrmann, *J. Mol. Catal.*, 1997, **116**, 49.

107. W. A. Herrmann, C. W. Kohlpaintner, H. Bahrmann and W. Konkol, *J. Mol. Catal.*, 1992, **73**, 191.

108. C. Deng, G. Ou, J. She and Y. Yuan, *J. Mol. Catal. A*, 2007, **270**, 76.

109. Q. Peng, X. Liao and Y. Yuan, *Catal. Commun.*, 2004, **5**, 447.

110. Y. Wang, J. Jiang, Q. Miao, X. Wu and Z. Jin, *Catal. Today*, 2002, **74**, 85.

111. Johnson-Matthey Co. (M. J. H. Russel, B. A. Murrer, US 4.399.312 (1983).

112. C. Liu, J. Jiang, Y. Wang, F. Cheng and Z. Jin, *J. Mol. Catal. A*, 2003, **198**, 23.

113. D. Baskakov and W. A. Herrmann, *J. Mol. Catal. A*, 2008, **283**, 166.

114. H. Klein, R. Jackstell and M. Beller, *Chem. Commun.*, 2005, 2283.

115. R. V. Chaudhari, B. M. Bhanage, R. M. Deshpande and H. Delmas, *Nature*, 1995, **373**, 501.

116. A. A. Dabbawala, D. U. Parmar, H. C. Bajaj and R. V. Jasra, *J. Mol. Catal. A*, 2008, **282**, 99.

117. N. Sieffert and G. Wipff, *Chem. Eur. J.*, 2007, **13**, 1978.

118. B. Fell, G. Papadogianakis and C. Schobben, *J. Mol. Catal. A*, 1995, **101**, 179.

119. B. Fell, D. Leckel and C. Schobben, *Fat. Sci. Technol.*, 1995, **97**, 219.

120. H. Chen, Y. Li, R. Li, P. Cheng and X. Li, *J. Mol. Catal. A*, 2003, **198**, 1.

121. H. Fu, M. Li, H. Mao, Q. Lin, M. Yuan, X. Li and H. Chen, *Catal. Commun.*, 2008, **9**, 1539.

122. G. Papadogianakis, L. Maat and R. A. Sheldon, *Chem. Commun.*, 1994, 2659.

123. G. Verspui, J. Feiken, G. Papadogianakis and R. A. Sheldon, *J. Mol. Catal. A*, 1999, **189**, 163.

124. M. Haumann and A. Riisager, *Chem. Rev.*, 2008, **108**, 1474.

125. P. Wasserscheid, *J. Ind. Eng. Chem.*, 2007, **13**, 325.

126. Y. Chauvin, L. Mussmann and H. Olivier, *Angew. Chem., Int. Ed.*, 1995, **34**, 2698.

127. C. P. Mehnert, R. A. Cook, N. C. Dispenziere and E. J. Mozeleski, *Polyhedron*, 2004, **23**, 2679.

128. B. Hamers, P. S. Baeuerlein, C. Muller and D. Vogt, *Adv. Synth. Catal.*, 2008, **350**, 332.

129. Y. Y. Wang, M. M. Luo, Q. Lin, H. Chen and X. J. Li, *Green Chem.*, 2006, **8**, 545.

130. C. C. Brasse, U. Englert, A. Salzer, H. Waffenschmidt and P. Wasserscheid, *Organometallics*, 2000, **19**, 3818.

131. F. Favre, H. Olivier-Bourbigou, D. Commereuc and L. Saussine, *Chem. Commun.*, 2001, 1360.

132. D. J. Brauer, K. W. Kottsieper, C. Liek, O. Stelzer, H. Waffenschmidt and P. Wasserscheid, *J Organomet. Chem.*, 2001, **630**, 177.

133. P. Wasserscheid, H. Waffenschmidt, P. Machnitzki, K. W. Kottsieper and O. Stelzer, *Chem. Commun.*, 2001, 451.

134. R. P. J. Bronger, S. M. Silva, P. C. J. Kamer and P. W. N. M. van Leeuwen, *Chem. Commun.*, 2002, 3044.

135. C. Deng, G. Ou, J. She and Y. Yuan, *J. Mol. Catal. A*, 2007, **270**, 76.

136. S. Paganelli, A. Perosa and M. Selva, *Adv. Synth. Catal.*, 2007, **349**, 1858.

137. X. Zhao, H. Alper and Z. Yu, *J. Org. Chem.*, 2006, **71**, 3988.

138. G. Rangits and L. Kollár, *J. Mol. Catal. A*, 2006, **246**, 59.

139. F. Shi, J. Peng and Y. Deng, *J. Catal.*, 2003, **219**, 372.

140. D. Koch and W. Leitner, *J. Am. Chem. Soc.*, 1998, **120**, 13398.

141. (a) G. Franciò and W. Leitner, *Chem. Commun.*, 1999, 1663; (b) G. Franciò, K. Wittmann and W. Leitner, *J. Organomet. Chem.*, 2001, **621**, 130.

142. Z. Jingchang, W. Hongbin, L. Hongtao and C. Weiliang, *J. Mol. Catal. A*, 2006, **260**, 95.

143. M. Solinas, J. Jiang, O. Stelzer and W. Leitner, *Angew. Chem., Int. Ed.*, 2005, **44**, 2291.

144. B. P. Mason, K. E. Price, J. L. Steinbacher, A. R. Bogdan and D. T. McQuade, *Chem. Rev.*, 2007, **107**, 2300.

145. P. B. Webb and D. J. Cole-Hamilton, *Chem. Commun.*, 2004, 612.

146. C. P. Mehnert, *Chem. Eur. J.*, 2005, **11**, 50.

147. (a) A. Kirschning, W. Solodenko and K. Mennecke, *Chem. Eur. J.*, 2006, **12**, 5972; (b) A. Riisager, R. Fehrmann, M. Haumann and P. Wasserscheid, *Top. Catal.*, 2006, **40**, 91.

148. (a) A. Riisager, R. Fehrmann, M. Haumann and P. Wasserscheid, *Eur. J. Inorg. Chem.*, 2006, 695; (b) A. Riisager, R. Fehrmann, M. Haumann, B. S. K. Gorle and P. Wasserscheid, *Ind. Eng. Chem. Res.*, 2005, **44**, 9853; (c) A. Riisager, R. Fehrmann, S. Flicker, R. van Hal, M. Haumann and P. Wasserscheid, *Angew. Chem. Int. Ed.*, 2005, **44**, 815.

149. A. Riisager, R. Fehrmann, M. Haumann, M. Jakuttis, J. Joni, P. Wasserscheid, *DGMK Tagungsbericht, 2007, Preprints of the DGMK/SCI-Conference "Opportunities and Challenges at the Interface between Petrochemistry and Refinery"*, 2007, 2, 139.

150. M. Haumann, K. Dentler, J. Joni, A. Riisager and P. Wasserscheid, *Adv. Synth. Catal.*, 2007, **349**, 425.

151. U. Hintermair, G. Zhao, C. C. Santini, M. J. Muldoon and D. J. Cole-Hamilton, *Chem. Commun.*, 2007, 1462.

152. A. Riisager, B. Jørgensen, P. Wasserscheid and R. Fehrmann, *Chem. Commun.*, 2006, 994.

153. A. G. Panda, S. R. Jagtap, N. S. Nandurkar and B. M. Bhanage, *Ind. Eng. Chem. Res.*, 2008, **47**, 969.

154. Yong Yang, Changxi Deng and Youzhu Yuan, *J. Catal.* 2005, **232**, 108.

155. R. A. Grubbs, *Tetrahedron*, 2004, **60**, 7117.

156. (a) A. Fürstner and P. W. Davies, *Chem. Commun.*, 2005, 2307; (b) A. Fürstner, *Angew. Chem. Int. Ed.*, 2000, **39**, 3013.

157. (a) A. H. Hoveyda and A. R. Zhugralin, *Nature*, 2007, **450**, 243; (b) K. C. Nicolaou, P. G. Bulger and D. Sarlah, *Angew. Chem. Int. Ed.*, 2005, **44**, 4490.

158. (a) P. Schwab, R. H. Grubbs and J. W. Ziller, *J. Am. Chem. Soc.*, 1996, **118**, 100; (b) M. Scholl, S. Ding, C. W. Lee and R. H. Grubbs, *Org. Lett.*, 1999, **1**, 953; (c) J. S. Kingsbury, J. P. A. Harrity, P. J. Bonitatebus and A. H. Hoveyda, *J. Am. Chem. Soc.*, 1999, **121**, 791; (d) S. B. Carber, J. S. Kingsbury, B. L. Gray and A. H. Hoveyada, *J. Am. Chem. Soc.*, 2000, **122**, 8168.

159. (a) D. M. Lynn, B. Mohr, L. M. Henling, M. W. Day and R. H. Grubbs, *J. Am. Chem. Soc.*, 2000, **122**, 6601; (b) S. J. Connon, M. Rivard, M. Zaja and S. Blechert, *Adv. Synth. Catal.*, 2003, **345**, 572.

160. G. C. Fu, S. T. Nguyen and R. H. Grubbs, *J. Am. Chem. Soc.*, 1993, **115**, 9856.

161. P. Schwab, M. B. France, J. W. Ziller and R. H. Grubbs, *Angew. Chem. Int. Ed.*, 1995, **34**, 2039.

162. M. Scholl, S. Ding, C. W. Lee and R. H. Grubbs, *Org. Lett.*, 1999, **1**, 953.

163. J. Huang, E. D. Stevens, S. P. Nolan and J. L. Petersen, *J. Am. Chem. Soc.*, 1999, **121**, 2674.

164. J. S. Kingsbury, J. P. A. Harrity, P. J. Bonitatebus and A. H. Hoveyda, *J. Am. Chem. Soc.*, 1999, **121**, 791.

165. S. B. Garber, J. S. Kingsbury, B. L. Gray and A. H. Hoveyda, *J. Am. Chem. Soc.*, 2000, **122**, 8168.

166. H. Wakamatsu and S. Blechert, *Angew. Chem. Int. Ed.*, 2002, **41**, 794.

167. K. Grela, S. Harutyunyan and A. Michrowska, *Angew. Chem. Int. Ed.*, 2002, **41**, 4038.

168. (a) H. Clavier, K. Grela, A. Kirschning, M. Mauduit and P. S. Nolan, *Angew. Chem. Int. Ed.*, 2007, **6**, 6786; (b) P. H. Deshmukh and S. Blechert, *Dalton Transact.*, 2007, 2479.

169. (a) J. Donohoe, A. J. Orr and M. Bingham, *Angew. Chem. Int. Ed.*, 2006, **45**, 2664; (b) I. Nakamura and Y. Yamamoto, *Chem. Rev.*, 2004, **104**, 2127.

170. H. Clavier, K. Grela, A. Kirschning, M. Mauduit and S. P. Nolan, *Angew. Chem. Int. Ed.*, 2007, **46**, 6786.

171. J. Horn, F. Michalek, C. C. Tzschucke and W. Bannwarth, *Top. Curr. Chem.*, 2004, **242**, 43.

172. A. Kirschning, K. Mennecke, U. Kunz, A. Michrowska and K. Grela, *J. Am. Chem. Soc.*, 2006, **128**, 13261.

173. A. Fürstner, L. Ackermann, K. Beck, H. Hori, D. Koch, K. Langemann, M. Liebl, C. Six and W. Leitner, *J. Am. Chem. Soc.*, 2001, **123**, 9000.

174. J. B. Binder, J. J. Blank and R. T. Raines, *Org. Lett.*, 2007, **9**, 4885.

175. B. H. Lipshutz, G. T. Aguinaldo, S. Ghorai and K. Voigtritter, *Org. Lett.*, 2008, **10**, 1325.

176. Ł. Gulajski, P. Sledz, A. Lupa and K. Grela, *Green Chem.*, 2008, **10**, 271.

177. (a) Ł. Gułajski, A. Michrowska, J. Naroznik, Z. Kaczmarska, L. Rupnicki and K. Grela, *ChemSusChem*, 2008, **1**, 103; (b) A. Michrowska, L. Gulajski, Z. Kaczmarska, K. Mennecke, A. Kirschning and K. Grela, *Green Chem.*, 2006, **8**, 685.

178. A. Michrowska, K. Mennecke, U. Kunz, A. Kirschning and K. Grela, *J. Am. Chem. Soc.*, 2006, **128**, 13261.

179. J. P. Jordan and R. H. Grubbs, *Angew. Chem. Int. Ed.*, 2007, **46**, 5152.

180. S. H. Hong and R. H. Grubbs, *J. Am. Chem. Soc.*, 2006, **128**, 3508.

181. J. P. Gallivan, J. P. Jordan and R. H. Grubbs, *Tetrahedron Lett.*, 2005, **46**, 2577.

182. D. M. Lynn, B. Mohr, R. H. Grubbs, L. M. Henling and M. W. Day, *J. Am. Chem. Soc.*, 2000, **122**, 6601.
183. X. Ding, X. Lv, B. Hui, Z. Chen, M. Xiao, B. Guo and W. Tang, *Tetrahedron Lett.*, 2006, **47**, 2921.
184. A. L. Miller II and N. B. Bowden, *Chem. Commun.*, 2007, 2051.
185. K. G. Mayo, E. H. Nearhoof and J. J. Kiddle, *Org. Lett.*, 2002, **4**, 1567.
186. H. Clavier, N. Audic, M. Mauduit and J-C. Guillemin, *Chem. Commun.*, 2004, 2282.
187. D. Sémeril, H. Olivier-Bourbigou, C. Bruneau and P. H. Dixneuf, *Chem. Commun.*, 2002, 146.
188. C. Thurier, C. Fischmeister, C. Bruneau, H. Olivier-Bourbigou and P. H. Dixneuf, *J. Mol. Catal. A*, 2007, **268**, 127.
189. (a) H. Clavier, N. Audic, J-C. Guillemi and M. Mauduit, *J. Organomet. Chem.*, 2005, **690**, 3585; (b) N. Audic, H. Clavier, M. Mauduit and J-C. Guillemin, *J. Am. Chem. Soc.*, 2003, **125**, 9248.
190. (a) Q. Yao and M. Sheets, *J. Organomet. Chem.*, 2005, **690**, 3577; (b) Q. Yao and Y. Zhang, *Angew. Chem. Int. Ed.*, 2003, **42**, 3395.
191. E. Negishi, *Bull. Chem. Soc. Jpn.*, 2007, **80**, 233.
192. A. B. Dounay and L. E. Overman, *Chem. Rev.*, 2003, **103**, 2945.
193. P. Espinet and A. M. Echavarren, *Angew. Chem. Int. Ed.*, 2004, **43**, 4704.
194. A. Suzuki, *Chem. Commun.*, 2005, 4759.
195. B. M. Trost and X. Ariza, *J. Am. Chem. Soc.*, 1999, **121**, 10727.
196. R. Chinchilla and C. Nájera, *Chem. Rev.*, 2007, **107**, 874.
197. D. Alberico, M. E. Scott and M. Lautens, *Chem. Rev.*, 2007, **107**, 174.
198. L. F. Tietze, H. Ila and H. P. Bell, *Chem. Rev.*, 2004, **104**, 3453.
199. L. Yin and J. Liebscher, *Chem. Rev.*, 2007, **107**, 133.
200. Y. Tsuji and T. Fujihara, *Inorg. Chem.*, 2007, **46**, 1895.
201. (a) M. Weck and C. W. Jones, *Inorg. Chem.*, 2007, **46**, 1865; (b) V. Farina, *Adv. Synth. Catal.*, 2004, **346**, 1553.
202. V. Calò, A. Nacci and A. Monopoli, *Eur. J. Org. Chem.*, 2006, 3791.
203. E. Assen, B. Kantchev, C. J. O'Brien and M. G. Organ, *Angew. Chem. Int. Ed.*, 2007, **46**, 2768.
204. L. Xu, W. Chen and J. Xiao, *Organometallics*, 2000, **19**, 1123.
205. F. McLachlan, C. J. Mathews, P. J. Smith and T. Welton, *Organometallics*, 2003, **22**, 5350.
206. S. Liu, T. Fukuyama, M. Sato and I. Ryu, *Synlett.*, 2004, 1814.
207. A. Perosa, P. Tundo, M. Selva, S. Zinovyev and A. Testa, *Org. Biomol. Chem.*, 2004, **2**, 2249.
208. D. Tsvelikhovsky and J. Blum, *Eur. J. Org. Chem.*, 2008, 2417.
209. L. Zhou and L. Wang, *Synthesis*, 2006, 2653.
210. D. Zhao, Z. Fei, T. J. Geldbach, R. Scopelliti and P. J. Dyson, *J. Am. Chem. Soc.*, 2004, **126**, 15876.
211. D. Zhao, Z. Fei, R. Scopelliti and P. J. Dyson, *Inorg. Chem.*, 2004, **43**, 2197.
212. Z. Fei, D. Zhao, D. Pieraccini, W. H. Ang, T. J. Geldbach, R. Scopelliti, C. Chiappe and P. J. Dyson, *Organometallics*, 2007, **26**, 1588.

213. C. Chiappe, D. Pieraccini, D. Zhao, Z. Fei and P. J. Dyson, *Adv. Synth. Catal.*, 2006, **348**, 68.
214. D. Astruc, F. Lu and J. R. Aranzaes, *Angew. Chem. Int. Ed.*, 2005, **44**, 7852.
215. (a) M. Weck and C. W. Jones, *Inorg. Chem.*, 2007, **46**, 1865; (b) N. T. S. Phan, M. van der Sluys and C. W. Jones, *Adv. Synth. Catal.*, 2006, **348**, 609.
216. (a) P. Migowski and J. Dupont, *Chem. Eur. J.*, 2007, **13**, 32; (b) J. A. Dahl, B. L. S. Maddux and J. E. Hutchison, *Chem. Rev.*, 2007, **107**, 2228.
217. C. C. Cassol, A. P. Umpierre, G. Machado, S. I. Wolke and J. Dupont, *J. Am. Chem. Soc.*, 2005, **127**, 3298.
218. J. Durand, E. Teuma, F. Malbosc, Y. Kihn and M. Gómez, *Catal. Commun.*, 2008, **9**, 273.
219. (a) D. Astruc, *Inorg. Chem.*, 2007, **46**, 1884; (b) M. T. Reetz and J. G. de Vries, *Chem. Commun.*, 2004, 1559.
220. X. Yang, Z. Fei, D. Zhao, W. H. Ang, Y. Li and P. J. Dyson, *Inorg. Chem.*, 2008, **47**, 3292.
221. S. Li, Y. Lin, J. Cao and S. Zhang, *J. Org. Chem.*, 2007, **72**, 4067.
222. (a) K. H. Shaughnessy, *Eur. J. Org. Chem.*, 2006, 1827; (b) K. H. Shaughnessy and R. B. DeVasher, *Curr. Org. Chem.*, 2005, **9**, 585.
223. A. L. Casalnuovo and J. C. Calabrese, *J. Am. Chem. Soc.*, 1990, **112**, 4324.
224. K. H. Shaughnessy and R. S. Booth, *Org. Lett.*, 2001, **3**, 2757.
225. R. Huang and K. H. Shaughnessy, *Organometallics*, 2006, **25**, 4105.
226. N. Jiang and A. J. Ragauskas, *Tetrahedron Lett.*, 2006, **47**, 197.
227. N. E. Leadbeater and M. Marco, *Org. Lett.*, 2002, **4**, 2973.
228. R. K. Arvela, N. E. Leadbeater, T. L. Mack and C. M. Kormos, *Tetrahedron Lett.*, 2006, **47**, 217.
229. G. A. Leeke, R. C. D. Santos, B. Al-Duri, J. P. K. Seville, C. J. Smith, C. K. Y. Lee, A. B. Holmes and I. F. McConvey, *Org. Proc. Res. Develop.*, 2007, **11**, 144.
230. K. C. Nicolaou and S. A. Snyder, *Classics in Total Synthesis II*, Wiley-VCH, Weinheim, 2003.
231. R. Noyori, *Chem. Commun.*, 2005, 1807.

CHAPTER 2

The Contribution of Photochemistry to Green Chemistry

STEFANO PROTTI, SIMONE MANZINI, MAURIZIO FAGNONI AND ANGELO ALBINI

Department of Organic Chemistry, University of Pavia, Viale Taramelli 10, 27100 Pavia, Italy

2.1 Introduction

2.1.1 Photochemistry and Green Chemistry

At the beginning of the last century, Giacomo Ciamician, a professor of chemistry in Bologna, Italy, observed[1] that the art of synthesis had reached an important target by making available various compounds identical to those of natural origin in the laboratory, yet it remained one step behind with respect to nature, in the sense that the use of a high temperature or of aggressive reagents was always required, whereas green plants appeared to be able to prepare the same compounds under mild conditions. To our knowledge, this is the first formulation of the concept of green chemistry.[2,3] The next question was, what gave plants this ability? Ciamician thought that, apart from the role of enzymes, the key was the fact that plants absorbed light from the sun. Therefore, he proceeded to test whether also in the laboratory one may exploit light for making ("green") chemical reactions. This surmise resulted in a series of papers published over the first 15 years of the 20th century that lay the

RSC Green Chemistry Book Series
Eco-Friendly Synthesis of Fine Chemicals
Edited by Roberto Ballini
© Royal Society of Chemistry 2009
Published by the Royal Society of Chemistry, www.rsc.org

foundations of organic photochemistry. Thus, photochemistry and green chemistry have been closely connected since their birth, over a century ago, although both disciplines have been long forgotten, with synthetic photochemistry surfacing again after 1950 and green chemistry only in the last decade of the 21st century.

Nowadays photochemistry is usually considered among the environmentally advantageous methods for synthesis and the large number of reactions induced by light that have been characterized in the meantime offer useful synthetic paths.[4-6] As briefly outlined below, a photochemical reaction does not "automatically" classify as "green", *e.g.*, because a considerable amount of energy from non-renewable sources has to be supplied for the irradiation, unless solar light is used, but it certainly has some peculiar advantages.[7] The use of solar irradiation for preparative reactions has been tested and found to be perfectly viable and susceptible to scaling up. The only problem is that the light flux changes with the weather conditions and thus the time required for reaction is variable.[8] With handier artificial lamps, the environmental burden relative to the production of electric energy and its relatively inefficient conversion into light must be considered.

The key point, however, is that the *only reagent* used in photochemical reactions is *light*. No reagent is added, either stoichiometric or catalytic, making the atom economy of the process intrinsically better and discounting any trouble with the use of toxic, delicate or expensive additives as well as for the recovery of spent reagents.

This chapter is subdivided in two parts, the first briefly introduces the green aspect of photochemistry in synthesis, while the second outlines a series of representative reactions, subdivided according to the process involved.

Photochemical reactions are not common in industry, and therefore the environmental aspect has been not considered in detail, since in most cases only small-scale explorative studies have been carried out. An important point that concerns the photochemical literature in general is that a large fraction of it is devoted to mechanistic studies. In many such studies no attempt to develop the preparative issue has been considered. This does not mean, however, that many such reactions are not suited for organic synthesis;[7,9] it does mean that often one has to elaborate the experimental part to make it better suited for a synthetic application rather than using the method as reported in the literature.

Two important questions encountered when evaluating the suitability of a photochemical method for synthesis, and in particular for green synthesis are the choice of the lamp and that of the solvent.

Apart for the use of lasers, which may be advantageous in a suitably built apparatus,[10] irradiation is usually done by means of a mercury arc. A worry when considering a photoreaction is the high price of the lamp and the (monetary and environmental) cost incurred for the electrical power required. As one can see from the literature, high-pressure mercury arcs are often satisfactorily general-purpose sources and many laboratories use 400–500 W lamps of this type. These are in fact somewhat expensive and with a limited lifetime (hundreds of hours). However, 125–150 W mercury arcs are often almost as effective and much less expensive. More importantly, low-pressure, phosphor-coated lamps (15 W, typically available for 300–320, 340–380 and 350–450 nm and other wavelength ranges, as well as uncoated, emitting at 254 nm) are less expensive, quite long-lived (thousands of hours) and more effective, minimizing the costs incurred. For dye-sensitized oxygenations, lamps emitting in the visible are used. High-pressure sodium arcs emit efficiently and can be used in an immersion well apparatus. However, the common "quartz-halogen" lamps, for external use, are generally as convenient and much less expensive. For chain processes (radical halogenations, $S_{RN}1$ arylation) a simple tungsten bulb is usually sufficient. Finally, an unparalleled opportunity with photochemical reactions of course is running them by using solar light. The amount of energy falling on the Earth in the form of solar light is tremendous, although the fraction absorbed by common organic molecules is small. The slow flux can be compensated for by using mirrors that achieve a "moderate" concentration of the light flux.[11,12]

Likewise, explorative studies are conveniently carried out in very simple apparatus. This may be built with the lamp(s) outside a reaction vessel (made of a material transparent to the desired wavelength, *e.g.*, Pyrex absorbs below *ca.* 300 nm), including test tubes, and irradiating by means of some external low-pressure lamps (10–20 W, see above). Alternatively, with the lamp inside, a vessel fitted with a refrigerated immersion well (of the correct material) is used in which is inserted a mercury arc (125 or 400 W – the former is sufficient for the great majority of cases).

The other important question concerns the solvent. The high molecular absorptivity of organic molecules makes it desirable to carry out irradiations in dilute solutions, typically 1 to 5×10^{-2} M or even less. This avoids complete absorption of light in the first layer, which would cause secondary photo-reactions to occur in the first layers and/or leave unchanged the fraction of starting materials present in layers furthest from the light source. Indeed, to have a "clean" reaction, many explorative studies are carried out at $\leq 1 \times 10^{-3}$ M concentration, implying that too large an amount of solvent must be used. An efficient recovery must thus be planned unless any "green" value of the process is lost. Apart from the amount, also the quality of the solvent may be unfavorable from the environmental point of view. Indeed, in view of the high reactivity of excited states, a reaction with the solvent is a likely possibility. Thus, the solvent(s) used for explorative studies are often chosen within a small group of compounds that have been shown to be (relatively) inert in photo-chemical reactions, not taking into account the green aspect. As an example, chloroform is often used in photosensitized oxygenation because singlet oxygen has a long lifetime under these conditions, but it is certainly not a good choice for the green characteristics of the process.

In conclusion, to satisfy green chemistry criteria, photochemical syntheses as reported in the literature usually require further work, aimed at minimizing the use of electricity for the lamps and diminishing the amount of solvent involved. Of course, it is important not only that light is produced as economically as possible but also that it is effectively absorbed. Apart from avoiding some naïve mistakes, such as using a solvent that absorbs the desired wavelength range, this implies maximizing the effective use of the radiation, again an aspect that is rarely considered in explorative papers. However, if industrial application is seriously considered, somewhat more expensive, but distinctly better performing irradiation apparatuses can be used. Practically all of these are based on the idea that light absorption by a thin layer of solution, from a few mm to some μm, leads to a more uniform absorption and thus both to a more efficient use of light and a higher yield of the primary product(s), since only the starting material absorbs, with less unreacted starting material remaining (because some inner parts of the solution receive no light) and fewer secondary photoproduct(s), which are due to competitive absorption by the primary product. The principle is applied in "falling film" reactors, where the solution is sprayed on the outer side of the lamp well, or is simply circulated in a tubing wrapped around the lamp well,[13] or alternatively in microreactors,[14–17] where higher concentrations can be used and different light sources such as LED or lasers may be advantageous. Both home-made and commercial equipments of this type have been tested in a significant number of cases and marked improvements of yields and/or selectivity have been reported. The use of such experimental set-ups aims to optimize the use of emitted light and/or the selectivity of the reaction and thus helps in planning the scaling up of the process.

Likewise, the choice of the solvent should be reconsidered. It is usually possible to shift to less environmentally unfriendly solvents, *e.g.*, from halogenated hydrocarbons to esters, with no substantial lowering of the yield. Furthermore, good results have been obtained in the solid phase. At first sight this would seem the worst possible choice, since light-penetration beyond the crystal surface is minimal, but apart from cases where the product is transparent in the wavelength region where the reagent absorbs, and thus light penetrates more and more with the progress of the reaction, here again going to smaller dimensions can be advantageous. Thus, excellent results have been obtained by irradiating a suspension of nanocrystals in water.[18] Another solvent-free choice that has been explored has the reagent embedded on polystyrene beads (for the case of photooxygenations).[19]

Actually, most of the limitations listed above for photoreactions reported in the literature (or similar ones) apply to the generality of the reported "research" papers, where some reactions, whether thermal or photochemical, are reported for the sake of demonstrating new synthetic possibilities, with little care for optimizing the process (and even less for the green aspect). Clearly, nobody would dream of proposing such processes for an industrial application (or simply for scaling them up to some grams) before reconsidering the experimental procedure in detail. At the very least, one can state photochemical reactions bear no negative connotation in this respect. As mentioned above, attention to this point is growing, however, and better experimental set-ups can be bought or built. There is no reason why specific aspects, such as a large amount of solvent to recovery and the cost of the photon, can not be improved.

2.2 Examples of Green Photochemical Reactions: Formation of a Carbon–Carbon Bond

It is obviously not possible to give a fair account of the synthetic significance of photochemical reactions in a single chapter. The reader is referred to any text of photochemistry for a more balanced account.[20–23] In choosing the few examples discussed below it has been attempted to present a cross section of most common photochemical reactions, while preferably selecting procedures that have been tested at least at the gram scale and where the authors had taken in consideration the environmental aspect of their syntheses. The reactions are classed on the basis of the bond formed. Attention to the mechanism is limited to what is essential for understanding the type of process occurring. Photochemical reactions involve either a chemical reaction of the excited state directly to the product or the generation of an active intermediate, such as a radical or an ion, *e.g.*, by fragmentation of the excited state (Scheme 2.1). In the latter case, an alternative procedure where the same intermediate is prepared thermally is often available, but the photochemical alternative is generally more versatile in terms of structure of the precursor of the active intermediate and reaction conditions.

A large variety of reactions forming carbon–carbon bonds are available. For the synthesis of open-chain compounds important processes occur *via* radical intermediates, such as the (photocatalyzed) alkylation of alkenes and alkynes and radical coupling by elimination of an interposed group. Convenient arylation methods *via* aryl radicals or cations ($ArS_{RN}1$[24] and ArS_N1,[25,26] respectively) as well as acylation reactions have likewise been reported (Sections 2.2.1–2.2.4). Cyclization processes occurring directly from the excited state include the important di-π-methane rearrangement (as well as the oxa-analogous process) that lead to a three-membered ring,[27] as well as 6π cyclizations leading to five- or six-membered rings. Cyclizations that occur *via* a discrete intermediate include the synthesis of hydroxycyclobutanes *via* intramolecular γ-hydrogen abstraction by ketones followed by cyclization (Sections 2.2.5–2.2.8). Photochemical cycloaddition reactions are well known, ranging from the (formal) [2π + 2π] cycloaddition to form cyclobutanes or oxetanes from carbonyls[28] to the various modes of benzene + alkene cycloaddition.

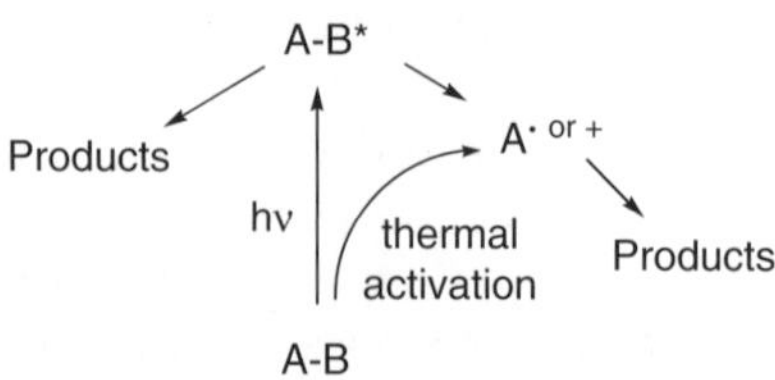

Scheme 2.1

2.2.1 Formation of C–C Bonds *via* Coupling Reactions

Formation of a single C–C bond from a precursor of structure R–X–R′ by elimination of an interposed function, according to Reaction (2.1), has found satisfactory application in photochemistry, with typical examples being the elimination of CO from ketones and of nitrogen from azo compounds.[29–31]

$$R\text{-}X\text{-}R' \xrightarrow{h\nu} R\text{-}X\cdot + \cdot R' \longrightarrow R\text{-}R'$$

$$X = CO, N_2$$
$$R, R' = \text{Alkyl, Aryl}$$

$$(2.1)$$

Again, carrying out the reaction in the solid state is all important in determining the reaction course, in particular for controlling side reactions. Thus, Garcia-Garibay found that irradiation of dicumyl ketone in solution gave dicumene in 60% yield, isopropylbenzene and α-methylstyrene in *ca.* 10% yield each, and several other unidentified products (20%). In contrast, finely powdered crystals of the ketone in aqueous suspension react in a very clean solid-to-solid fashion to give dicumyl as the only product in >99% yield. This reaction was conveniently scaled up to 10 g. To obtain a homogeneous light absorption, a nanocrystalline suspension of the starting ketone, obtained by adding a saturated acetone solution to water containing a tensioactive agent (sodium dodecyl sulfate), was used. Irradiation by means of a 400 W medium-pressure Hg lamp led to conversion at a rate of $0.5\text{--}1\,\mathrm{g\,h^{-1}}$.[18]

Photodecarbonylation in the solid state allows to maintain stereochemical control. Thus, this highly stereospecific process can be exploited for the synthesis of enantiomerically pure natural products. As an example, the photodecarbonylation of an aqueous suspension of the diastereomeric difluorodioxaborinane complexes of β-keto-(S)-α-methylbenzylamide (**1** and **2**) was exploited for the synthesis of both enantiomers of α-cuparenone (**3** and **4**) with 100% stereoselectivity and in 80% yield (Scheme 2.2).[32]

The last examples demonstrated that the solid-to-solid photoinduced decarbonylation of crystalline ketones can have an important role in natural product synthesis and in green chemistry due to the high yield, the fewer steps involved and the easy scale-up of such reactions.

Another group of reactions involves coupling of radicals arising by hydrogen transfer, the most investigated case being the reductive coupling of aromatic ketones to give pinacols. This is one of the longest known photochemical reactions, but recently it has been found that it takes place conveniently in room-temperature ionic liquids (RTILs), a new class of environmentally friendly solvents. The strong absorption by ketones in the near-UV overcomes that of the heterocycles often present in RTILs and ensures selective excitation. Moreover, the photoreduction mode of the starting ketone is determined simply by the appropriate choice of the RTIL used. A nice example is shown in Scheme 2.3.[33] When the reaction is carried out in *sec*-butylammonium trifluoroacetate (sBATFA), the photoinduced electron transfer (PET) process in

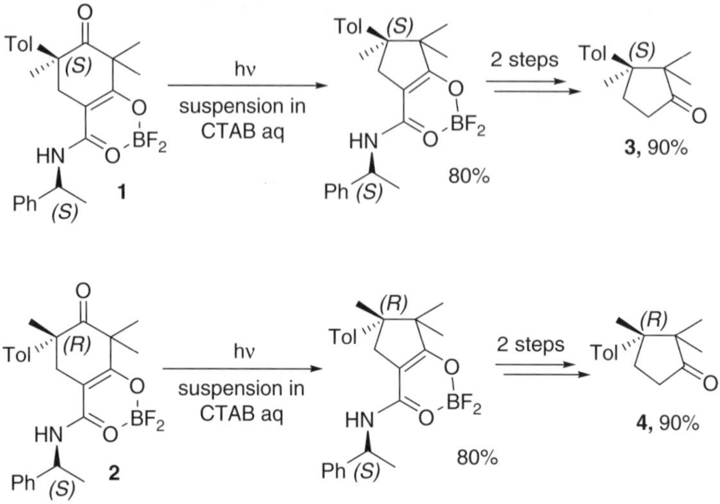

Scheme 2.2

the presence of amines forms the radical pair **5/6** and pinacol **7** from it, *via* radical coupling, in 100% yield. Conversely, a neat two-electron photoreduction to benzhydrol, rather than reductive coupling, occurs when the equilibrium between the radical pair **5/6** and ion pair **8/9** is shifted toward the latter species, as was found to be the case when using 1-butyl-3-methylimidazolium tetrafluoroborate $BMI(BF_4)$ as the ionic liquid.[33] Under these conditions 4,4'-dichlorobenzophenone was quantitatively reduced by using triethylamine (at least 40 equiv) as the reducing agent (Scheme 2.3). Notably, $BMI(BF_4)$ can be reused after elimination *in vacuo* of the solvent employed for the extraction of the product during the work up and further drying in a vacuum oven at 120 °C.

2.2.2 Photoinduced Alkylation of Alkenes and Alkynes by Alkanes, Alcohols and Alkyl Halides

The direct and selective activation under mild conditions of C–H bonds in organic molecules (especially in alkanes) still remains a "Holy Grail in chemistry." To reach this target, photocatalytic reactions have been developed[34-36] where a homolytic hydrogen abstraction occurred as a result of the absorption of a photon by a suitable photocatalyst, often a ketone. The nucleophilic carbon radical generated in this way undergoes Michael-type addition to an electron-deficient alkene or alkyne and afford various alkylated products. Photocatalysis is suitable for the efficient generation of alkyl radicals directly from alkanes. Accordingly, β-cycloalkylketones were prepared starting from the corresponding cycloalkanes through the photocatalyzed radical addition reaction onto enones. In this case, the alkyl radicals were generated *via* hydrogen abstraction by using an inorganic (tetrabutylammonium decatungstate, TBADT) photocatalyst. This

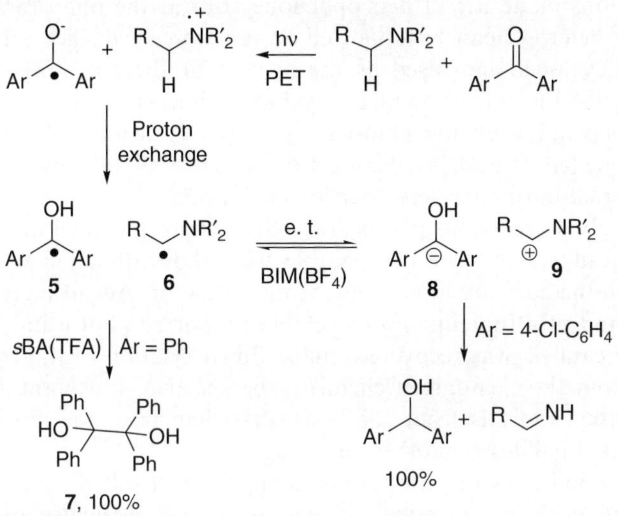

Scheme 2.3

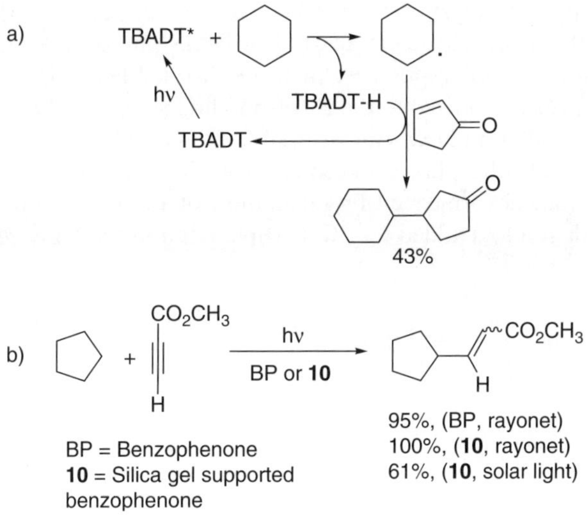

Scheme 2.4

was found to be active for at least 50 cycles and to give isolated yields ranging from 30% to 80%.[37] The alkane was used in a five-fold excess with respect to the enone. In this way, 3-cyclohexylcyclopentanone (*ca.* 0.7 g, 43% yield) was obtained by irradiation in an immersion-well apparatus (Scheme 2.4a)

Reactions involving cycloalkanes as hydrogen donors and electron-poor alkynes (*e.g.*, methyl propiolate or dimethyl acetylenedicarboxylate) were carried out both in a photochemical reactor (350 nm) and under solar irradiation.

These reactions made use of benzophenone (**BP**) as the photocatalyst, either dissolved or heterogeneously supported on functionalized silica gel (**10**).[38] The reaction of cyclopentane (used as the solvent of the reaction) with methyl propiolate gave the corresponding β-cyclopentyl-acrylate in a satisfying yield (60–95%) depending on the photocatalyst used (Scheme 2.4b). When using benzophenone (*ca.* 15 mol.%), the reaction yield was slightly lower than for the reaction carried out under heterogeneous conditions.

Interestingly, in reactions photocatalyzed by benzophenone the use of solar light gave results that were comparable with those obtained when using a conventional mercury arc in a photochemical reactor. Accordingly, the use of solar light and of a (potentially) recyclable supported photocatalyst (actually most of the catalyst was recovered) made this alkylation method particularly attractive from the clean/green chemistry perspective. Furthermore, the generation of alkyl radicals from alkanes rather than from the more toxic and expensive alkyl halides is another advantage.

The mild conditions of photocatalysis appear well suited for the conjugate alkylation of sensitive substrates such as unsaturated aldehydes, in contrast to the high susceptibility of such substrates to 1,2 rather than 1,4 attack in thermal methods. In this case, γ-butyrolactols (or lactones) have been synthesized. The organic solvent could in part be replaced by water, thus allowing the use of a hydrosoluble photocatalyst (disodium benzophenonedisulfonate (BPSS) or the sodium salt of 4-benzoylbenzoic acid). The BPSS photocatalyzed reaction of 2-hexenal in a methanol–water 1 : 1 mixture afforded lactol **11** (*ca.* 2.6 g, 57% yield) that was easily oxidized to the corresponding γ-lactone (**12**, Scheme 2.5).[39]

Furthermore, alkyl radicals can be easily obtained from aliphatic iodides or bromides by means of a photoinduced electron transfer reaction with a tertiary amine.[40] The radical formed by fragmentation of the radical anion with concomitant halide ion loss added to a C–C triple bond in the *5-exo-dig* cyclization

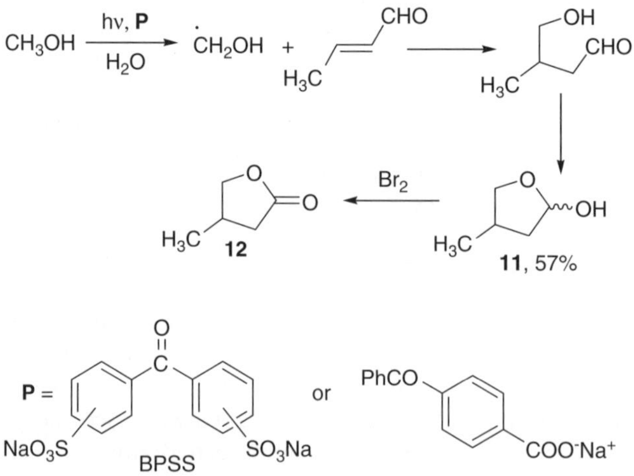

Scheme 2.5

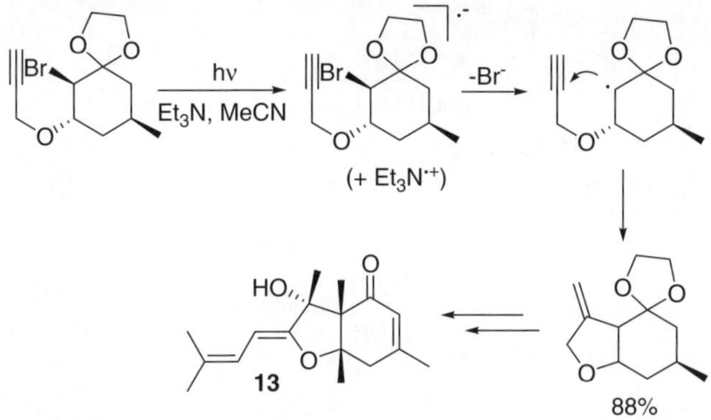

Scheme 2.6

mode to afford the bicyclic core of ($\pm$)-bisabolangelone (**13**), a sesquiterpene with antifeeding properties (Scheme 2.6). Notably, neither toxic tin derivatives (*i.e.*, Bu$_3$SnH) nor radical initiators were required to generate the radical intermediate.

2.2.3 Acylation Reactions

Aldehydes have been likewise employed as a source of acyl radicals *via* photochemical hydrogen exchange reactions. Acylated hydroquinones were readily obtained in a high yield by irradiation of an alcoholic solution of 1,4-quinones and aldehydes. This photo-Friedel–Crafts-like process relies on the ability of the excited state of quinones to abstract the formyl group hydrogen. Radical–radical coupling of the thus-formed radical pair allowed the synthesis of the final hydroquinones. Since quinones absorb at wavelengths higher than 350 nm the reaction can be carried out by using solar irradiation.[41] Thus, exposing a solution of **14** in a *tert*-butanol–acetone mixture to the concentrated solar light in a PROPHIS reactor[41] in the presence of an excess of butyraldehyde for three days afforded **15** in 90% yield (Scheme 2.7). Notably, the reaction was scaled-up to 500 g of **14** in 80 L of solvent.

The easy access to acyl radicals by photocatalyzed selective abstraction of the carbonyl hydrogen atom in aldehydes has originated further convenient applications. For example, trapping of these nucleophilic radicals by electrophilic olefins gave rise to unsymmetrical ketones. A low amount (2 mol.%) of tetrabutylammonium decatungstate (TBADT) as the photocatalyst was sufficient. Scheme 2.8 shows as an example the synthesis of β-ketoester **16** in 63% yield upon photolysis of an acetonitrile solution of heptaldehyde and dimethyl maleate.[42]

This method made use of equimolar amounts of the reagents (while an excess of aldehyde was required in the corresponding thermally initiated reactions) and avoided the use of foul-smelling auxiliaries, such as thiols. These facts qualify the reaction as a "green" synthetic method. In addition, the atom

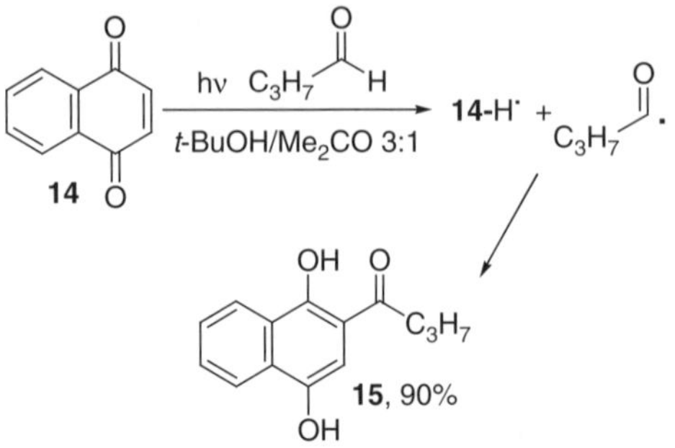

Scheme 2.7

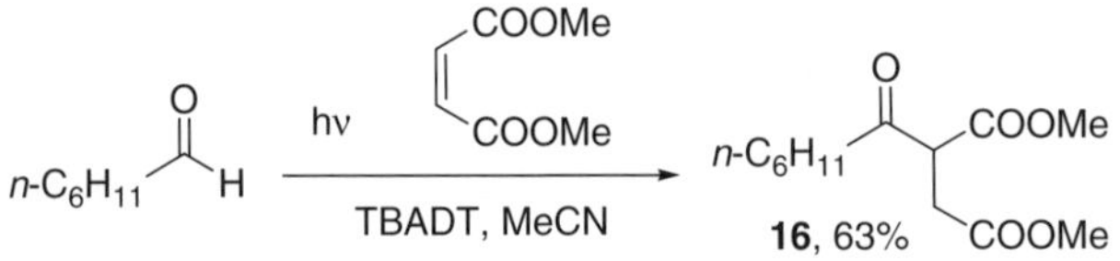

Scheme 2.8

economy of the process was improved by the fact that aldehydes are used rather than the most common precursors of thermal initiated reactions, such as acyl selenides or acylcobalt(III) derivatives, which are more toxic and elaborated.[43]

2.2.4 Aryl–Aryl and Aryl–Alkyl Bond Formation

One way of carrying out nucleophilic aromatic substitution reactions under mild conditions is the ArS_{RN}1 process, which is initiated by (usually, but not necessarily, photoinduced) electron transfer to an aryl halide, *e.g.*, from an enolate.[24] Cleavage of the resulting aryl radical anion with loss of a halide anion gives an aryl radical that combines with the enolate, thus forming the desired aryl–carbon bond.

An example is given in Scheme 2.9, where product **20** is formed (along with a small amount of aniline) by irradiation of **19** in DMSO under N_2 at 40 °C in the presence of ketone **17** and *t*-BuOK (2.0 mmol).[44]

An elegant photochemical formation of an aryl–carbon bond through a PET mechanism was recently reported in the total synthesis of the potent antimitotic polycycle (–)-diazonamide A.[45] The reaction was initiated by intramolecular electron transfer between the indole chromophore and the adjacent bromoarene (Scheme 2.10). Thus, compound **21** was treated with an aqueous-acetonitrile solution of LiOH and the resulting lithium phenoxide solution was degassed and photolyzed (Rayonet, 300 nm) to yield biaryl **22** (as a single atropodiastereomer) in a good yield. A radical–radical anion pair (**23**) was formed upon excitation, and

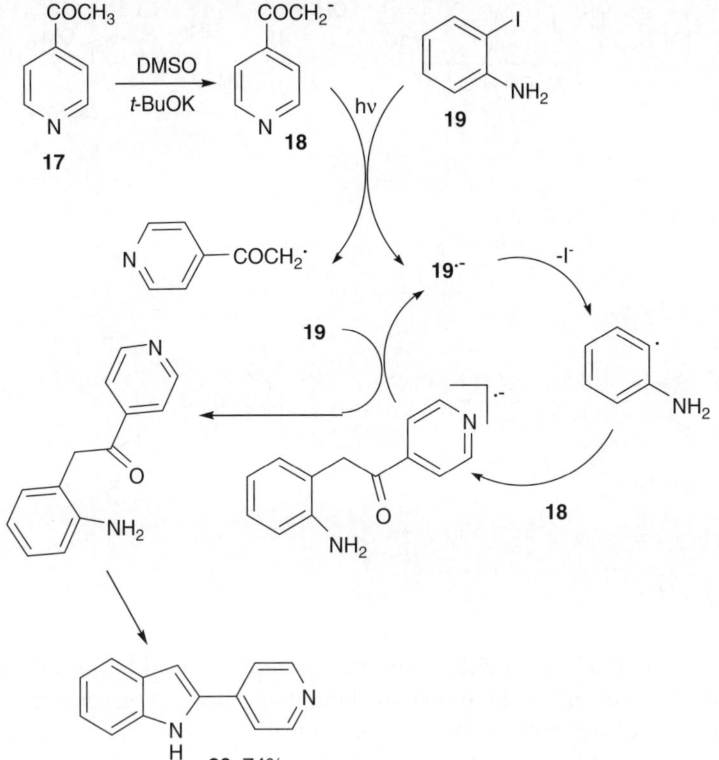

Scheme 2.9

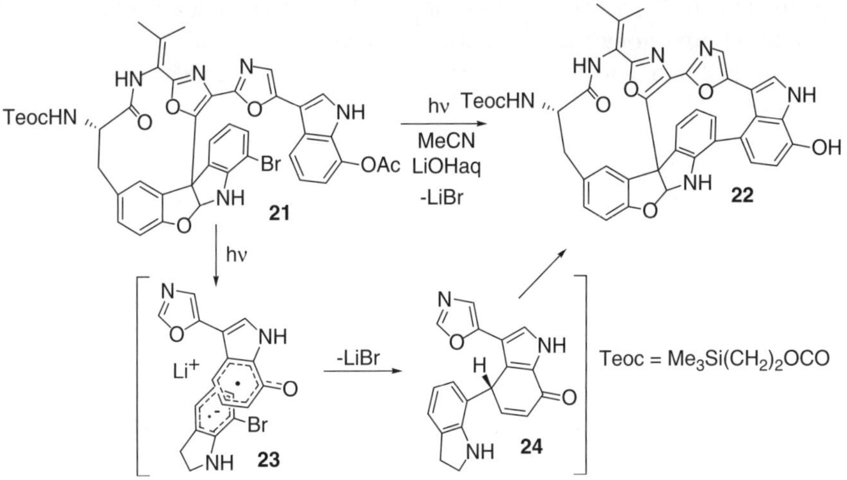

Scheme 2.10

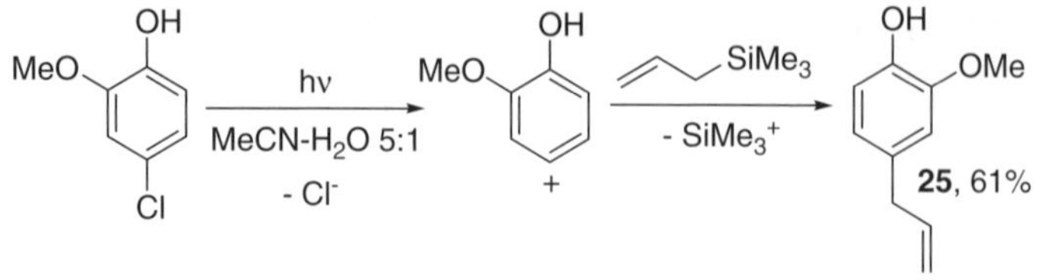

Scheme 2.11

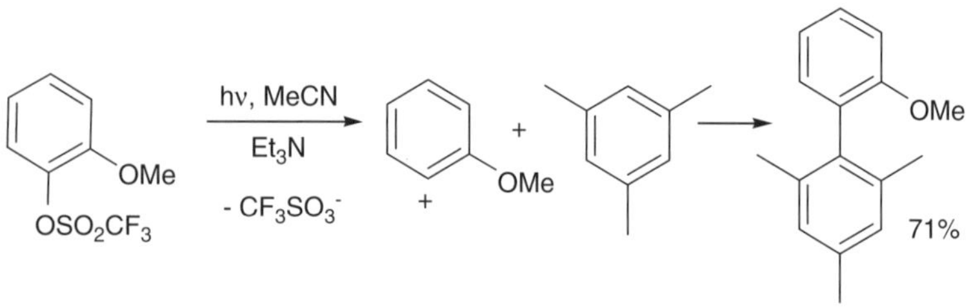

Scheme 2.12

after elimination of the bromide anion the biradical coupled to form the desired C–C bond in indolenone **24**, which finally tautomerized and generated **22**.[45]

A new metal-free method for photochemical arylations based on the intermediacy of aryl cations formed through the heterolytic fragmentation of substituted aryl halides or esters has recently emerged. A series of bioactive allylphenol and allylanisole derivatives present in several plants of the genus *Piper* have been synthesized in medium to high yields through this procedure.[46] As a representative example, Scheme 2.11 shows the single-step synthesis of eugenol (**25**, 61% yield), a compound used in perfumeries, flavorings and essential oils and in medicine as a local antiseptic and anesthetic.

This expeditious metal-free procedure involved the irradiation of the corresponding chloroaromatic in a polar solvent (*e.g.*, aqueous-acetonitrile) by using allyltrimethylsilane as allylating agent. Larger scale reactions (up to 1 g reported) can be performed in an immersion-well apparatus. Notably, in all cases the desired allylated products were obtained in an acceptable purity (>95%) through a simple work up by extraction with CH_2Cl_2 in a liquid–liquid extractor, or alternatively could be purified by bulb to bulb distillation.[46] The above synthesis avoided the protection–deprotection sequence of the OH group usually required in related metal-catalyzed allylation.

Aryl cation chemistry was likewise successful in the synthesis of biaryls, a common structural component of many pharmaceutically and biologically active compounds. In this way, sterically crowded biphenyls were smoothly synthesized by a direct Ar–H substitution in alkyl benzenes. The desired coupling took place satisfactorily, forming various bulky biaryl compounds in yields greater than 50% (mostly >70%, see Scheme 2.12).[47]

This remarkable reaction is relevant first because the process did not require an expensive and toxic metal catalyst and second because an aromatic hydrocarbon ArH (mesitylene in the reported example) was directly used as the nucleophile, in contrast to what happens with thermal reactions, where a nucleophilic organometallic derivative Ar–M is used, as in the Stille (M = SnR_3), Kumada (M = MgX) and Suzuki [M = $B(OH)_2$] reactions.

2.2.5 Cyclization *via* Atom and Energy Transfer

Excited carbonyls are known for their ability to abstract hydrogen intramolecularly, particularly when present in the gamma position with respect to the ketone function, thus yielding a 1,4-biradical intermediate. A four-membered ring thus results from radical–radical coupling (the reaction is known as the Norrish–Yang cyclization).[48] The stereochemical outcome of the reaction can be governed by carrying out the irradiation of the ketone in the solid state. An asymmetric photochemical synthesis of β-lactams has been reported by irradiating a suspension in hexane of crystalline *N,N*-dialkylarylglyoxamide chiral salt (**26**) in an immersion well apparatus.[49] Lactam **27** was smoothly formed in a high yield and with excellent enantiomeric excess (94% yield, 99% ee), with the further bonus of a negligible amount of a five-membered ring by-product (Scheme 2.13).[49]

The presence of a good leaving group X (*e.g.*, a sulfonyl ester) adjacent to the carbonyl group modified the course of the Norrish–Yang reaction. In the

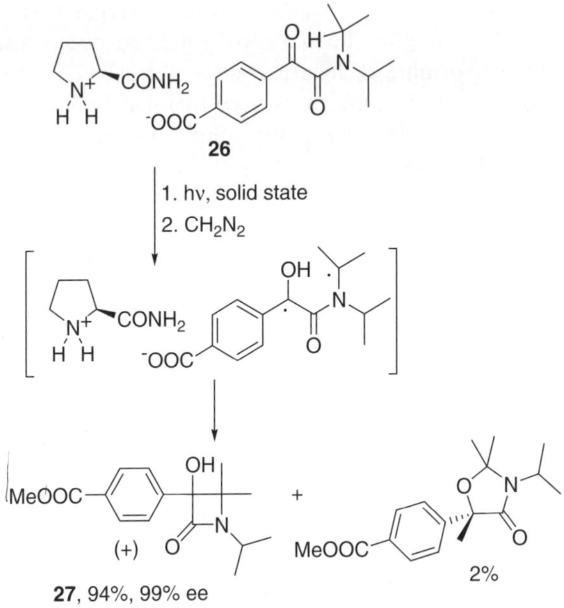

Scheme 2.13

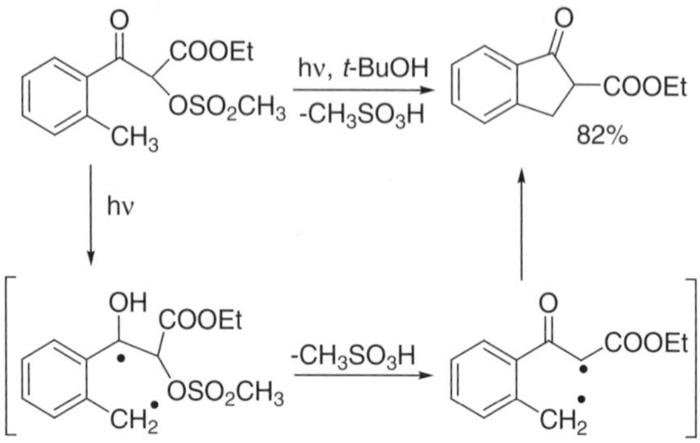

Scheme 2.14

example illustrated in Scheme 2.14, the initial 1,4-diradical was subjected to a spin center shift forming a 1,5-diradical *via* elimination of a CH_3SO_3H molecule (the acid scavenger *N*-methylimidazole was required in some cases). 1-Indanones were thus formed in a reasonable yield starting from *o*-alkylphenyl alkyl ketone precursors.[50]

The presence of an electron-donating group directs the reaction, by causing intramolecular electron transfer and deprotonation from the position to which it is bonded. As an example, when a suitable donating group is present in a δ position the radical is formed in that position and a five-membered ring results. The sequence of intramolecular electron transfer from donating group to the carbonyl (functioning as the acceptor) followed by proton transfer has been extended to the formation of larger rings. As a result, reduction of the carbonyl group to a hydroxy group was accompanied by formation of a C–C bond, as exemplified by the decarboxylative photocyclization of ω-phthalimido carboxylates to yield [1,4]benzodiazepines (Scheme 2.15).[51] An eco-friendly acetone–water mixture was adopted as the reaction medium. This reaction affords an interesting case where the concept of "memory of chirality" has been exploited. Thus, the reaction is not controlled by asymmetric induction by a chiral auxiliary but by the lifetime and the conformational flexibility of the intermediates. The diastereoselective control, however, strongly depended on the scale of the reaction; accordingly, a small-scale reaction led to the pure *trans* isomer, whereas a larger-scale irradiation (0.1 mol) using a 3 kW XeCl excimer as light source gave benzodiazepine **29** as a 1 : 1 mixture of diastereoisomers starting from phthalimide **28**.[51]

A similar strategy was developed for the construction of cyclic peptide analogues by using the PET-promoted photocyclization reactions of peptides having N-terminal phthalimides as light absorbing electron-acceptor groups and α-amidosilane or α-amidocarboxylate groups as electron donating C-terminal moieties (Scheme 2.16). Notably, both the length of the peptide chains separating the donor–acceptor centers and the nature of the electron

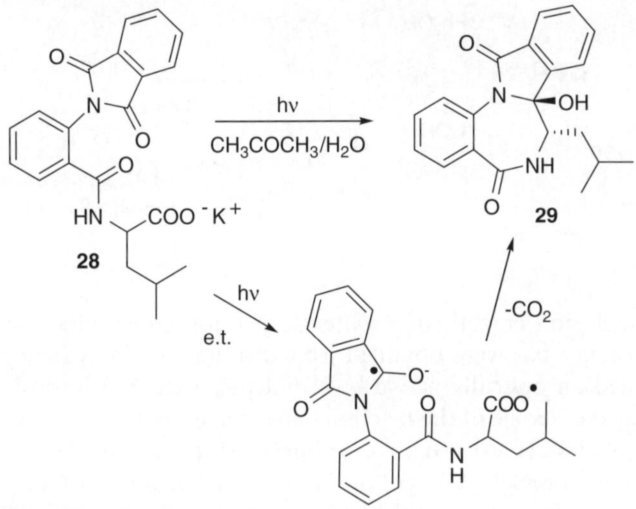

Scheme 2.15

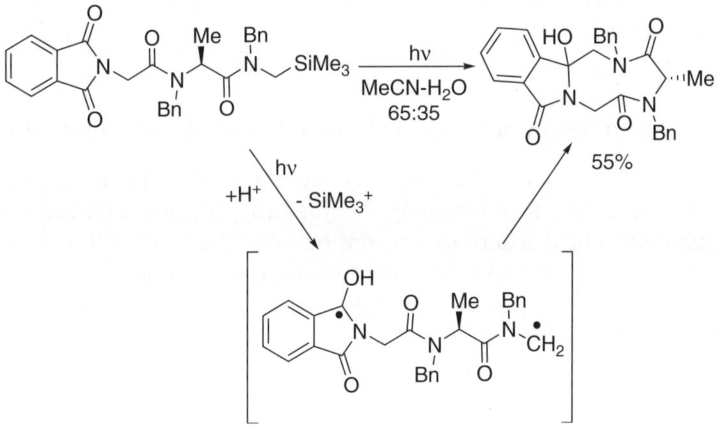

Scheme 2.16

donor did not significantly affect the chemical efficiency.[52] Moreover, in this process a significant amount of water (*ca.* 35%) was used to replace part of the toxic acetonitrile used as the solvent.

2.2.6 Electrocyclic Reactions

The 6π photocyclization of *cis*-stilbene to dihydrophenanthrene has been investigated in depth. A related reaction, noteworthy because of the high degree of enantioselectivity involved, has been reported to occur upon irradiation of

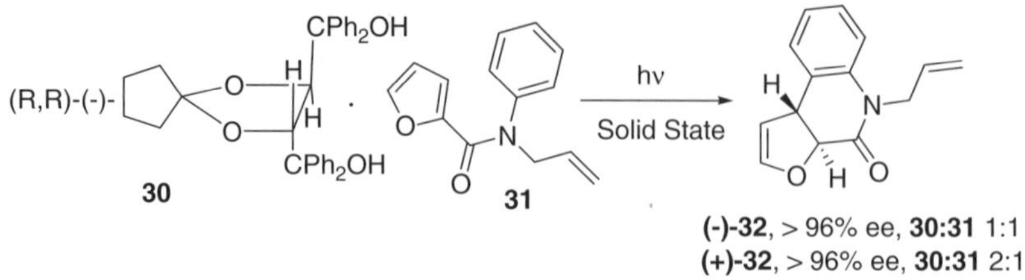

Scheme 2.17

powdered inclusion crystals in a water suspension containing a surfactant.[53] The inclusion crystals were obtained from a furan-2-carboxyanilide derivative and contained an optically active host diol prepared from tartaric acid. The stereochemical outcome of the reaction depended on the host-to-anilide ratio in the inclusion crystals (**30** : **31**). Accordingly, irradiation of the 1 : 1 complex gave the (–)-enantiomer (96% ee) of the *trans*-dihydrofuran (**32**) in 50% yield, while the 2 : 1 complex formed the (+)-enantiomer in 86% yield with a 98% ee (Scheme 2.17). Interestingly, the steric course of the reaction was also significantly affected by the preparation mode of the complex.[53]

Analogously, starting from *N*-methyl-*N*-phenyl-3-amino-2-cyclohexen-1-one derivatives, an enantioselective reaction took place forming *N*-methylhexahydro-4-carbazolones.[54]

2.2.7 Di-π-methane and Oxa-di-π-methane Rearrangements

1,5-Dienes are known for the characteristic rearrangement to vinylcyclopropanes (di-π-methane rearrangement). Analogously, β,γ-enones are known for a typical photochemical reaction, *i.e.*, the oxa-di-π-methane (ODPM) rearrangement *via* a diradical intermediate.[55] Acetophenone sensitization was conveniently used to furnish the triplet of the reagent and to induce the rearrangement to a cyclopropyl ketone. A typical example is the conversion of quinuclidinones (*e.g.*, 1-azabicyclo[2.2.2]octenones, **33**) into cyclopropane tricyclic photoproducts such as **34** (Scheme 2.18). This was formed in *ca.* 70% yield.[56] Subsequent reductive cleavage of **34** by using lithium dimethylcuprate or *via* hydrogenolysis caused a ring-opening to produce the pyrrolizidine skeleton, a structural motif often present among alkaloid derivatives.

2.2.8 Cycloaddition Reactions

The [2 + 2] photodimerization is a well-known photochemical reaction, for which many examples have been reported, in particular for arylalkenes and α,β-unsaturated carbonyl or carboxyl derivatives. In solution, *E/Z* isomerization competes with the bimolecular reaction; in the solid state, on the other hand, cycloaddition may be quite effective, as has long been known.[57] Relations between the arrangement of the molecules in the crystal lattice and the reaction occurring have

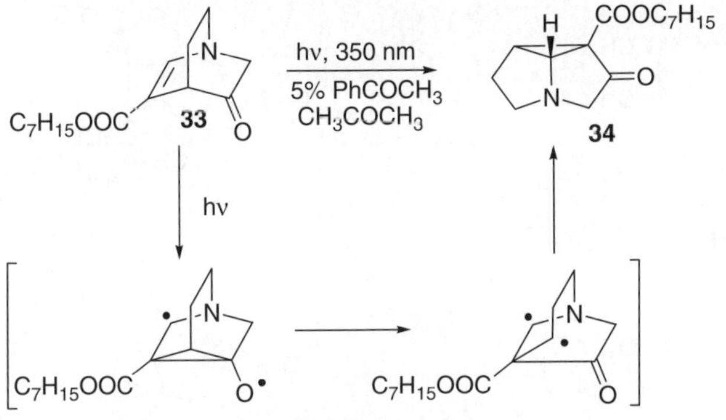

Scheme 2.18

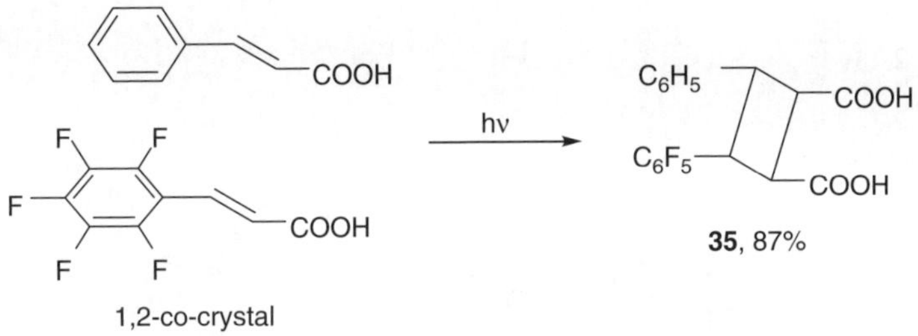

Scheme 2.19

been established. Furthermore, cross-dimerization in mixed crystals has also been developed.[58] As an example, co-crystals formed by crystallization in ethanolic solution of an equimolar mixture of *trans*-cinnamic acid and *trans*-2,3,4,5,6-pentafluorocinnamic acid were irradiated. As a result, a [2 + 2] cycloaddition to form **35** in a high yield (87%) was obtained (Scheme 2.19).[59] With a few exceptions, olefins arranged in parallel orientations that have center-to-center separations of 3.5–4.2 Å undergo the photocycloaddition reaction. More precisely, the reaction was successful when a stacked interaction of the aromatic groups between phenyl and perfluorophenyl groups was present in the lattice, but failed in the case of a slipped-stacked orientation.[59]

Tetrahydrophthalic anhydride and the corresponding imide were found to be efficient partners in intermolecular [2 + 2] photochemical cycloaddition reactions with alkenols (*e.g.*, *cis*-buten-2-ene-1,4-diol) and alkynols (Scheme 2.20). The corresponding cyclobutane adducts have been synthesized in high yields with levels of stereoselection (as high as 10 : 1) almost unprecedented in classical intermolecular cycloadditions.[60,61]

Only a moderate excess of allyl alcohol is required (*ca.* 1.5 equiv.) and the reaction is complete within 2 h with the concomitant formation of a lactone moiety in a stereoselective fashion.

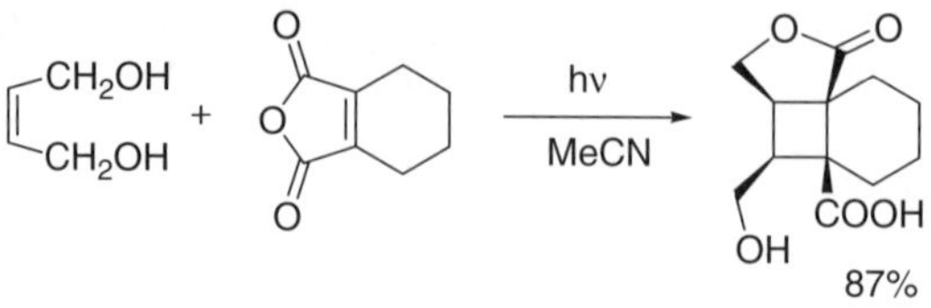

87%

Scheme 2.20

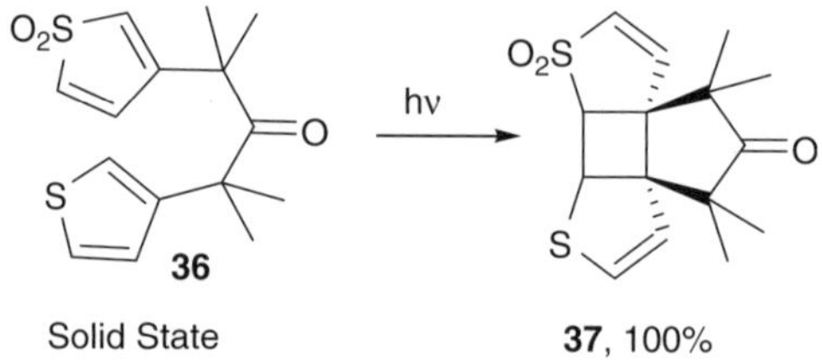

36

Solid State

37, 100%

Scheme 2.21

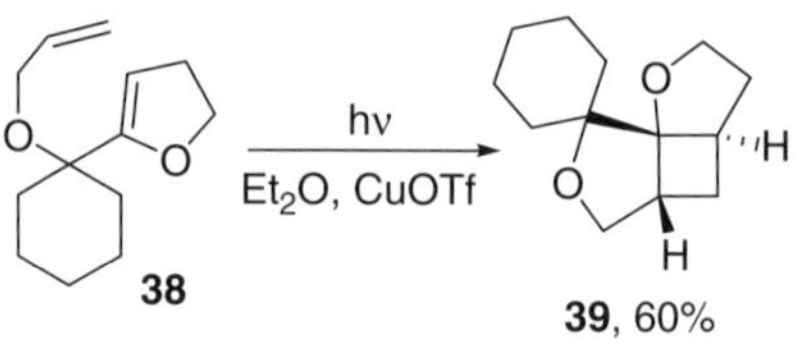

38

39, 60%

Scheme 2.22

Notably, in some cases photocycloaddition has been observed in the solid state in preference to other processes that are usually very fast. As an example, ketosulfone **36** did not undergo the expected photodecarbonylation reaction (see Section 2.2.1) but a very clean solid-to-solid intramolecular [2 + 2] cyclization to form the tetracyclic compound **37** as the only product in quantitative yield (Scheme 2.21). Small-scale reactions (*ca.* 50 mg) could be completed also with sunlight within 2 h and with no apparent changes in the aspect of the crystalline specimens.[62]

Functionalized substituted cyclopentanes and spirocyclopentanes were easily accessed by intramolecular photocycloaddition of dihydrofuran based dienes catalyzed by Cu(I) salts. Thus, tricyclic compound **39** was synthesized in 60% yield starting from **38** (Scheme 2.22). Notably, **39** can be easily transformed upon treatment with triflic acid into substituted cyclopentanones, which are important building blocks in the total synthesis of natural products.[63]

The *N*-pentenyl-substituted maleimide **40** undergoes an efficient formal [5 + 2] cycloaddition reaction upon UV irradiation and forms azepine **42**. The reaction was thought to proceed by initial [2 + 2] cycloaddition to form zwitterion **41**, which evolved to **42** *via* a fragmentation step (Scheme 2.23). This

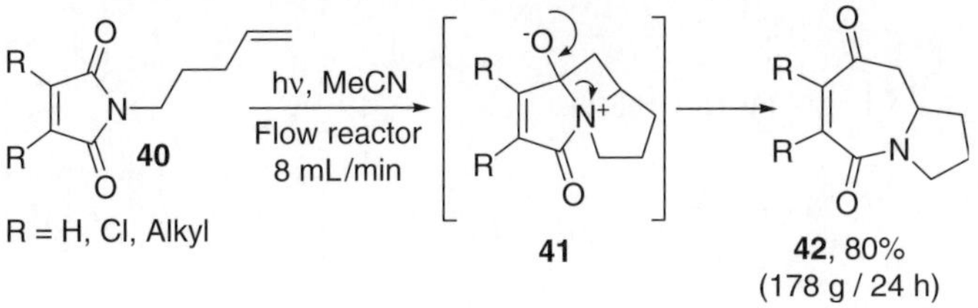

Scheme 2.23

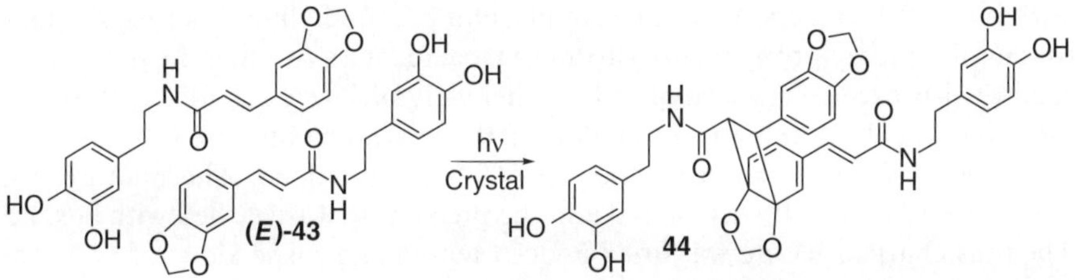

Scheme 2.24

kind of reaction has proved to be useful for the construction of the CDE ring skeleton of (−)-cephalotaxine.[64] Interestingly, this type of cycloaddition was also performed in a continuous flow photochemical reactor by passing the solution once in the irradiation compartment, in what is a nice example of organic photochemistry on a (relatively) large scale.[13] The reactor was assembled simply by winding three layers of fluorinated ethylenepropylene (FEP) tubing (a total length of tubing of 37 m with a total internal volume of 210 mL have been used) around the custom-made Pyrex immersion well. This system was capable of producing 175 g of adduct (R = methyl) in a continuous 24 h operation period thanks to the solvent resistant polymeric material used (FEP), which also has excellent UV transmission properties.[13]

Solvent-free conditions were found to be useful for inducing a synthetically useful photoreaction of the cinnamide derivatives of some catecholamines. This reaction is the first example of solid-state [2 + 2] photoaddition of an alkene to the benzene ring and is exemplified in Scheme 2.24 by the photodimerization of *N*-[(*E*)-3,4-methylenedioxycinnamoyl]dopamine [(*E*)-**43**] to afford the tricyclic product **44** exclusively. By contrast, photolysis in solution (methanol) resulted in *E–Z* isomerization in all cases.[65]

Another synthetically important photochemical reaction is the Paternò–Büchi reaction,[66] *i.e.*, the photocycloaddition of ketones and aldehydes to olefins. This is a milestone in organic photochemistry and involves attack of the n,π* triplet of the carbonyl compounds to an alkene in the ground state, mostly in the triplet multiplicity, although reactions *via* the singlet are well known.[67] With nucleophilic olefins, the reaction occurs through the initial formation of a C O bond, in the opposite case, formation of a C–C bond occurs first. The use

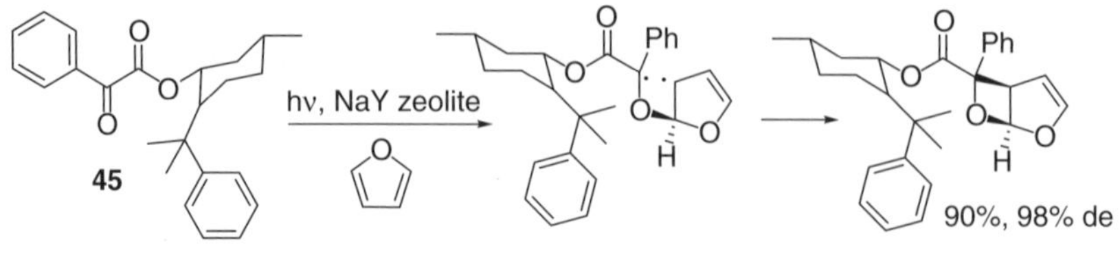

Scheme 2.25

of chiral phenylglyoxylate gave the best results with reference to the stereo-selectivity of the reaction. As an example, chiral 2-methylbutyl benzoylformate (or chiral 1-methylpropyl benzoylformate) reacted with furan in the presence of zeolite with excellent selectivity. The phenylglyoxylate ester (**45**, 0.26 g) was dissolved in a 1 : 4 anhydrous dichloromethane–hexane mixture (100 mL) and NaY zeolite (12 g, previously dried at 500 °C) was added. The mixture was stirred for 12 h, then it was filtered and the precipitate was washed with hexane. The thus-charged zeolite was dried under vacuum and suspended in furan. The mixture was then flushed with nitrogen for 1 h and irradiated in an immersion apparatus with a 125 W high-pressure mercury arc surrounded with a Pyrex water jacket for 7 h (Scheme 2.25). The solvent was evaporated and the residue extracted with diethyl ether. The most interesting point is probably that the diastereoselectivity obtained in Paternò–Büchi reactions can be rationalized by considering the different stabilities of the biradical intermediates.

2.3 Formation of a Carbon–Heteroatom Bond

Reactions involving the formation of carbon–heteroatom bonds include the industrially best known photochemical reactions. In fact, chlorination, bromination and sulfochlorination are major processes in industrial chemistry, and oxygenation has likewise an important role.[68] Due to the focus on fine chemistry of this chapter, the discussion below is limited to laboratory-scale preparations and in particular to some bromination and oxygenation reactions illustrating the advantage of the photochemical approach, as well as to some alkoxylation, hydroxylation and amination reactions.

2.3.1 Oxygenations

Reactions *via* singlet oxygen are one of the strongholds of organic photochemistry, often characterized by high yield and selectivity as well as by a simple experimental set-up, since oxygen activation is conveniently obtained by using various dye families as sensitizers and the inexpensive "quartz-halogen" lamps are generally quite effective. Of excellent synthetic value is the ene reaction with alkenes[69] and the [4 + 2] cycloaddition to dienes,[70] leading to allyl hydroperoxides and to cyclic peroxides, respectively. The latter compounds can be conveniently reduced to alcohols or otherwise elaborated. As an example, both reactions have been exploited in the synthesis of (±)-*proto*-quercitol and (±)-*gala*-quercitol from 1,4-cyclohexadiene, a reaction that has been carried out on a 1 g scale. Thus, the tetraphenylporphyrin-sensitized photooxygenation of 1,4-cyclohexadiene in methylene chloride at room temperature resulted in the formation of the two bicyclic endoperoxides **46** and **47** in a ratio of 88 : 12 (63 and 7% yield, respectively, Scheme 2.26). The two quercitols were then prepared from the respective endoperoxides after separation of the latter on a silica gel column.[71]

Similarly, a convenient synthesis of (±)-*talo*- and (±)-*vibo*-quercitol and other inositol derivatives has been accomplished, by adopting the addition of singlet oxygen onto a suitable cyclohexene precursor as the key photochemical step (Scheme 2.26).[72]

Oxygenations have also been carried out under solvent-free conditions with the starting material embedded in porphyrin-loaded polystyrene beads that serve as photosensitizer. In this way, work up is greatly simplified and the dye remains on the beads, so that contamination of the product is avoided, while the reaction proceeds as efficiently as in the best experiments in solution, which make use of chlorinated hydrocarbon that is certainly a poor choice in terms of green chemistry.[19] By using this protocol, the alcohol derived from sorbic acid (**48**) gave, in excellent yields, the singlet oxygen product **49**. Likewise, β-pinene (**50**) was transformed into the corresponding allylic hydroperoxide **51** in 84% yield (Scheme 2.27).

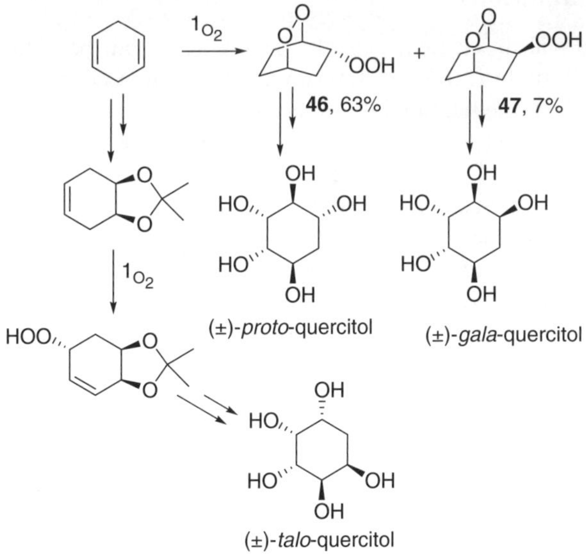

Scheme 2.26

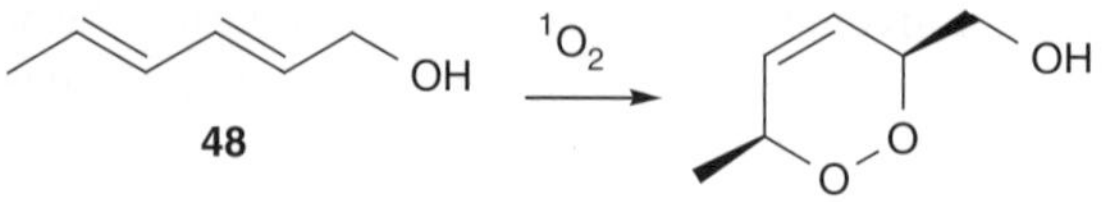

48

49, 85%

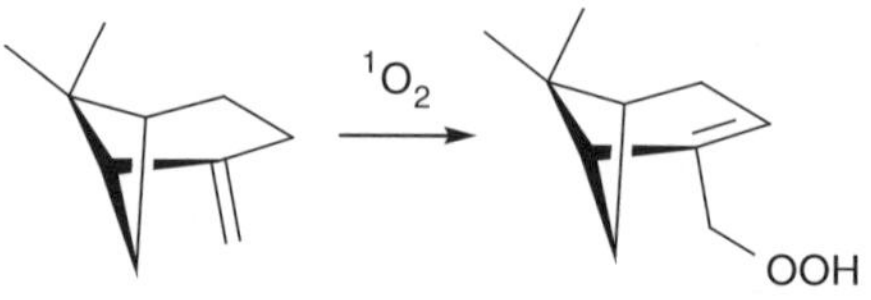

50

51, 84%

Scheme 2.27

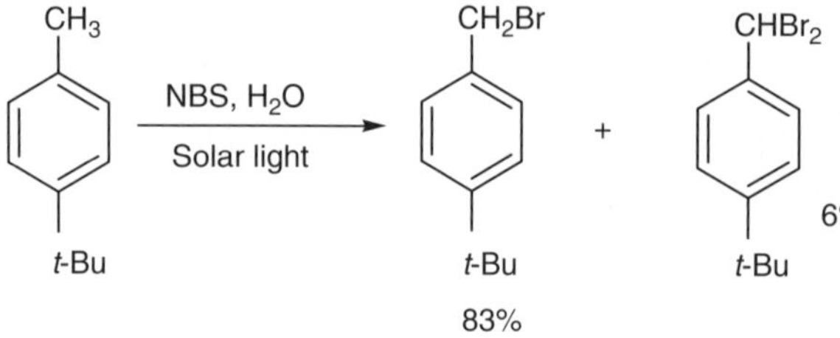

Scheme 2.28

2.3.2 Brominations

As is well known, an efficient benzylic bromination can be obtained by using
N-bromosuccinimide (NBS) as a brominating agent under visible light irradiation in aqueous media. Sunlight induced radical chain bromination of 4-*tert*-
butyltoluene afforded 4-*tert*-butylbenzyl bromide in 83% yield along with a
small amount (6%) of dibrominated derivative (Scheme 2.28).[73] Illumination
with a 40 W incandescent light-bulb was found likewise to be effective. The
starting methylbenzenes formed a layer "on water," but as soon as these were
brominated the specific weight increased and the organic phase sank to the
bottom of the flask. Water was found to be a suitable medium for this green
process also regarding the simple isolation protocol; in fact, the only by-product formed is succinimide. This is soluble in water, thus allowing an easy
separation (by phase separation or filtration) of the hydrophobic organic
products.

2.3.3 Synthesis of Heterocycles

(Aminoalkyl)stilbenes were employed as precursors for the photochemical
induced building of six- to eight-membered nitrogen-containing rings (*e.g.*, of

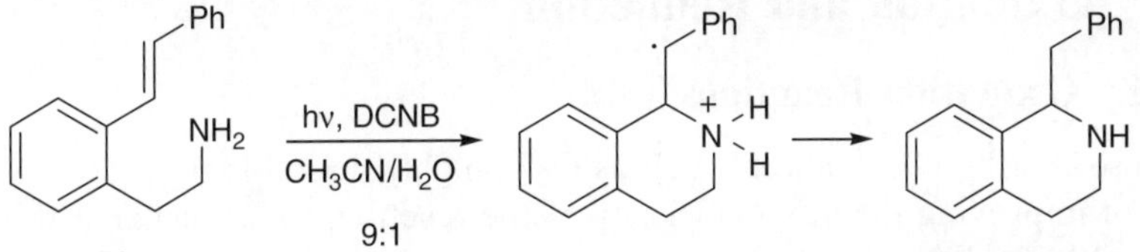

Scheme 2.29

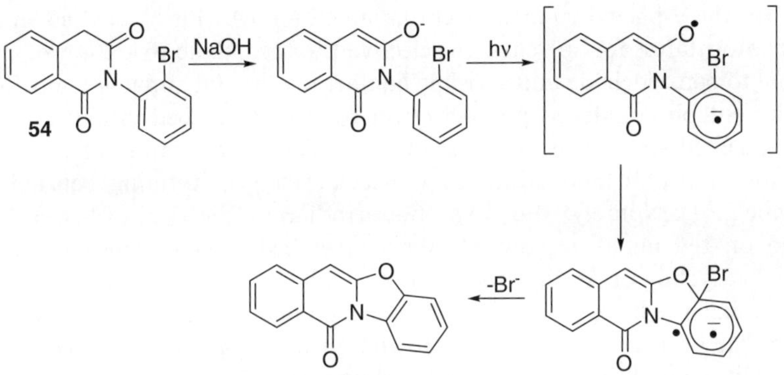

Scheme 2.30

tetrahydrobenzazepines).[74] The presence of a photocatalyst (a dicyanoarene) is crucial for the success of the reaction. Thus, although direct irradiation of the primary (aminoalkyl)stilbenes results only in *E–Z* isomerization, irradiation in the presence of the electron-acceptor dicyanobenzene resulted in a regioselective intramolecular N–H addition to the C=C double bond of the thus-formed radical cation. Scheme 2.29 demonstrates that the irradiation of a deoxygenated 9 : 1 acetonitrile–water solution of (aminoalkyl)stilbene **52** under *m*-dicyanobenzene (DCNB) photocatalyzed conditions furnishes 1-benzyltetrahydroisoquinoline (**53**), in 76% isolated yield.[74]

A photoinduced electron transfer between an enolate and an aromatic halide was likewise adopted for building the isoquinoline ring as a part of the skeleton of alkaloids largely found in antihypertensive and antidepressant agents. Scheme 2.30 shows the formation of an aryl–O bond by means of photolysis of tetrahydroisoquinoline (**54**) under basic conditions (CH$_3$CN/1 M NaOH) in a multilamp reactor, affording the benzoxazolo[3,2-b]isoquinolin-11-one (**55**) in *ca.* 50% yield.[75] The intramolecular displacement of the halogen of the haloarene moiety by the enolate oxygen involves either a reaction in the singlet excited state (*via* S_N2Ar^*) or, more probably, an intramolecular attack of the oxygen radical onto the radical anion of the haloarene [*via* $S_N(ET)Ar^*$].[75]

2.4 Oxidation and Reduction

2.4.1 Oxidation Reactions

The use of a "green oxidant" such as O_2 is highly desirable from the point of view of improving the atom efficiency of the reaction, in particular with reference to the increasing need to use environmentally benign oxidants in place of conventional stoichiometric oxidizing chemical reagents, which are usually hazardous and/or toxic products that generate a large amount of noxious by-products. Photochemistry is an elective method for achieving selective oxidations using molecular oxygen as an oxidant under mild conditions. As an example, the 9-phenyl-10-methylacridinium cation (AcrPh$^+$) acted as an effective photocatalyst for solvent-free selective photocatalytic oxidation of benzyl alcohol to benzaldehyde under visible light irradiation (at ambient temperature) *via* efficient photoinduced photoelectron transfer from benzyl alcohol to the singlet excited state of the acridinium ion.[76] The radical cation of the alcohol then lost a proton and added to molecular oxygen, forming benzaldehyde (Scheme 2.31). Notably, after 15 h photoirradiation, the yield of benzaldehyde (based on the initial amount of AcrPh$^+$) exceeds 800%, demonstrating an efficient recycling of AcrPh$^+$ in the photocatalytic oxidation step.[76]

A similarly efficient, selective and environmentally benign photocatalytic system has been developed for the oxidation by oxygen of activated benzylic and allylic alcohols into their corresponding carbonyls in moderate to excellent yields (Scheme 2.32). The process did not require a transition metal to occur,

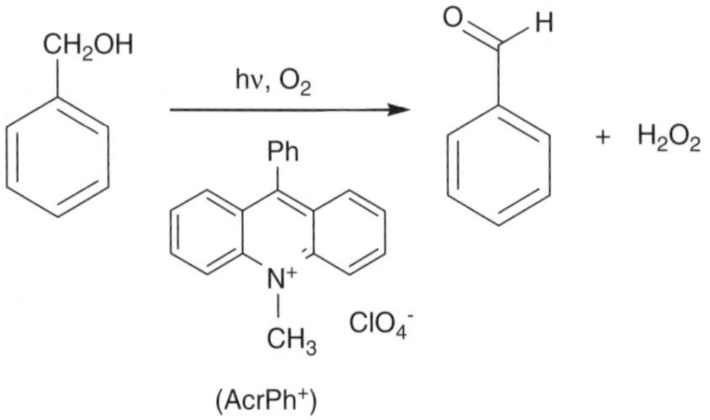

Scheme 2.31

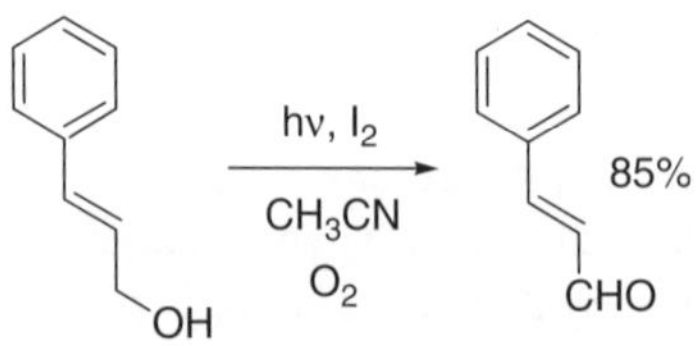

Scheme 2.32

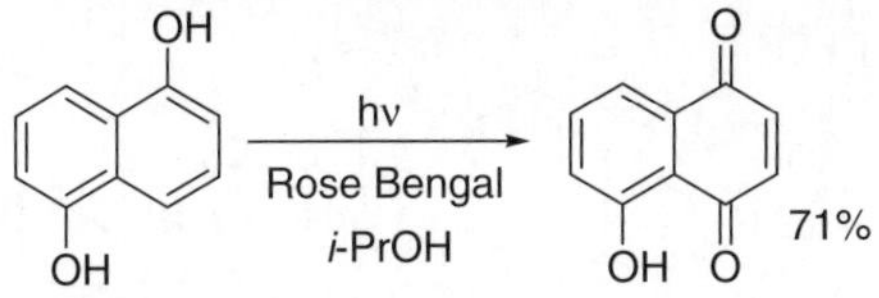

Scheme 2.33

but a catalytic amount of iodine under an O_2 atmosphere, which acted as the stoichiometric reoxidant of the catalyst.[77] Under these conditions, very high inter- and intramolecular chemoselectivities are observed when benzylic OH groups are oxidized in the presence of aliphatic (nonbenzylic) hydroxyls. Moreover, this photochemical process takes place under mild conditions, *viz.* room temperature and atmospheric pressure and, more importantly, it avoids the use of stoichiometric reagents, so that this method is particularly attractive for synthetic purposes from economic and environmental points of view. In one case, *e.g.*, the oxidation of benzhydrol (1 mmol) to benzophenone, only 0.05 mmol of I_2 was required.[77]

Dye-sensitized (*e.g.*, with rose bengal or methylene blue) photooxygenations allow the use of solar light due to the favorable absorption of most dyes within the visible spectrum. Accordingly, the 1,4-naphthoquinone derivative juglone was synthesized from the cheap and commercially available 1,5-dihydroxynaphthalene by taking advantage of this friendly and benign "green photochemical" protocol.[78] Irradiation (*ca.* 2 h) with solar light of an isopropanol solution of 1,5-dihydroxynaphthalene (*ca.* 500 mg) in the presence of a catalytic amount of rose bengal, while purging with a gentle stream of oxygen, was effective for the juglone synthesis (Scheme 2.33).[78]

2.4.2 Oxidative Cleavages

The conversion of stigmasterol into progesterone relies on a key photooxygenation step (Scheme 2.34). A 10% methanolic potassium hydroxide solution of aldehyde **56** (obtained in a high yield from stigmasterol) and rose bengal sensitizer was irradiated with a 1000 W tungsten lamp, under oxygen bubbling. Ketone **57** was then formed *via* a hydroperoxide intermediate and was easily converted into progesterone precursor pregn-5-en-3β-ol-20-one (**58**) under acidic conditions in 94% overall yield (based on **56**).[79]

A photooxygenation step was also found useful for the synthesis of dehydropregnenolone acetate (16-DPA, **61**) – an ideal starting material for the preparation of important steroidal drugs – from naturally occurring diosgenin (**59**, Scheme 2.35).[80]

Acetone, a solvent of low toxicity and low price, was chosen as the reaction media and hematoporphyrin as the photocatalyst. When the reaction was carried out in the presence of both acetic anhydride and pyridine the oxygenation yield reached 75%. Although a mixture of oxygenated products, namely a dioxetane and an allylic hydroperoxide (not shown) were formed, they can be

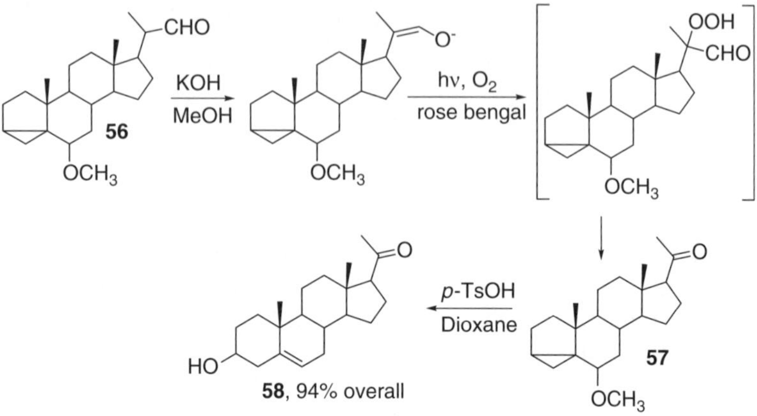

Scheme 2.34

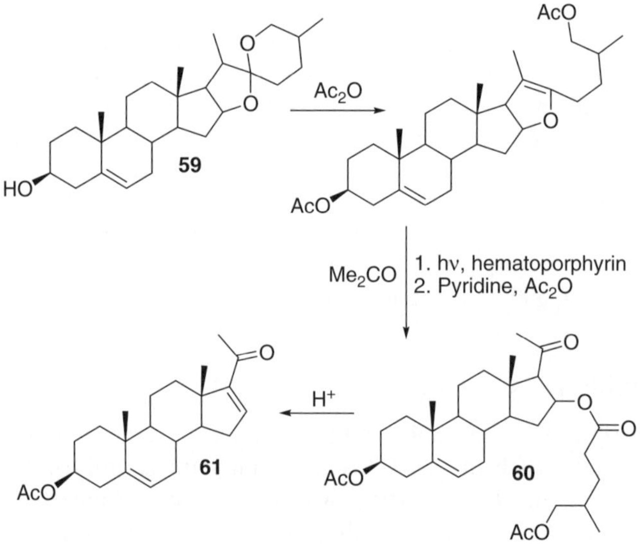

Scheme 2.35

easily converted into the dicarbonyl diosone **60**. This environmentally friendly approach allowed the use of photogenerated singlet oxygen in place of any inorganic oxidant (*e.g.*, CrO_3), thereby avoiding the handling or recovery of metal wastes during the purification step.

2.4.3 Reductions

Amines are convenient reducing agents in photochemistry, as seen in the reductive coupling of ketones (Section 2.2.1). This principle has also been

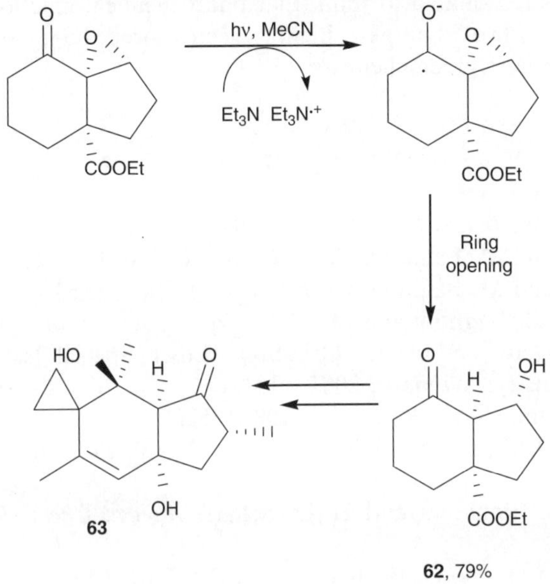

hv, MeCN

Et₃N Et₃N·⁺

Ring
opening

63

OH

COOEt

H

62, 79%

Scheme 2.36

applied to the photoreductive ring opening of αβ-epoxyketones.[81] In such a way, the bicyclic β-hydroxyketone **62** was formed in 79% yield upon 254 nm irradiation in MeCN in the presence of triethylamine (10 equiv.) (Scheme 2.36). Compound **62** was then used for the synthesis of ptaquilosin (**63**), the aglycone of the potent bracken carcinogenic ptaquiloside.[82]

2.5 Conclusions and Outlook

The limited and admittedly personal choice of reaction methods presented here should give a flavor of the variety of synthetic paths available among photochemical reactions. Although most of the cited reports were not conceived with green chemistry postulates in mind, there is an intrinsic advantage in employing a reagent such as the photon that is effective under unparalleled mild conditions and leaves no residue behind. Several of the above methods are satisfactory as reported and are among the best processes for fine chemicals; for the others, suitable changes can generally be devised (*e.g.*, changing solvents such as chloroform to a less noxious compound) so that their introduction makes the method acceptable, something that would be not as simple in a thermal process. The recent literature makes it apparent that the versatility of photochemistry has much to offer organic synthesis. To quote a single case, consider how easy is to take advantage of supramolecular interactions for selective reactions when the activation is caused by a photon, and thus no heating and no reagent are introduced that may disturb the supramolecular organization before reaction.

It thus appears reasonable to think that photochemical reactions will have a significant role in providing new, highly selective procedures, and will acquire an increasing role in green chemistry.

References

1. G. Ciamician, *Bull. Chim. Soc. Fr*, 1908, **3**, i.
2. A. Albini and M. Fagnoni, *Green Chem*, 2004, **6**, 1.
3. A. Albini and M. Fagnoni, *ChemSusChem*, 2008, **1**, 63.
4. A. Albini, M. Fagnoni and M. Mella, *Pure Appl. Chem*, 2000, **72**, 1321.
5. E. M. Carreira and A. G. Griesbeck (Eds.), Special Issue on Organic Photochemistry, *Synthesis*, 2001, 1111.
6. N. Hoffmann, *Chem. Rev.*, 2008, **108**, 1052.
7. S. Protti, D. Dondi, M. Fagnoni and A. Albini, *Pure Appl. Chem.*, 2007, **79**, 1929.
8. P. Esser, B. Pohlman and H.-D. Scharf, *Angew. Chem. Int. Ed.*, 1994, **33**, 2009.
9. C.-L. Ciana and C. G. Bochet, *Chimia*, 2007, **61**, 650.
10. A. G. Griesbeck, N. Maptue, S. Bondock and M. Oelgemöller, *Photochem. Photobiol. Sci.*, 2003, **2**, 450.
11. M. Oelgemöller, C. Jung and J. Mattay, *Pure Appl. Chem.*, 2007, **79**, 1939.
12. C. Jung, K.-H. Funken and J. Ortner, *Photochem. Photobiol. Sci.*, 2005, **4**, 409.
13. B. D. A. Hook, W. Dohle, P. R. Hirst, M. Pickworth, M. B. Berry and K. I. Booker-Milburn, *J. Org. Chem.*, 2005, **70**, 7558.
14. Y. Matsushita, T. Ichimura, N. Ohba, S. Kumada, K. Sakeda, T. Suzuki, H. Tanibata and T. Murata, *Pure Appl. Chem.*, 2007, **79**, 1959.
15. H. Maeda, H. Mukae and K. Mizuno, *Chem. Lett.*, 2005, **34**, 66.
16. P. Watts and S. J. Haswell, *Drug Discovery Today*, 2003, **8**, 586.
17. C. H. Tung, L.-Z. Wu, L.-P. Zhang, H.-R. Li, X.-Y. Yi, K. Song, M. Xu, Z.-Y. Yuan, J.-Q. Guan, H.-W. Wang, Y.-M. Ying and X.-H. Xu, *Pure Appl. Chem.*, 2000, **72**, 2289.
18. M. Veerman, M. J. E. Resendiz and M. A. Garcia-Garibay, *Org. Lett.*, 2006, **8**, 2615.
19. A. G. Griesbeck and A. Bartoschek, *Chem. Commun.*, 2002, 1594.
20. W. Horspool and D. Armesto, *Organic Photochemistry. A Comprehensive Treatment,* Ellis Horwood, Chichester, 1992.
21. J. Mattay and A. G. Griesbeck, *Photochemical Key Steps in Organic Synthesis,* VCH, New York, 1994.
22. W. Horspool and F. Lenci (Eds.), *Organic Photochemistry and Photobiology, 2nd Edition* CRC Press, Boca Raton, 2004.
23. A. G. Griesbeck and J. Mattay (Eds.), *Synthetic Organic Photochemistry,* Marcel Dekker, New York, 2005.
24. R. A. Rossi, A. B. Pierini and A.-B. Peñéñory, *Chem. Rev.*, 2003, **103**, 71.
25. M. Fagnoni and A. Albini, *Acc. Chem. Res.*, 2005, **38**, 713.

26. V. Dichiarante and M. Fagnoni, *Synlett*, 2008, 787.
27. H. E. Zimmerman and D. Armesto, *Chem. Rev.*, 1996, **96**, 3065.
28. See for example: J. Iriondo-Alberdi and M. F. Greaney, *Eur. J. Org. Chem.*, 2007, 4801.
29. W. Adam and A. V. Trofimov, Photomechanistic aspects of bicyclic azoalkanes: triplet states, photoreduction, and double inversion, Chapter 93 in ref. 22.
30. T. Shinmyozu, R. Nogita, M. Akita and C. Lim, Photochemical routes to cyclophanes involving decarbonylation reactions and related process, Chapter 51 in ref. 22.
31. M. A. Garcia-Garibay and L. M. Campos, Photochemical decarbonylation of ketones: recent advances and reactions in crystalline solids, Chapter 48 in ref. 22.
32. A. Natarajan, D. Ng, Z. Yang and M. A. Garcia-Garibay, *Angew. Chem. Int. Ed.*, 2007, **46**, 6485.
33. J. L. Reynolds, K. R. Erdner and P. B. Jones, *Org. Lett.*, 2002, **4**, 917.
34. M. Fagnoni, D. Dondi, D. Ravelli and A. Albini, *Chem. Rev.*, 2007, **107**, 2725.
35. A. Maldotti, A. Molinari and R. Amadelli, *Chem. Rev.*, 2002, **102**, 3811.
36. G. Palmisano, V. Augugliaro, M. Pagliaro and L. Palmisano, *Chem. Commun.*, 2007, 3425.
37. D. Dondi, A. M. Cardarelli, M. Fagnoni and A. Albini, *Tetrahedron*, 2006, **62**, 5527.
38. R. A. Doohan and N. W. A. Geraghty, *Green Chem.*, 2005, **7**, 91.
39. D. Dondi, I. Caprioli, M. Fagnoni, M. Mella and A. Albini, *Tetrahedron*, 2003, **59**, 947.
40. J. Cossy, V. Bellosta, J.-L. Ranaivosata and B. Gille, *Tetrahedron*, 2001, **57**, 5173.
41. C. Schiel, M. Oelgemöller, J. Ortner and J. Mattay, *Green. Chem.*, 2001, **3**, 224.
42. S. Esposti, D. Dondi, M. Fagnoni and A. Albini, *Angew. Chem. Int. Ed.*, 2007, **46**, 2531.
43. C. Chatgilialoglu, D. Crich, M. Komatsu and I. Rvu, *Chem. Rev.*, 1999, **99**, 1991.
44. S. M. Barolo, A. E. Lukach and R. A. Rossi, *J. Org. Chem.*, 2003, **68**, 2807.
45. A. W. G. Burgett, Q. Li, Q. Wei and P. G. Harran, *Angew. Chem. Int. Ed.*, 2003, **42**, 4961.
46. S. Protti, M. Fagnoni and A. Albini, *Org. Biomol. Chem.*, 2005, **3**, 2868.
47. V. Dichiarante, M. Fagnoni and A. Albini, *Angew. Chem. Int. Ed.*, 2007, **46**, 6495.
48. P. Wessig, Regioselective photochemical synthesis of carbo- and heterocyclic compounds: the Norrish/Yang reaction, Chapter 57 in ref. 22.
49. J. R. Scheffer and K. Wang, *Synthesis*, 2001, 1253.
50. P. Wessig, C. Glombitza, G. Müller and J. Teubner, *J. Org. Chem.*, 2004, **69**, 7582.

51. A. G. Griesbeck, W. Kramer and J. Lex, *Angew. Chem. Int. Ed.*, 2001, **40**, 577.

52. U. C. Yoon, Y. X. Jin, S. W. Oh, C. H. Park, J. H. Park, C. F. Campana, X. Cai, E. N. Duesler and P. S. Mariano, *J. Am. Chem. Soc.*, 2003, **125**, 10664.

53. F. Toda, H. Miyamoto, K. Kanemoto, K. Tanaka, Y. Takahashi and Y. Takenaka, *J. Org. Chem.*, 1999, **64**, 2096.

54. F. Toda, H. Miyamoto, T. Tamashima, M. Kondo and Y. Ohashi, *J. Org. Chem.*, 1999, **64**, 2690.

55. V. Singh, Photochemical rearrangements in β,γ-unsaturated enones: the oxa-di-π-methane rearrangement, Chapter 78 in ref. 22.

56. C. K. McClure, A. J. Kiessling and J. S. Link, *Org. Lett.*, 2003, **5**, 3811.

57. D. M. Bassani, The dimerization of cinnamic acid derivatives, Chapter 20 in ref. 22.

58. Y. Sonoda, [2 + 2]-Photocycloadditions in the solid state, Chapter 73 in ref. 22.

59. G. W. Coates, A. R. Dunn, L. M. Henling, J. W. Ziller, E. B. Lobkovsky and R. H. Grubbs, *J. Am. Chem. Soc.*, 1998, **120**, 3641.

60. K. I. Booker-Milburn, J. K. Cowell, A. Sharpe and F. Delgado Jiménez, *Chem. Commun.*, 1996, 249.

61. K. I. Booker-Milburn, J. K. Cowell, F. Delgado Jiménez, A. Sharpe and A. J. White, *Tetrahedron*, 1999, **55**, 5875.

62. M. J. E. Resendiz, J. Taing, S. I. Khan and M. A. Garcia-Garibay, *J. Org. Chem.*, 2008, **73**, 638.

63. A. Haque, A. Ghatak, S. Ghosh and N. Ghoshal, *J. Org. Chem.*, 1997, **62**, 5211.

64. K. I. Booker-Milburn, L. F. Dudin, C. E. Anson and S. D. Guile, *Org. Lett.*, 2001, **3**, 3005.

65. Y. Ito, S. Horie and Y. Shindo, *Org. Lett.*, 2001, **3**, 2411.

66. A. G. Griesbeck and S. Bondock, Oxetane formation: intermolecular additions, Chapter 60 in ref. 22.

67. M. D'Auria, L. Emanuele and R. Racioppi, *Photochem. Photobiol. Sci.*, 2003, **2**, 904.

68. A. M. Braun, M.-T. Maurette and E. Oliveros, in *Photochemical Technology*, John Wiley & Sons, Chichester, 1991.

69. E. L. Clennan, Photo-oxygenation of the ene-type, Chapter 12, p 365, in ref. 23.

70. M. R. Iesce, Photo-oxygenation of the [4 + 2] and [2 + 2] type, Chapter 11, p 299 in ref. 23.

71. E. Salamci, H. Seçen, Y. Sütbeyaz and M. Balci, *J. Org. Chem.*, 1997, **62**, 2453.

72. A. Maraş, H. Seçen, Y. Sütbeyaz and M. Balci, *J. Org. Chem.*, 1998, **63**, 2039.

73. A. Podgoršek, S. Stavber, M. Zupan and J. Iskra, *Tetrahedron Lett.*, 2006, **47**, 1097.

74. F. D. Lewis, D. M. Bassad, E. L. Burch, B. E. Cohen, J. A. Engleman, G. Dasharatha Reddy, S. Schneider, W. Jaeger, P. Gedeck and M. Gahr, *J. Am. Chem. Soc.*, 1995, **117**, 660.
75. A. Senthilvelan, D. Thirumalai and V. T. Ramakrishnan, *Tetrahedron*, 2005, **61**, 4213.
76. K. Ohkubo, K. Suga and S. Fukuzumi, *Chem. Commun.*, 2006, **20**, 2018.
77. S. Farhadi, A. Zabardasti and Z. Babazadeh, *Tetrahedron Lett.*, 2006, **47**, 8953.
78. O. Suchard, R. Kane, B. J. Roe, E. Zimmermann, C. Jung, P. A. Waske, J. Mattay and M. Oelgemöller, *Tetrahedron*, 2006, **62**, 1467.
79. R. K. Boeckman Jr., J. P. Bershas, J. Clardy and B. Solheim, *J. Org. Chem.*, 1977, **42**, 3630.
80. Y. Zhang, Y.-Y. Liu, X.-X. Cheng, X.-S. Wang and B.-W. Zhang, *Chinese J. Chem.*, 2005, **23**, 753.
81. J. Cossy, A. Bouzide, S. Ibhi and P. Aclinou, *Tetrahedron*, 1991, **47**, 7775.
82. J. Cossy, S. Ibhi, P. H. Khan and L. Tacchini, *Tetrahedron Lett.*, 1995, **36**, 7877.

CHAPTER 3

Supported Organic Bases: A Green Tool for Carbon–Carbon Bond Formation

GIOVANNI SARTORI, RAIMONDO MAGGI, CHIARA ORO AND LAURA SOLDI

"Clean Synthetic Methodologies Group", Dipartimento di Chimica Organica e Industriale, Università degli Studi di Parma, and Consorzio Interuniversitario "La Chimica per l'Ambiente" (INCA), UdR PR2, Viale G.P. Usberti 17A, I-43100 Parma, Italy

3.1   Introduction

The demands of modern society for new functional chemical entities has driven the development of novel methods that are revolutionising the way we think, plan and optimize chemical processes. In future, the impact will be even greater by influencing both materials discovery and catalysts design.[1] Heterogeneous catalysis represents a powerful tool for the application of clean technologies in chemical synthesis with the emphasis on reduction of waste at source. The preparation of fine chemicals and pharmaceuticals is frequently accompanied by the production of large amounts of waste, reaching values 25–100 times higher than those of the target compounds.[2] This worrying situation is partly because the production of fine chemicals and specialties generally involves multi-step syntheses, and partly to the widespread use of stoichiometric reagents rather than catalytic methods. Consequently, the worldwide interest in problems related to sustainable development and waste minimization has prompted many

RSC Green Chemistry Book Series
Eco-Friendly Synthesis of Fine Chemicals
Edited by Roberto Ballini
© Royal Society of Chemistry 2009
Published by the Royal Society of Chemistry, www.rsc.org

academic and industrial research groups to study the possibility of replacing multi-step, expensive and environmentally harmful processes with single step, multicomponent, solvent-free and benign ones.[3] The replacement of stoichiometric reactions by catalytic ones represents, without doubt, a great instrument with which to make synthetic procedures more eco-efficient. Moreover, the use of solid and recyclable catalysts represents a substantial improvement, especially if their homogeneous counterparts are particularly expensive. In fact, one of the major problems related to the use of homogeneous catalysts is the difficulty in their recovery, and consequently they are often destroyed in the work-up procedures; of consequence, large amounts of waste, typically salts and contaminated aqueous solutions, are produced.[4] Recently, various methodologies have been developed to immobilize metal as well as molecular homogeneous catalysts on the surface of heterogeneous organic and inorganic polymeric materials.[5] This approach allowed many new applications of catalysis in fine chemical preparation since the heterogeneous catalysis supplies the opportunity for easy separation and recycling of the catalyst, easy purification of the products and, possibly, continuous or multiple processing of organic compounds.[6]

Here we review recent articles concerning the preparation of basic organic supported catalysts and their use in fine chemical synthesis. The review is split up in two general sections: the first dealing with monofunctional heterogeneous catalysts, and the second showing the applicability of polyfunctional catalysts able to promote multi-step sequential reactions or to enhance the reaction rates. Three fundamental classes of carbon–carbon forming reactions will be considered in both sections, namely the aldol condensation, the Michael condensation and the Knoevenagel reaction. We particularly focussed our attention on the synthetic aspects underlining the practical advantages connected with the application of supported organic bases. Techniques for synthesising the supported catalysts, that are well documented in the literature, are also considered.

3.2　Monofunctional Catalysts

Catalysts and reagents immobilized upon a range of insoluble supports have been utilized and reported since the late 1960s.[7]

The monofunctional catalysts were obviously the first supported organic catalysts prepared. The advantages offered are a direct consequence of the support matrix. In addition to the easy separation from the reaction mixture, simply by filtration, other advantages of the supported base catalysts are due to the unique microenvironment created for the reactants within the support, which is responsible for improved catalyst stability, increased selectivity due to steric hindrance and superior activity due to support surface cooperation.

Generally, immobilization may be divided up into four distinct methodologies: (i) formation of a covalent bond with the organic catalyst, (ii) adsorption or ion-pair formation, (iii) encapsulation and (iv) entrapment.

In any case of catalyst immobilization, the support material needs to be thermally, chemically and mechanically stable during the reaction process. Moreover, the structure of the support needs to be such that the active sites are well dispersed on its surface and that these sites are easily accessible. Also, the support must have a reasonably high surface area (typically $> 100\,\mathrm{m^2\,g^{-1}}$) and appropriate pore size (*i.e.* $> 20\,\text{Å}$) to allow easy diffusion of the reactants to the active sites.

There are, however, some issues of critical importance in the production and use of supported base catalysts, namely the correct evaluation of the recycling efficiency, which is frequently nullified by the leaching phenomena. Unfortunately, these fundamental problems are frequently either ignored or barely considered in the synthetic studies.

3.2.1 Aldol Reactions

The aldol condensation of trichlorosilyl enol ethers and aldehydes was reported to be efficiently catalysed by polystyrene-supported hexamethylphosphoric triamide (HMPA) catalyst.[8] HMPA has been extensively used in organic synthesis because of its superior solvation and coordination properties.[9] Unfortunately, HMPA is also toxic and a potential carcinogenic reagent; this limits its broad utility in organic synthesis. To address the toxicity problems of HMPA, while maintaining its excellent synthetic utility, a polystyrene-supported HMPA-type derivative (PS-HMPA) was prepared by a multi-step methodology involving the treatment of methylamino polystyrene with tetramethylphosphorodiamidic chloride in the presence of potassium carbonate, dimethylaminopyridine and triethylamine (Scheme 3.1).

This supported triamide was utilized in the model reaction between the trichlorosilyl enol ether of cyclohexanone and benzaldehyde (Scheme 3.2), carried out at $-78\,^{\circ}\mathrm{C}$: after 3 h the aldol product was obtained in 80% yield and 1:1 *syn/anti* diastereoselectivity; when the temperature was increased to $-23\,^{\circ}\mathrm{C}$,

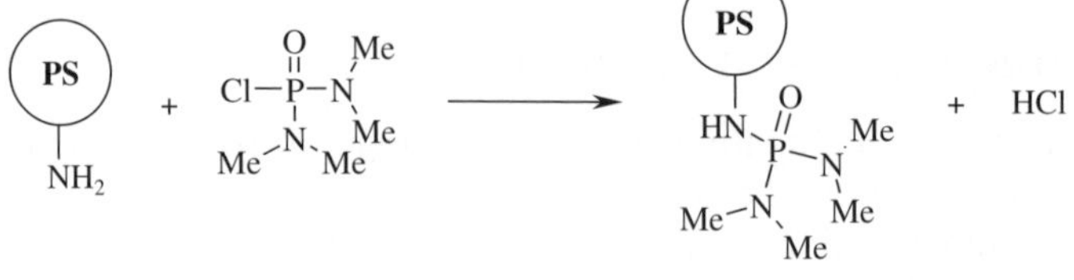

Scheme 3.1

Scheme 3.2

the reaction time for complete conversion was shortened to 2 h, and the diastereoselectivity was improved to 9:1 (*syn/anti*).

The aldol products in the present reaction showed an opposite configuration to that of the products obtained with homogeneous, chiral phosphoramide catalysts.[10] This observation suggests that the reaction can proceed through two possible pathways: the first involves the formation of a pre-aldol complex responsible for an open-chain mechanism, the second involves a closed, boat-like transition state. The authors proposed the first hypothesis as more likely, due to the bulky environment on the polymer; this hypothesis was indirectly supported by the fact that the use of Denmark's bulky chiral phosphoramide catalyst resulted in prevalent *syn* selectivity.[11] The polymer could be reused in a second run but, however, its effectiveness was somewhat diminished (about 15% lower then the original reaction). The diminished catalytic activity was presumably due to slow destruction of the polymer backbone when subjected to the vigorous magnetic stirring.

The self-aldol condensation of unmodified aldehydes in toluene catalyzed by propylamines supported on mesoporous silica FSM-16 (surface area $881 \, m^2 g^{-1}$) has been reported.[12] The aminopropyl groups (propylamine, *N*-methylpropylamine and *N,N*-diethylpropylamine) were anchored on FSM–16 silica by post-modification methodology by using 3-aminopropyl-, *N*-methyl-3-aminopropyl- and *N,N*-diethyl-3-aminopropyltriethoxysilane respectively (Scheme 3.3).[13]

Nucleophilic reactions of unmodified aldehydes are usually difficult to control, affording complex mixture of products, often due to the high reactivity of the formyl group under either basic or acidic reaction conditions. The activity order of the supported amines was secondary > primary > tertiary, which may suggest the intervention of an enamine pathway;[14] the enals were exclusively obtained as (*E*) isomers. Notably, FSM-16–(CH$_2$)$_3$–NHMe exhibited higher activity than conventional solid bases such as MgO and Mg-Al-hydrotalcite [hexanal self-aldol condensation: FSM-16–(CH$_2$)$_3$–NHMe 97% conversion and 85% yield in 2 h, MgO 56% conversion and 26% yield in 20 h, Mg-Al-hydrotalcite 22% conversion and 11% yield in 24 h].

The leaching test allowed exclusion of the possible migration of any active catalytic species in solution. The catalyst on recycling showed a decrease in

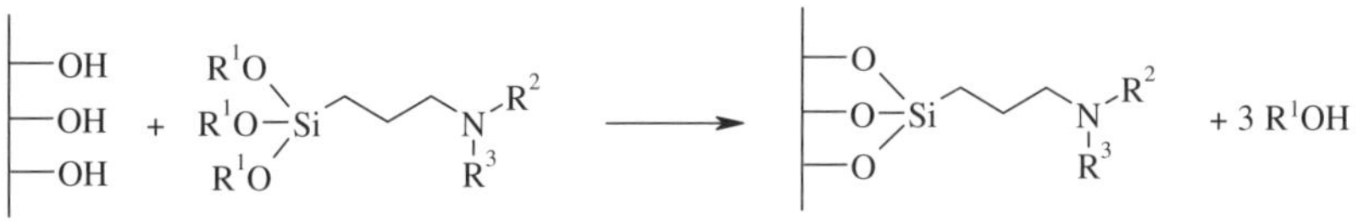

$$R^1 = Me, Et$$
$$R^2 = H, Me, Et$$
$$R^3 = H, Et$$

Scheme 3.3

catalytic activity (60% yield after 2 h). However, the activity was comparable to that observed for the first run when the filtered catalyst was simply dispersed in a diluted aqueous solution of potassium carbonate followed by washing with distilled water and subsequent drying. Moreover, the catalyst supported on mesoporous silica showed higher activity than that supported on amorphous silica: the initial rate of the former catalyst was 2.2 times higher than that of the latter. In addition, FSM-16–$(CH_2)_3$–NHMe exhibited much higher activity than the homogenous amine catalyst in terms of both initial rate and yield. The authors ascribed this behaviour to the enrichment of the reactants inside the well-ordered pores of mesoporous sili*ca*. The reaction could be more efficiently promoted by using an ionic liquid ([bmim]PF_6) instead of toluene.[15]

Guanidines are strong bases that find applications in numerous reactions widely employed in organic synthesis, including carbon–carbon bond formation.[16] 1,2,3-Tricyclohexylguanidine (TCG) **15**, synthesized as shown in Scheme 3.4, was heterogenized onto hydrophobic zeolite Y by the "ship in a bottle" method[17] and supported onto mesoporous silica MCM-41 (**14**) by reaction of the guanidine with 3-glycidyloxypropyl MCM-41 silica prepared by the post-modification method from MCM-41 silica and 3-glycidyloxypropyltrimethoxysilane (Scheme 3.5).[18]

The catalytic activity of this material was tested in the condensation of acetone with benzaldehyde to give the α,β-unsaturated ketone **19** (Scheme 3.6).[19]

In the presence of TCG (10 mol.%) under homogeneous conditions benzaldehyde was quantitatively converted after 2 h, giving compound **19** with 94% yield, whereas 10 mol.% TCG encapsulated in zeolite Wessalith® (**12**) afforded **19** with 8% yield accompanied by compound **18** in 48% yield.

By using TCG supported on MCM-41 silica the total yield (**18** + **19** products) ranged from 31% to 89%, and the amount of the addition product **18** depended on the alcohol solvent utilized (**19/18**: MeOH 89/0, Pr^{i}OH 40/22, Bu^{t}OH 20/11).

The strongly hindered guanidine base 1,5,7-triazabicyclo[4.4.0]dec-5-ene (TBD) was supported to MCM-41 silica by the same post-modification methodology[18] and the catalyst was utilized in the condensation reaction between benzaldehyde and heptanal to produce jasminaldehyde **21** (Scheme 3.7).[20]

This reaction could also afford the heptaldehyde self-condensation product, some benzyl alcohol and benzoic acid (from the Cannizzaro reaction) as by-products; authors reported that a higher temperature favours the jasminaldehyde formation. By slowly adding heptaldehyde, to keep its concentration low, a 99%

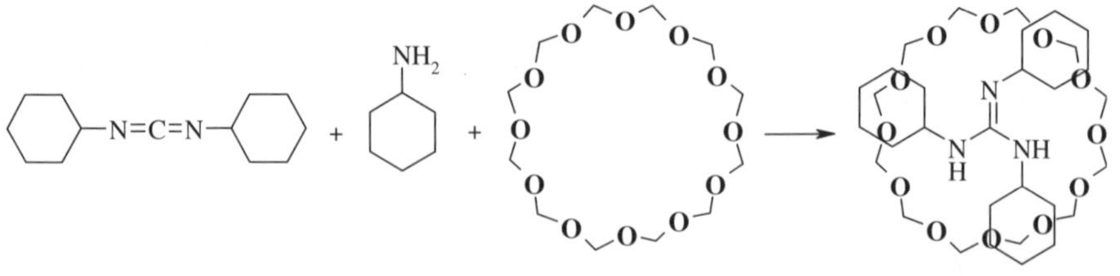

Scheme 3.4

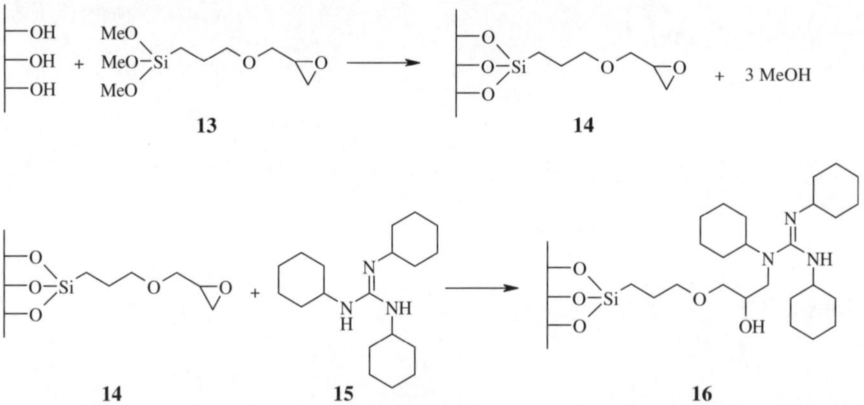

13 14

14 15 16

Scheme 3.5

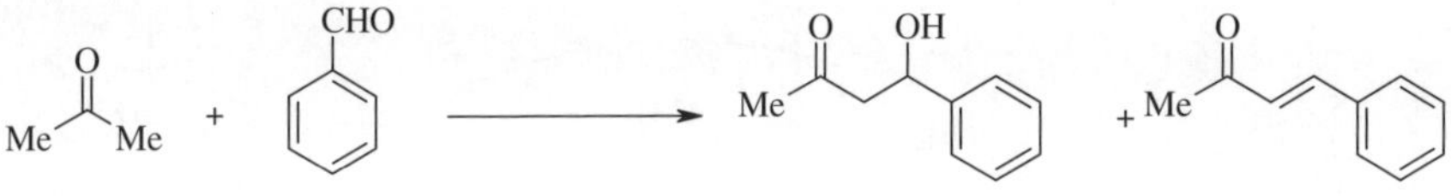

17 5 18 19

Scheme 3.6

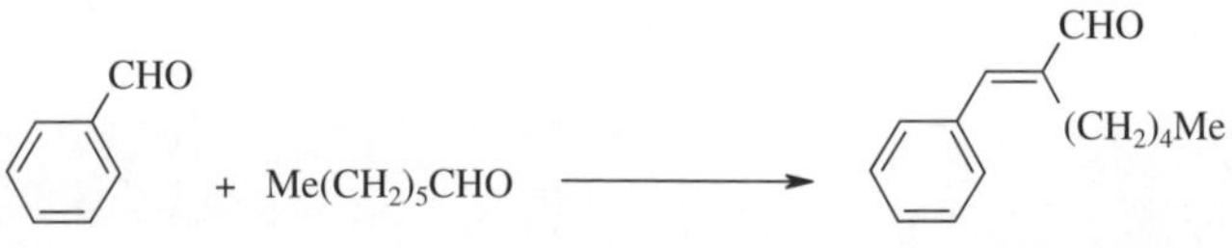

5 20 21

Scheme 3.7

conversion accompanied by a 91% selectivity could be reached; conversely, when the heptaldehyde was added in one portion, a much lower selectivity was obtained (49%) at the same conversion (95%). Moreover, the catalyst suffered from severe deactivation and consequently it could not be successfully recycled.

The nitroaldol condensation between nitroalkanes and aromatic aldehydes to give (*E*)-nitrostyrenes **24**[21] (Scheme 3.8), routinely performed in the presence of a wide range of base catalysts, has been studied in the presence of aliphatic amines supported on MCM-41 silica.[13]

The inert nature of the amine-free material suggested that the supported amines were responsible for the activity. Their efficiency follows the trend primary ≫ secondary > tertiary, in disagreement with the tabulated basicity order of aliphatic amines in polar solvents.[22] These features, along with the reported remarkable ease with which aromatic aldehydes gave rise to the supported imines by reaction with aminopropylsilica,[23] allowed the authors to

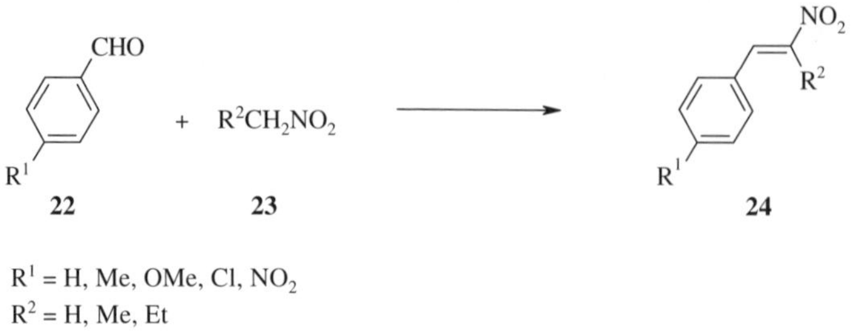

$R^1 = H, Me, OMe, Cl, NO_2$
$R^2 = H, Me, Et$

Scheme 3.8

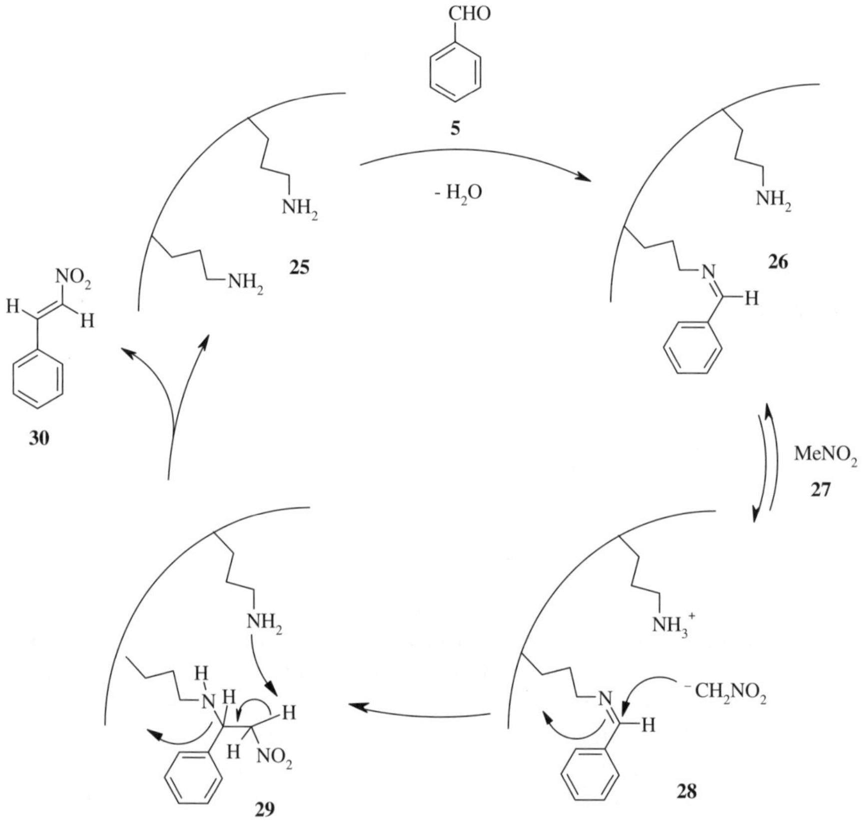

Scheme 3.9

formulate a mechanistic hypothesis involving a supported imine intermediate (**26**, Scheme 3.9), according to that already reported for the reaction of benzaldehyde and ethyl cyanoacetate in the presence of supported primary amines **25**.[24]

Repeated condensation tests in the reaction between benzaldehyde and nitromethane were successively performed for five cycles, affording nitrostyrene

Table 3.1 Condensation reaction between nitroalkanes and aromatic aldehydes in the presence of MCM-41–$(CH_2)_3$–NH_2.

R^1	R^2	Time (h)	24 Yield (%)	24 Selectivity (%)
H	H	1	98	99
Me	H	1	97	98
OMe	H	1	96	99
NO_2	H	1	98	99
Cl	H	1	97	98
H	Me	6	95	97
OMe	Me	6	95	98
Cl	Me	6	98	99
Cl	Et	6	88	98
NO_2	Et	6	92	97

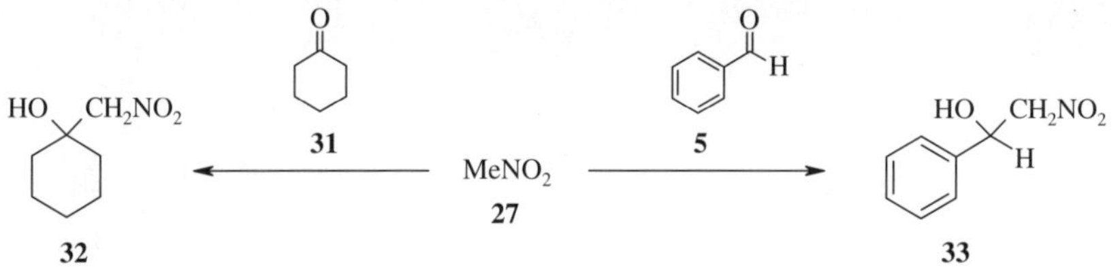

Scheme 3.10

in successive % yields of 98, 95, 95, 90 and 84. The reaction was extended to different aromatic aldehydes and nitroalkanes, giving the corresponding nitrostyrenes in high yield and excellent selectivity. In all cases the (*E*)-stereoisomer was the sole product detected (Table 3.1).

Commercially available TBD supported on polystyrene (PS-TBD) was utilized to promote the Henry reaction (Scheme 3.10).[25]

The addition of nitromethane to cyclohexanone was completed in 1 h at 0 °C, affording product **32** in 82% yield; as expected the reaction with benzaldehyde afforded the corresponding product **33** in 95% yield in only 5 min at the same temperature. However, the reaction was not generally applicable as other aliphatic ketones and acetophenone did not react.

The direct asymmetric aldol reaction between unmodified aldehydes and ketones plays an important role in nature as a source of carbohydrates[26] and it is used for the synthesis of chiral β-hydroxycarbonyl compounds. This reaction was performed by using (*S*)-proline/poly-(diallyldimethylammonium) hexafluorophosphate heterogeneous catalytic system **36**.[27] The catalyst was simply prepared by mixing a suspension of the commercially available polyelectrolyte **34** in methanol with a solution of (*S*)-proline (**35**) in the same solvent (Scheme 3.11).

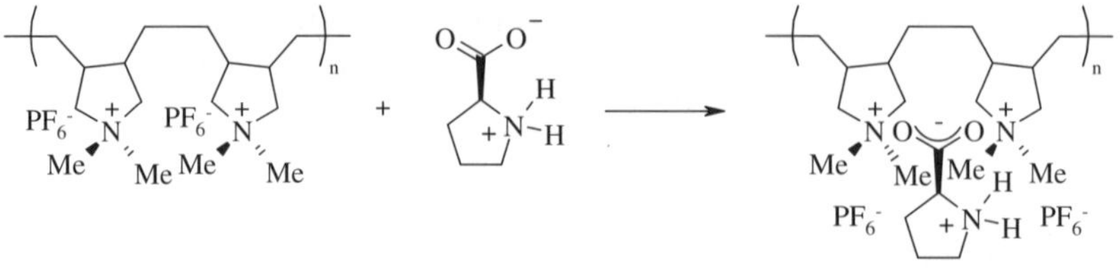

34   **35**   **36**

Scheme 3.11

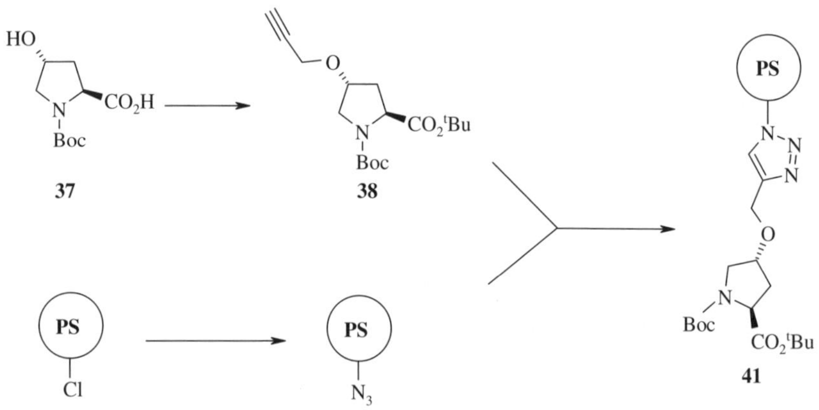

37   **38**

39   **40**   **41**

Scheme 3.12

In the model reaction of benzaldehyde (0.5 mmol) with acetone (1 mL) (Scheme 3.6) at 25 °C for 15 h, catalyst **36** gave a mixture of aldol (**18**) and elimination product (**19**) in 98:<1 molar ratio, with (*S*)-**18** obtained in 97% yield and 72% ee. The catalyst could be recovered and reused for six cycles, giving **18** with similar yield and ee values. The reaction could be applied to variously substituted aromatic aldehydes, furnishing the aldol products in 58–98% yield and 62–76% ee. The reaction could also be carried out with cyclic ketones such as cyclohexanone and cyclopentanone (18–80% yield and 20–91% ee).

The same reaction was performed by using polystyrene-supported hydroxyproline **37**.[28] The solid catalyst **41** was prepared according to the procedure depicted in Scheme 3.12 by 1,3-dipolar cycloaddition of the azide-substituted Merrifield resin **40** with the *O*-propargyl hydroxyproline **38**, which in turn was synthesized by a well-established method.[28,29]

The asymmetric aldol reaction was extended to different aldehydes and ketones in water (Scheme 3.13), catalysed by resin **41** and in the presence of a catalytic amount of the water-soluble DiMePEG (MW ~ 2000) which facilitates the diffusion of reactants to resin.

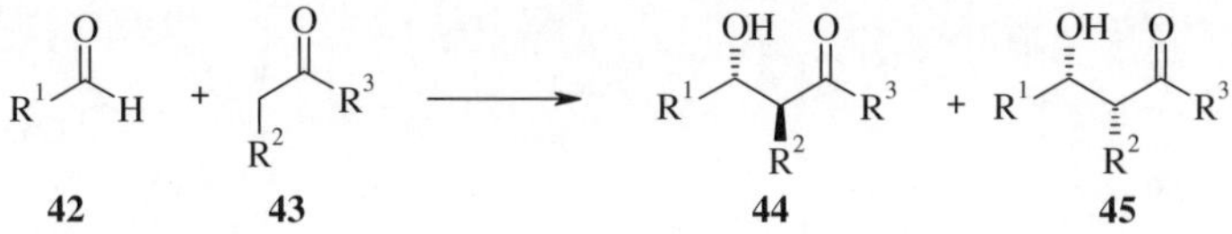

R^1 = Ph, 4-MeOC$_6$H$_4$, 4-NO$_2$C$_6$H$_4$, 4-CF$_3$C$_6$H$_4$, 2-ClC$_6$H$_4$, 2-naphthyl, 2-furfuryl
R^2 = H,OH
R^3 = Me
R^2R^3 = (CH$_2$)$_3$, (CH$_2$)$_4$

Scheme 3.13

Table 3.2 Asymmetric aldol reactions shown in Scheme 3.13, catalysed by resin **41**.

R^1	R^2	R^3	Time (h)	Yield (%)	**44 : 45** ratio
Ph	–(CH$_2$)$_4$–		60	96	96 : 4
2-Naphthyl	–(CH$_2$)$_4$–		84	95	95 : 5
4-BrC$_6$H$_4$	–(CH$_2$)$_4$–		84	95	96 : 4
2-Furfuryl	–(CH$_2$)$_4$–		65	94	82 : 18
4-FC$_6$H$_4$	–(CH$_2$)$_4$–		60	96	97 : 3
2-ClC$_6$H$_4$	Me	OH	144	45	58 : 42
Ph	–(CH$_2$)$_3$–		60	87	83 : 17

Some results are summarized in Table 3.2.

Interestingly, aromatic aldehydes react with cyclohexanone with higher diastereoselectivities than those recorded for the monomeric proline derivative in water. In addition, the robust polymeric catalyst **41** can be reused for at least three times with similar performance after filtration, washing with ethyl acetate and drying.

3.2.2 Michael Reactions

SiO$_2$–(CH$_2$)$_3$–NEt$_2$, prepared by post-modification methodology,[13] has been employed for the Michael addition of nitroalkanes **46** to electron-poor olefins **47** (Scheme 3.14).[30]

A wide range of nitroalkanes underwent Michael addition to electron poor olefins by this procedure, affording products **48**; some results are reported in Table 3.3.

The different reactivity was ascribed to the different activation of the electron-poor carbon–carbon double bond by the electron-withdrawing group EWG. The catalyst could be utilized with similar results for at least two further cycles in the model reaction between 1-nitropropane and methyl vinyl ketone (1st cycle: 80% yield, 2nd cycle: 78% yield, 3rd cycle: 79% yield).

$R^1 = $ Me, Et, Pr, C_5H_{11}
$R^2 = $ H,Me
EWG = MeCO, EtCO, $PhSO_2$, CN, MeOCO

Scheme 3.14

Table 3.3 Michael addition of different nitroalkanes to α,β-unsaturated compounds catalysed by SiO_2–$(CH_2)_3$–NEt_2 (see Scheme 3.14).

R^1	R^2	EWG	Time (h)	48 Yield (%)
Me	H	MeCO	3	75
Et	H	MeCO	5	80
Me	Me	MeCO	5	82
Pr	H	MeCO	5	85
Pr	H	EtCO	6	83
Me	H	$PhSO_2$	25	81
Me	Me	$PhSO_2$	25	90
Pr	H	$PhSO_2$	28	65
Et	H	CO_2Me	26	61
Pr	H	CO_2Me	30	62
Pr	H	CN	15	58
C_5H_{11}	H	CN	16	56

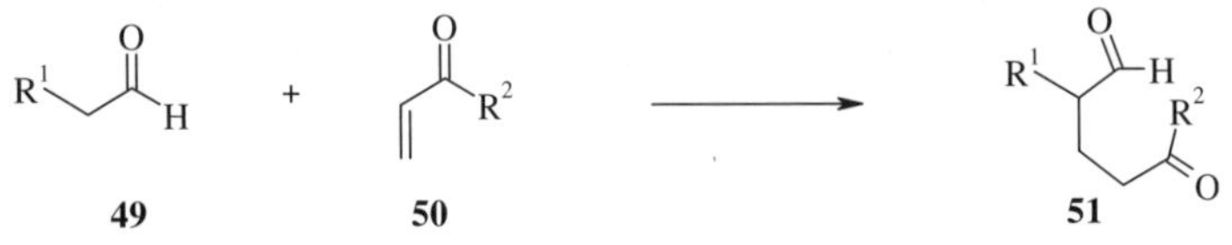

$R^1 = $ Et, Bu, C_6H_{13}, Bn
$R^2 = $ Me, Et

Scheme 3.15

Secondary amines, immobilized on mesoporous FSM-16 silica by the same methodology, were utilized to obtain substituted 5-ketoaldehydes, which are important synthons for the preparation of natural products such as terpenoids, by direct 1,4-conjugate addition of unmodified aldehydes to vinyl ketones (Scheme 3.15).[31]

The reaction, which could not be catalyzed by ordinary basic metal oxides such as MgO and Mg-Al-hydrotalcite, was efficiently promoted by FSM 16–$(CH_2)_3$–NHMe, suggesting the intervention of the enamine pathway.

Elemental analysis showed that the nitrogen content of the catalyst did not decrease after the reaction, confirming that the leaching of amino groups during the reaction was negligible. The catalyst could be recycled, showing a decrease in activity in the reaction between decanal and methyl vinyl ketone (from 70% to 43% yield). However, the activity of the recovered catalyst was restored by treatment with a dilute aqueous solution of potassium carbonate followed by washing with water. The reaction could be applied even with a small amount of catalyst (1 mol.%) to different aldehydes and vinyl ketones, giving the corresponding products in 37–93% yield.

The Michael 1,4-addition of methylene activated compounds to various α,β-unsaturated carbonyl compounds[32] in water was promoted by amines grafted on silica gel by a post-modification method.[13] Since the reaction was carried out in a triphasic system, due to low solubility of organic substrates, internal heating by microwave irradiation (100 W) was exploited. In the model reaction between ethyl 2-oxocyclopentanecarboxylate (**52**) and 3-buten-2-one (**53**) (Scheme 3.16), the temperature was ramped for 2 min from room temperature to 80 °C.

Without catalyst or in the presence of silica gel alone, the starting materials were recovered unchanged; among supported amine catalysts investigated, *N,N*-diethylaminopropylated silica gel (5 mol.%) provided the best result (92% yield of product **54**). Concerning the solvent effect, while in THF or xylenes the reaction did not proceed, without solvent product **54** could be isolated in 33% yield. Satisfactory Michael additions of various cyclic and acyclic 1,3-dicarbonyl compounds to various α,β-unsaturated carbonyl compounds were obtained (Table 3.4). The catalyst could be easily recycled at least four times with a slight decrease in activity; the original activity can be retrieved after washing with dilute aqueous sodium carbonate.

The commercially available polystyrene-supported iminophosphorane (PS-BEMP) **55** (Figure 3.1) has been utilized to promote the Michael addition of acyclic- and cyclic-1,3-dicarbonyl compounds, including those substituted at the active methylene, to electron-poor olefins.[33]

The reaction did not require dry solvents or inert atmosphere and afforded the desired adducts in 42–98% yield. As expected, the more reactive acrolein and methyl vinyl ketone gave very good results with both acyclic and cyclic β-ketoesters and β-diketones and, surprisingly, acrylonitrile and methyl acrylate, reported to be totally inactive under Lewis acid catalysis,[34] afforded the corresponding adducts with α-acetylbutyrolactone in high yield (77 and 98% respectively). In the reaction with crotonaldehyde, a 1:1 mixture of

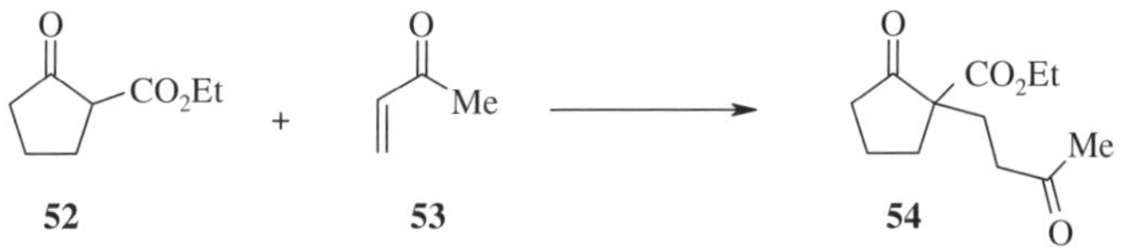

Scheme 3.16

Table 3.4 Michael addition of 1,3-dicarbonyl compounds to various α,β-unsaturated carbonyl compounds catalysed by SiO_2–$(CH_2)_3$–NEt_2.

Product	Time (min)	Yield (%)
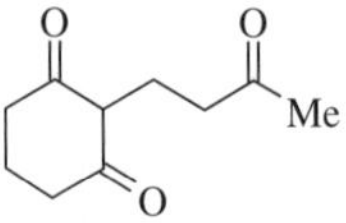 (2-substituted cyclohexane-1,3-dione; Me)	47	85
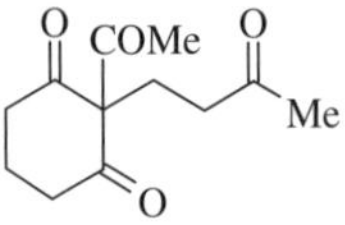 (2-COMe cyclohexane-1,3-dione; Me)	15	58
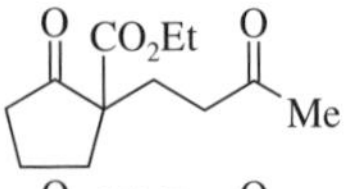 (2-CO₂Et cyclopentanone; Me)	10	92
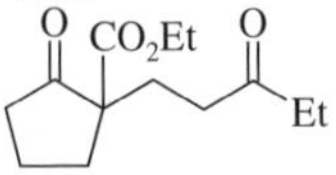 (2-CO₂Et cyclopentanone; Et)	10	94
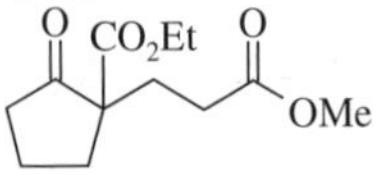 (2-CO₂Et cyclopentanone; OMe)	6	68
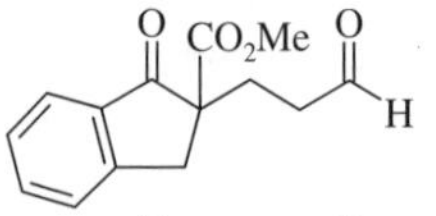 (indanone; CO₂Me; H)	20	75
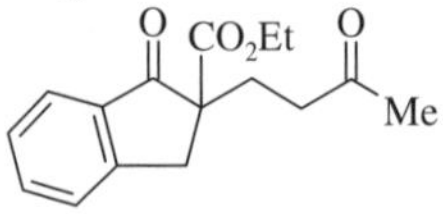 (indanone; CO₂Et; Me)	40	93
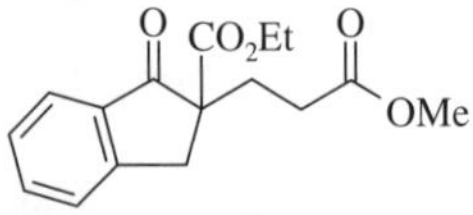 (indanone; CO₂Et; OMe)	120	87
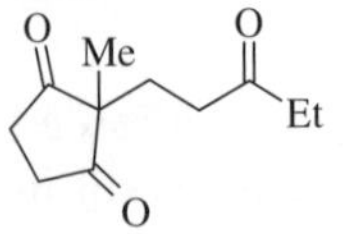 (2-Me cyclopentane-1,3-dione; Et)	62	82
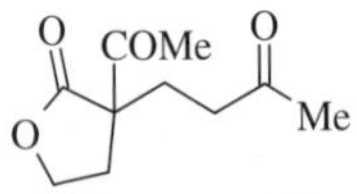 (COMe butyrolactone; Me)	14	84
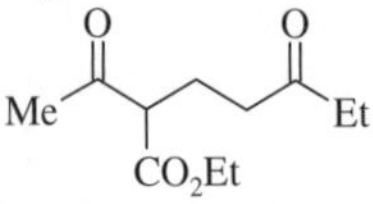 (Me; Et; CO₂Et)	10	79

diastereoisomers was detected. The catalyst could be recycled three times in the reaction of acetylbutyrolactone and methyl acrylate without significant lowering of product yield, although the reaction time increased from 4 to 7 and to 16 h; however, after three cycles the catalyst beads were crushed by mechanical stirring, leading to an inactive powder.

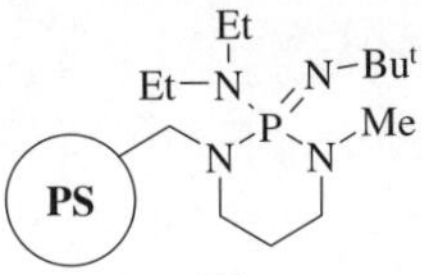

55

Figure 3.1

56   **57**   **58**

Scheme 3.17

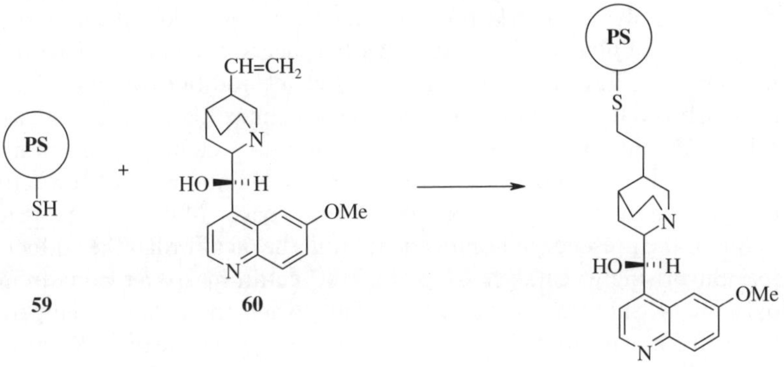

59   **60**   **61**

Scheme 3.18

MCM-41-TBD catalyst has been used in the Michael reaction between 2 cyclopenten-1-one and ethyl cyanoacetate (Scheme 3.17).[35]

Product **58** was obtained in 52% yield and 100% selectivity. Differences in reactivity were observed, related to the nature of the reagents: thus diethyl malonate reacted with methyl vinyl ketone, furnishing the Michael adduct in 34% yield after 30 min, whereas the adduct with 2-cyclohexen-1-one was obtained in 15% yield after 60 h.

The catalytic activity of cinchonidine, cinchonine, quinine and quinidine, bonded to crosslinked polystyrene bearing thiole residues in the presence of azobisisobutyronitrile, as exemplified in Scheme 3.18,[36] was compared in the enantioselective addition of thiophenols to 2-cyclohexen-1-one (Scheme 3.19).[37]

R = Me, Cl

Scheme 3.19

The reaction did not occur in the presence of unfunctionalized polystyrene and, as expected, cinchonidine and quinine gave predominantly the (*R*) enantiomer, whilst cinchonine and quinidine afforded the (*S*) one. The enantiomer obtained with each polymer-supported alkaloid was the same as that obtained under homogeneous conditions; nevertheless, the optical yields were always less and the best result was achieved with supported cinconine (81% yield, 45% ee). The catalyst could be recovered and reused in successive reactions for at least three times, affording chemical and optical yields similar to those obtained with the fresh catalyst.

As the reactivity of supported alkaloids was often lower than that of their homogeneous counterparts, a spacer was introduced between the catalytic centre and the polymer matrix to overcome this drawback. To this end polymer-supported quinine **60** derivatives with spacers of different length were prepared (Figure 3.2) and tested in the Michael addition of methyl 1-oxoindane-2-carboxylate to methyl vinyl ketone (Scheme 3.20).[38]

Catalysts **65** and **66** with spacer groups containing eight and fifteen atoms, respectively, showed comparable stereoselectivity (62 and 65% ee respectively) that is higher than that showed by **68** with no spacer (49% ee); the reaction carried out in the presence of homogeneous quinine **60** afforded the adduct with 76% ee. Surprisingly, catalyst **67**, with a seven-atom spacer containing an acryloyl moiety, showed a lower activity but, when the reaction temperature was raised from -48 to $25\,^{\circ}$C, the product was isolated with 92% yield and 45% ee, confirming that the configurational environment of the alkaloid moiety was retained after the polymerization reaction despite the low catalytic activity.

When the recovered polymers **65** and **66** were utilized in successive reactions, they exhibited lower catalytic activity but the stereoselectivity was comparable to that of the first use.

3.2.3 Knoevenagel Reactions

Primary amines anchored to silica have been utilized as catalysts for Knoevenagel condensation reactions. For example, silica gel functionalized with a propylamino moiety [SiO_2–$(CH_2)_3$–NH_2] was employed in the reaction of different methylene active compounds with aldehydes and ketones for the production of electron-poor olefins **73** (Scheme 3.21).[39]

The process was performed by flowing the reagent mixture through a catalytic bed charged in a vertical double jacket glass column; even after many runs the

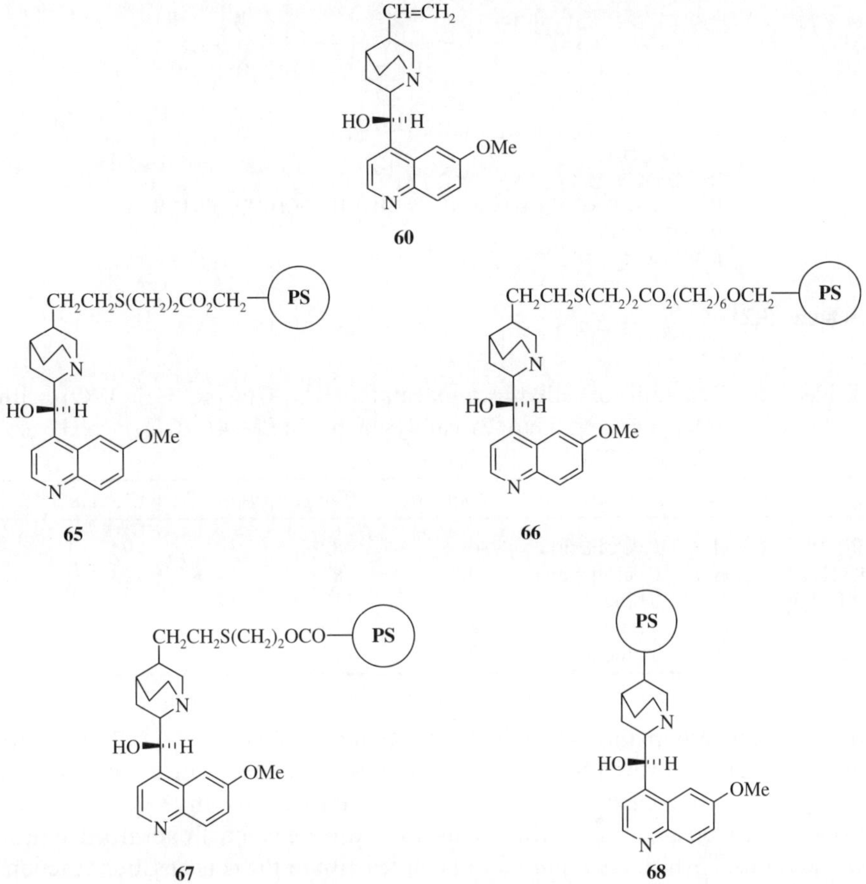

Figure 3.2

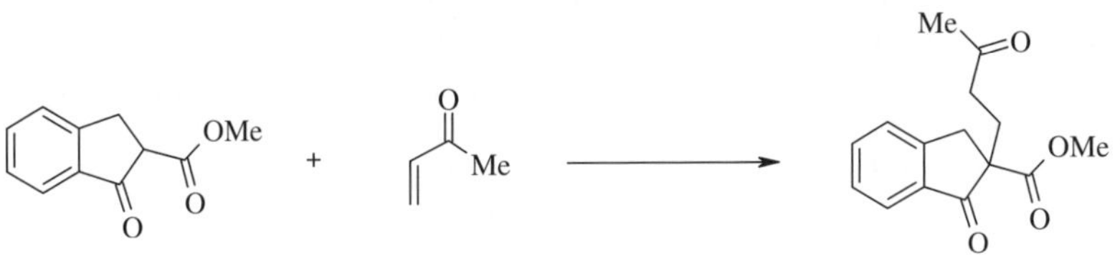

Scheme 3.20

activity of the catalyst remained unchanged. The products were obtained in 66–98% yields – comparable to those achieved under homogeneous catalysis. The authors reported that the catalytic process could involve the immobilized amino groups and the residual silanol groups of the silica surface: the amino group extracts a proton from the methylene active compound while the silanol activates

71 + **72** ⟶ **73**

R^1 = MeCO, CN
R^2 = EtOCO, MeCO, CN
R^3 = Ph, 4-OMeC$_6$H$_4$, 4-NO$_2$C$_6$H$_4$, PhCH=CH, PhCH$_2$CH$_2$, Et, C$_7$H$_{15}$
R^4 = H, Me, Et
R^{3}R^4 = (CH$_2$)$_5$

Scheme 3.21

Table 3.5 Reaction of ethyl cyanoacetate (R^1 = CN, R^2 = EtOCO) with aldehydes and ketones catalysed by MCM-41–(CH$_2$)$_3$–NH$_2$ (see Scheme 3.21).

R^3	R^4	Solvent	Time (h)	Temperature (°C)	73 Yield (%)
Ph	H	Cyclohexane	36	82	94
C$_7$H$_{15}$	H	Cyclohexane	0.5	82	93
–(CH$_2$)$_5$–		Toluene	2	110	92
Et	Et	Toluene	18	110	95
Me	Ph	Toluene	72	110	49

the carbonyl group through a hydrogen bond (Si OH···O=C). The matrix influence was also proved by the low reactivity of β-diketones with respect to other methylene active compounds, which was in disagreement with the trend of pK_a values: in fact on a silica gel surface β-diketones are preferentially adsorbed in their enolic forms,[40] which are reported to be unreactive in the condensation reaction.

SiO$_2$–(CH$_2$)$_3$–NH$_2$ was compared to MCM-41–(CH$_2$)$_3$–NH$_2$ in the model reaction of ethyl cyanoacetate with aldehydes and ketones (Scheme 3.21, R^1 = CN, R^2 = EtOCO).[41] In the reaction catalyzed by MCM-41–(CH$_2$)$_3$–NH$_2$ the yields of compounds **73** were generally high (Table 3.5) and the overall reactivity showed some similarity to that of SiO$_2$–(CH$_2$)$_3$–NH$_2$.

A point of divergence between the two catalysts was their behaviour in different solvents: whereas the amorphous silica-based catalyst was effective only in hydrocarbon solvents the activity of the MCM-41-based catalyst depended on the reflux temperature of the solvent; however, more polar solvents were clearly disadvantageous. This behaviour was explained by considering that the reaction rate was influenced by the partitioning of the reactants (polar) between the catalyst surface (polar) and the bulk medium (non-polar). The silica-based catalysts have a significantly lower polarity, and partitioning away from the catalyst is pronounced even in moderately polar solvents such as toluene; conversely, the much more polar MCM-41 catalyst can compete more effectively for the substrate, thus effectively extending the range of useful solvents. Another advantage of the MCM-41-based catalyst is the possibility of increasing the loading, promoting a rise in the reaction rate.

The mode of deactivation was also different between the two catalysts: the silica catalyst underwent a slow, irreversible formation of surface-bound amide groups (arising from the reaction of the surface-bound primary amine and the ester function of the ethyl cyanoacetate); in contrast, the MCM-41 recovered spent catalyst did not display bands for nitrile or amide groups, but rather the presence of some adsorbed complex organic compounds.

To improve the activity of the MCM-41-based catalyst, materials containing the aminopropyl catalytic centre and a second catalytically inert group (designed to modify the polarity of the catalyst surface) were prepared.[42] In particular, aminopropyl/phenyl and aminopropyl/methyl materials were synthesized and their activity evaluated in a model reaction between ethyl cyanoacetate and cyclohexanone [Scheme 3.21, $R^1 = CN$, $R^2 = EtOCO$, $R^3R^4 = (CH_2)_5$]. The phenyl-containing material was substantially more active than the parent aminopropyl material (the reaction took 30 min to go to completion whereas with the simple aminopropyl material 2 h were needed to achieve the same result); conversely, with the methyl-containing material no substantial change in reactivity was observed. This behaviour was ascribed to the increased lipophilicity of the catalyst surface. The catalyst reusability was also improved with the insertion of the phenyl group on the surface as, usually, the deactivation mechanism was related to the strong adsorption of the ketone, which was reduced by lowering the catalyst surface polarity.

An interesting comparison of the activity of primary and tertiary amino groups linked to MTS silicas in the reaction of benzaldehyde with ethyl cyanoacetate (Scheme 3.21, $R^1 = CN$, $R^2 = EtOCO$, $R^3 = Ph$, $R^4 = H$) was reported.[24] The results showed that catalysis induced by tertiary amine was relevant to classical base activation of the methylene group followed by nucleophilic attack to the carbonyl function, whereas primary amines could activate the carbonyl group by imine formation followed by Mannich-like nucleophilic attack by the activated ethyl cyanoacetate, as shown in Scheme 3.9.

The Knoevenagel reaction between benzaldehydes **22** and malononitrile or ethyl cyanoacetate **74** followed by condensation of the resulting α,β-unsaturated nitriles **75** with dimedone (**76**) was performed in the presence of SiO_2–$(CH_2)_3$–NEt_2 (Scheme 3.22).[43]

The one-portion addition of the reagents in the model reaction between benzaldehyde, ethyl cyanoacetate and dimedone resulted in exclusive formation of the unwanted dimeric product **78** (Figure 3.3), as a consequence of the preferential condensation of benzaldehyde and dimedone due to the higher acidity of dimedone (pK_a 12.0) than that of ethyl cyanoacetate (pK_a 13.1).

For this reason dimedone was added sequentially. A mixture of benzaldehyde and ethyl cyanoacetate in water–ethanol was irradiated at 100 W in the presence of SiO_2–$(CH_2)_3$–NEt_2 followed by addition of the dicarbonyl compound and subsequent heating: the desired tetrahydrobenzo[*b*]pyran (**77**) was isolated in 91% yield. The heterogeneous catalyst could be recycled at least three times, though a certain decrease in catalytic activity was observed. The methodology was then applied to various aromatic aldehydes bearing both electron-withdrawing and electron-donating groups; the reaction proceeded

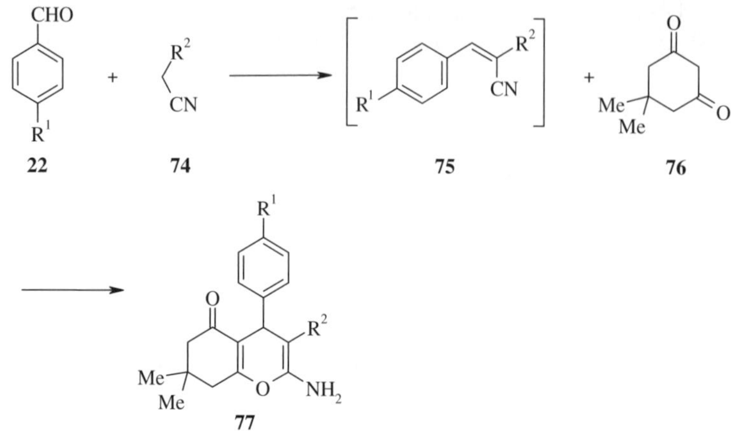

R^1 = H, Me, OMe, NO$_2$, Cl, OH, OAc, OTHP, CO$_2$H
R^2 = CN, CO$_2$Et

Scheme 3.22

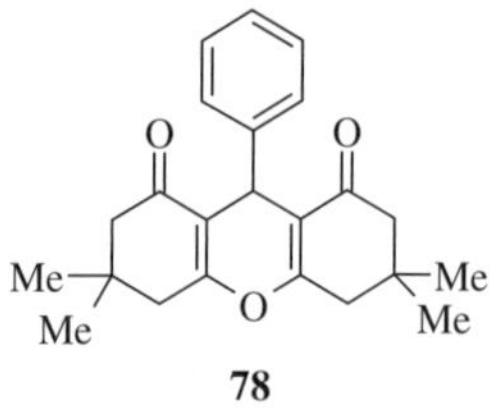

Figure 3.3

without the protection of acidic substituents (*i.e.* when 4-hydroxybenzaldehyde and 4-formylbenzoic acid were employed).

When malononitrile was utilized, due to its higher acidity (pK_a 11.5) with respect to dimedone (pK_a 12.0), the reaction could be carried out in a three-component manner by adding the reagents all at once – the corresponding tetrahydrobenzo[*b*]pyrans were isolated in 77–99% yield.

The one-pot synthesis of γ-hydroxy-α,β-unsaturated nitriles **82** starting from aldehydes **49** and arylsulfonylacetonitriles **79** has been carried out in the presence of SiO$_2$–(CH$_2$)$_3$–NEt$_2$ (Scheme 3.23).[44]

The reaction proceeded *via* Knoevenagel condensation followed by Mislow–Evans rearrangement,[45] affording the alcohol through the breaking of the oxygen–sulfur bond; the target products were obtained with *trans* geometry. Water was utilized as solvent due to its ability to promote the rate-determining Knoevenagel condensation. The reaction was applied to various aromatic and aliphatic aldehydes (Table 3.6); since the reaction conditions are very mild, acid- or base-sensitive protecting groups remained intact in the final products.

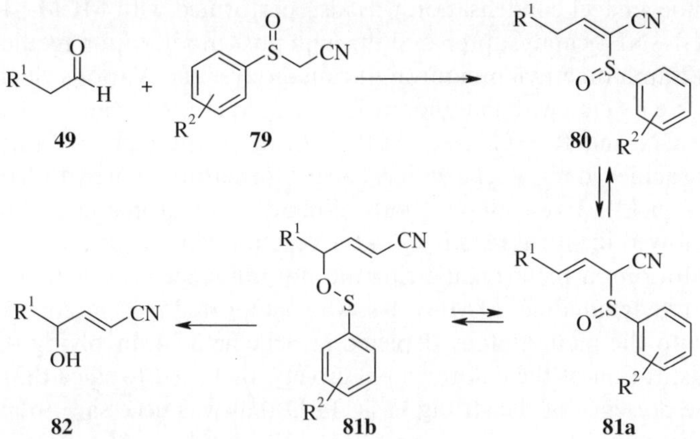

$R^1 = Bu, C_5H_{11}, C_8H_{17}, AcO(CH_2)_8, THPO(CH_2)_8, Ph$
$R^2 = H, 3\text{-}Cl, 4\text{-}Cl, 4\text{-}NO_2$

Scheme 3.23

Table 3.6 Sequential Knoevenagel/Mislow–Evans rearrangement with various aldehydes catalysed by SiO_2–$(CH_2)_3$–NEt_2 (see Scheme 3.23).

Product	Time (h)	Yield (%)
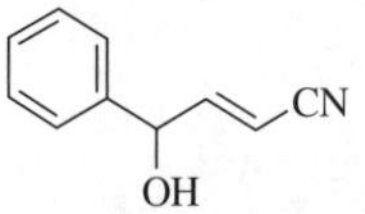 (1-phenyl-3-hydroxy butenenitrile)	6	82
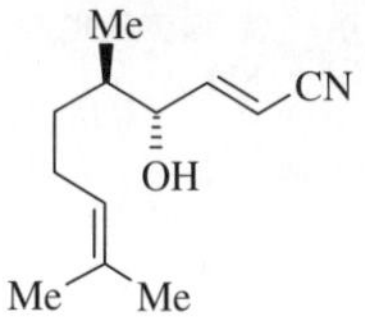 (Me-branched, Me₂C= terminal)	6	80
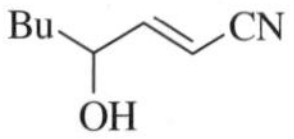 Bu	6	64
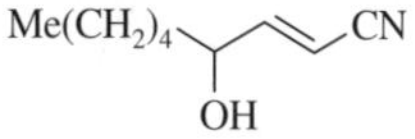 $Me(CH_2)_4$	6	61
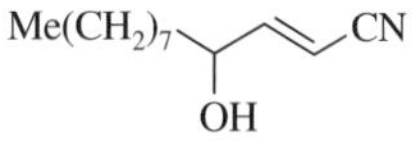 $Me(CH_2)_7$	6	61
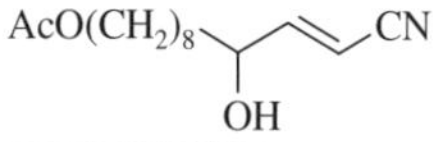 $AcO(CH_2)_8$	16	43
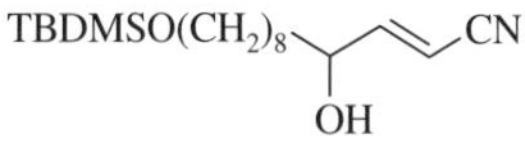 $TBDMSO(CH_2)_8$	24	39
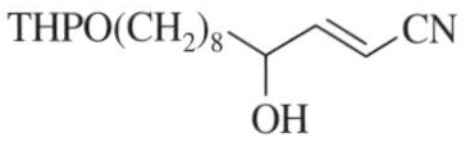 $THPO(CH_2)_8$	24	52

The Knoevenagel condensation was also performed with MCM-41–$(CH_2)_3$–NH–$(CH_2)_2$–NH_2 catalyst prepared through post-modification methodology,[13] utilizing (2-aminomethylaminopropyl)trimethoxysilane. Various aldehydes and ketones were reacted with malononitrile and ethyl cyanoacetate[46] (Scheme 3.21, $R^1 = R^2 = CN$ and $R^1 = CN$, $R^2 = EtOCO$): in all the reactions total conversions were achieved in toluene with exclusive formation of dehydrated products (75–100% yield). Interestingly, both aliphatic and aromatic carbonyl compounds showed identical reactivity in the reaction with ethyl cyanoacetate and the substitution on the aromatic ring did not influence the reactivity.

Tetramethylguanidine (TMG) **84** was supported on mesoporous silica according to the methodology depicted in Scheme 3.24, involving the nucleophilic substitution of the chloropropyl moiety anchored to silica (**83**) by TMG (**84**) in the presence of the strong basic TBD that was necessary to remove the formally produced hydrochloric acid.

Catalyst **85** was studied to promote the reaction between malonic acid and heptanal[47] (Scheme 3.25).

Decarboxylation of adduct **88** surprisingly led to 3-nonenoic acid (**89**). The selectivity of the β,γ-unsaturated product was quite moderate in all cases due to the side reaction involving autocondensation of heptanal. Reuse of the recovered catalyst still indicated a significant decrease of activity.

An interesting comparison between the catalytic activity of commercially available **PS-BEMP** and silica- and polystyrene-supported-TBD in the production of substituted aza-heterocycles **93** and **96** was reported (Scheme 3.26).[48]

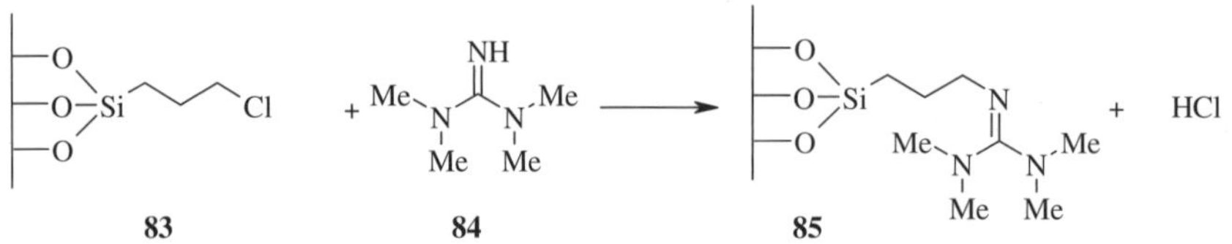

Scheme 3.24

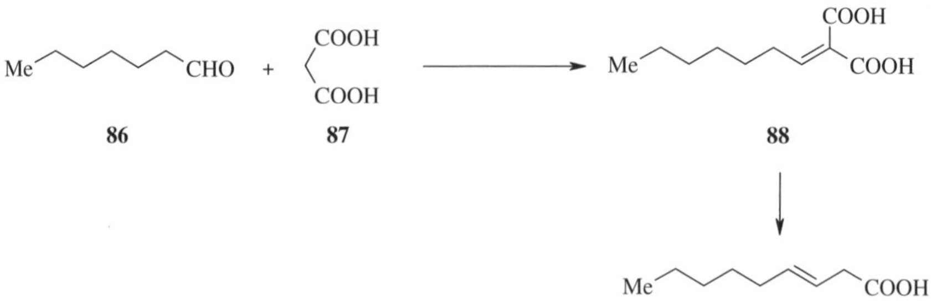

Scheme 3.25

R^1 = CO_2Me, CO_2Et, CO_2Ph, CN
R^2 = Me, Et, CH_2CO_2Et, Bn
R^3 = CO_2Me, CO_2Et, CO_2Pr^i, CN
HET = indole, 4(1H)-quinolone, pyridazin-3(2H)-one

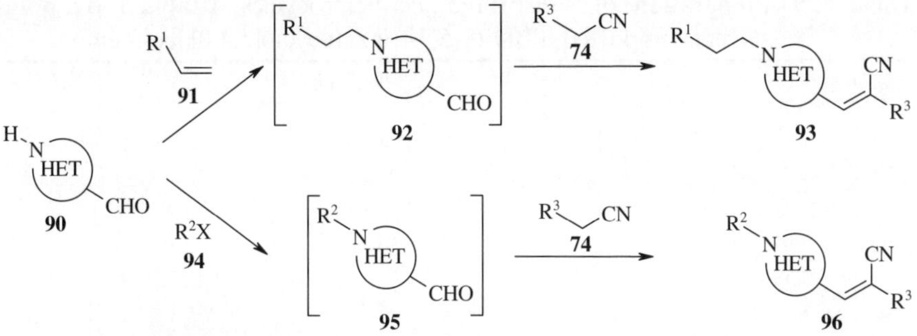

Scheme 3.26

PS-BEMP was able to promote the first step (the alkylation process) in higher yield but failed in the Knoevenagel condensation (very long reaction times, 48–72 h, were needed); the best results were obtained by employing the supported TBD catalysts. No significant differences between SiO_2-TBD and PS-TBD were observed, probably because the latter catalyst can swell sufficiently in THF. Consequently, the less expensive PS-TBD was employed for the preparation of a library of substituted aza-heterocycles; Table 3.7 gives some examples.

3.2.4 Other Reactions

MCM-41–$(CH_2)_3$–NH–$(CH_2)_2$–NH_2 was utilized for the cyanosilylation of carbonyl compounds to afford cyanohydrin trimethylsilyl ethers **98** (Scheme 3.27),[49] which are industrially valuable intermediates in the synthesis of cyanohydrins, β-amino alcohols, α-amino acids and other biologically active compounds.[50]

Various aldehydes and ketones were treated with Me_3SiCN to furnish the corresponding cyanohydrin trimethylsilyl ethers (Table 3.8).

The hypervalent silicon derivative **100**, formed from the interaction of the more basic secondary amine group of the catalyst **99** with Me_3SiCN (Scheme 3.28), was assumed by the authors to produce the transition state **101** by interaction with the carbonyl compound (since the nucleophilicity of the cyano group of **100** is enhanced by electron donation from the pentavalent silicon); immediate silylation gives the corresponding cyanohydrin trimethylsilyl ether **98** and restores the catalyst.

The reusability of the catalyst was checked and a consistent activity was observed after several cycles in the model reaction between *p*-chlorobenzaldehyde and Me_3SiCN.

Polymer-supported formamides **103** and **105** (Scheme 3.29), prepared through the functionalization of the chlorobenzyl Merrifield resin **39**, were utilized in the allylation of aldehydes.[51]

Table 3.7 Preparation of substituted aza-heterocycles catalysed by poly-styrene-supported-TBD (1,5,7-triazabicyclo[4.4.0]dec-5-ene).

Product	Time (h)	Yield (%)
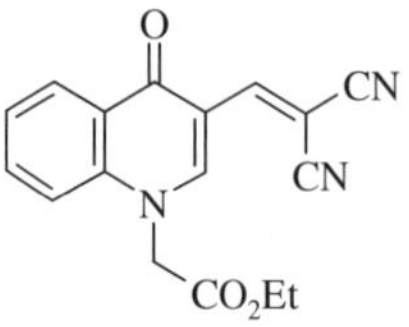	12	72
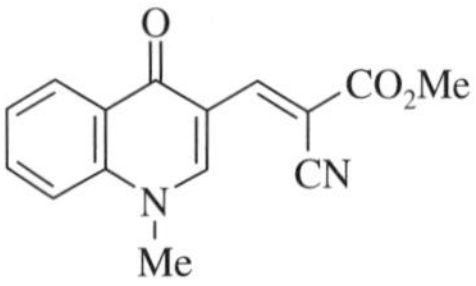	12	77
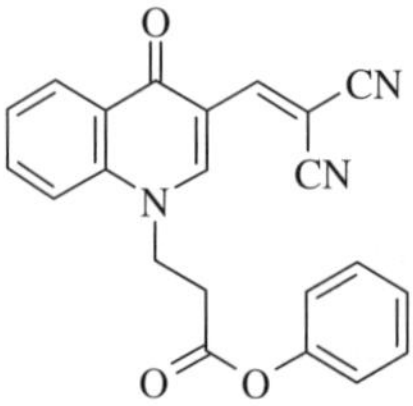	8	54
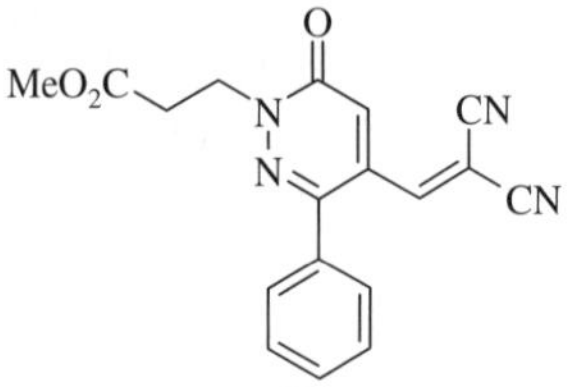	4	65
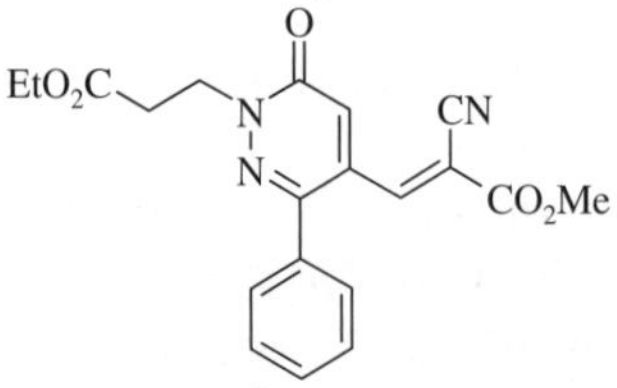	4	62
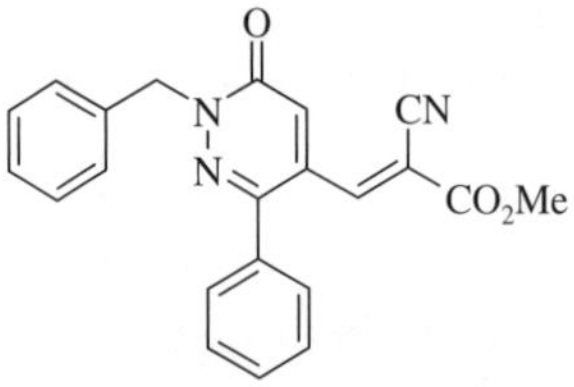	4	72
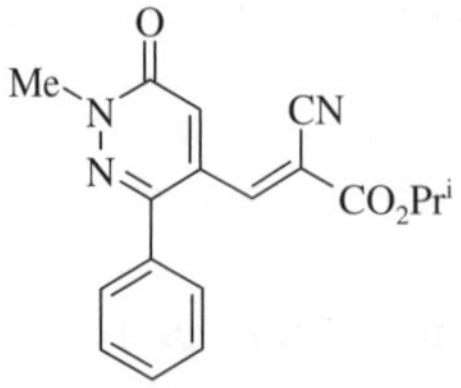	4	57

Table 3.7   (*Continued*).

Product	Time (h)	Yield (%)
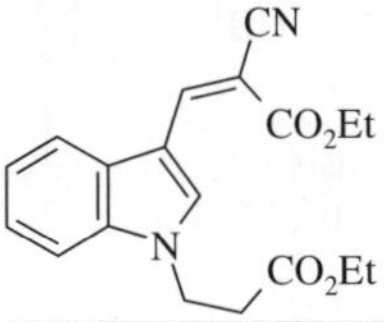 (indole derivative with CN, CO$_2$Et, CO$_2$Et substituents)	12	78

$$R^1\text{-CO-}R^2 \quad + \quad Me_3SiCN \quad \longrightarrow \quad \underset{R^1 \; R^2}{\overset{NC \; OSiMe_3}{\diagup\!\diagdown}}$$

72   **97**   **98**

R^1 = H, Me
R^2 = Bu^i, Ph, 4-ClC$_6$H$_4$, 4-OMeC$_6$H$_4$, 4-MeC$_6$H$_4$, 4-OHC$_6$H$_4$, 2-NO$_2$C$_6$H$_4$, PhCH=CH
R^1R^2 = (CH$_2$)$_4$, (CH$_2$)$_5$

Scheme 3.27

Table 3.8   Cyanosilylation of carbonyl compounds with TMSCN catalysed by MCM-41–(CH$_2$)$_3$–NH–(CH$_2$)$_2$–NH$_2$ (see Scheme 3.27).

R^1	R^2	Time (h)	**98** Yield (%)
4-ClC$_6$H$_4$	H	12	100
4-MeOC$_6$H$_4$	H	12	60
4-MeC$_6$H$_4$	H	12	80
4-OHC$_6$H$_4$	H	8	80
2-NO$_2$C$_6$H$_4$	H	12	100
PhCH=CH	H	12	63
2-Furfuryl	H	6	88
Ph	Me	12	28
–(CH$_2$)$_5$–		2	75
Me$_2$CHCH$_2$	H	6	100
–(CH$_2$)$_4$–		2	80

In the model reaction between 3-phenylpropanal and allyltrichlorosilane (**106**, Scheme 3.30, R = PhCH$_2$CH$_2$), catalyst **103** was more reactive than **105**; polymers with higher loadings showed higher activity and even a catalytic amount of **103** (10 mol.%) was shown to be effective at a longer reaction time.

As physical destruction of the polymer matrix by stirring might deactivate the catalyst, an automatic shaker instead of a magnetic stirrer was utilized, allowing preservation of the catalyst activity, even after multiple use, to afford the homoallylic alcohol in high yields (1st cycle: 90% yield, 2nd cycle: 86% yield, 3rd cycle: 86% yield).

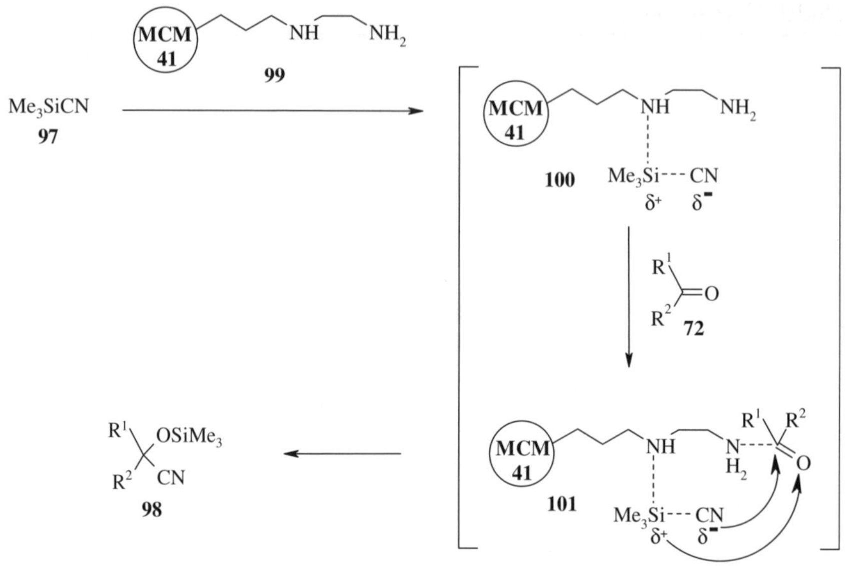

Scheme 3.28

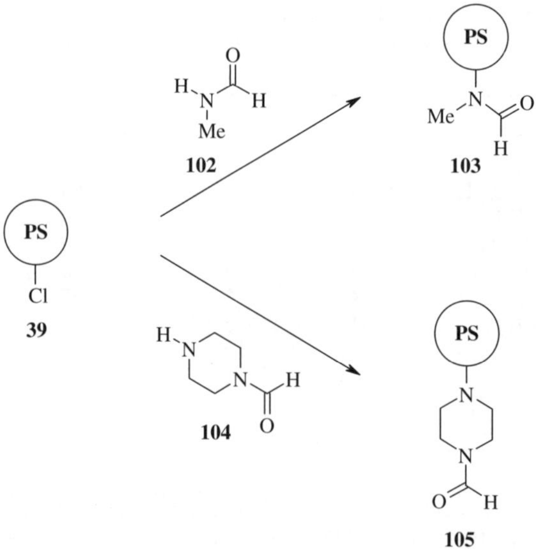

Scheme 3.29

The reaction could be extended to different aldehydes to afford the corresponding products in good yields (66–95%), although 300 mol.% of catalyst **103** and long reaction times were required to complete the reactions in some cases.

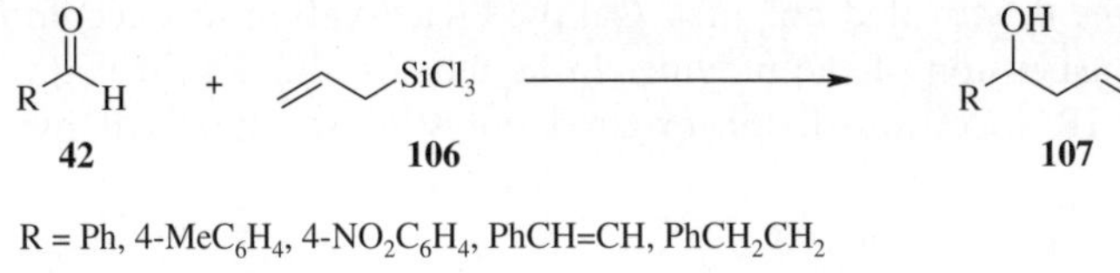

R
H
O
+
SiCl3
OH
R
42
106
107
R = Ph, 4-MeC6H4, 4-NO2C6H4, PhCH=CH, PhCH2CH2

Scheme 3.30

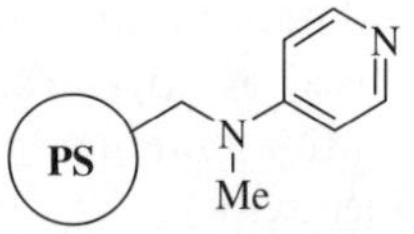

PS
N
N
Me
108

Figure 3.4

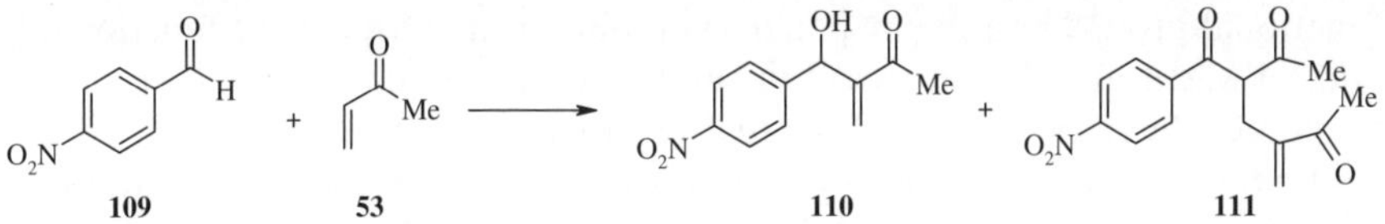

O2N
O
H
+
O
Me
O2N
OH
O
Me
+
O2N
O
O
Me
Me
O
109
53
110
111

Scheme 3.31

The polymer-supported 4-(*N*-methylamino)pyridine (Figure 3.4), prepared by reaction of 4-chloromethylstyrene and 4-(methylamino)pyridine sodium salt and successive copolymerization with styrene and divinyl benzene,[52] was utilized as reusable catalyst for promoting the Baylis–Hillman reaction;[53] usually this kind of reaction uses stoichiometric or overstoichiometric amounts of an organic amine as a base catalyst.

As a test reaction to determine the activity of this catalyst, the coupling of 4-nitrobenzaldehyde with methyl vinyl ketone, a paradigmatic example of the Baylis–Hillman reaction, was selected. The major product was the corresponding β-hydroxyketone **110** accompanied by minor amounts of the Michael adduct **111** that appeared at high conversions when an excess of methyl vinyl ketone was utilized (Scheme 3.31).

Concerning the influence of the solvent, while the reaction afforded low conversions in protic solvents, high conversions were obtained in dimethylformamide or dichloromethane. Under these conditions the catalyst was truly heterogeneous, but its activity decreased gradually upon successive reuses (from 75% to 51% yield of product **110**). The main cause of deactivation could be the covalent attachment of methyl vinyl ketone to the pyridine nitrogen of the polymer: IR spectra of the catalyst before and after extensive deactivation showed that reuse of the catalyst caused the appearance of bands from 1750 to 1650 cm^{-1}, which are attributable to the presence of several types of carbonyl groups in the deactivated polymer. Catalyst reactivation was accomplished by stirring a suspension of the polymer in aqueous sodium hydroxide; in accordance, the IR spectrum of the original polystyrene supported pyridine was restored.

3.3   Polyfunctional Catalysts

Much effort has been made in the last few years and continues to be devoted to the development of catalysts containing multiple types of active sites. Polyfunctional catalysts may be used to perform several steps in a reaction sequence, working in a cooperative manner to alter the characteristics of a single reaction, *e.g.* rate and selectivity.

Performing multiple reactions simultaneously in a single step offers possibilities for reduced waste and increased safety as well as manipulation of equilibria.[54] This approach was inspired by the action of enzymes, which constitute interesting examples of multifunctional catalysts as they can promote multi-step reactions. In fact, enzymes immobilize mutually incompatible functional groups in a manner that maintains their independent functionality and, as such, are able to carry out multi-step reaction sequences with functionalities that would not be tolerated together in solution.

Organic reactions have been performed through the combination of incompatible basic and acid catalysts by two general methodologies, namely by using a mixture of two supported catalysts with different active centres on separate supports or multifunctional single catalysts where incompatible reagents are rendered mutually inactive by confinement effect.

3.3.1   Aldol Reactions

An interesting approach based on the use of a combination of catalysts that can be recovered by magnetic and gravimetric methods has been reported;[55] the use of magnetically separable materials allows the creation of various versatile catalysts that can be easily recovered without the need for specialized equipment. This concept was demonstrated by combining base catalysts that are magnetically recoverable with acid catalysts that are recovered gravimetrically. Superparamagnetic spinel ferrite nanoparticles[56] were functionalized through silane chemistry with *N*-[3-(trimethoxysilyl)propyl]ethylenediamine to create surface base sites. The solid basic nanoparticles were then used in conjunction with a sulfonic acid polymer resin in the tandem deacetalization–Knoevenagel reaction (Scheme 3.32).

Both catalysts and all the reagents were added to a reaction vessel at time zero, and the mixture was stirred for 30 min: complete conversion of **112** into **114** was observed. After reaction, the catalysts were separated by sonicating the reaction vessel and affixing a small permanent magnet externally to one wall of the vessel; the nonmagnetic resin catalyst was removed from the reaction vessel by decantation while the magnetic nanoparticle base catalyst was held

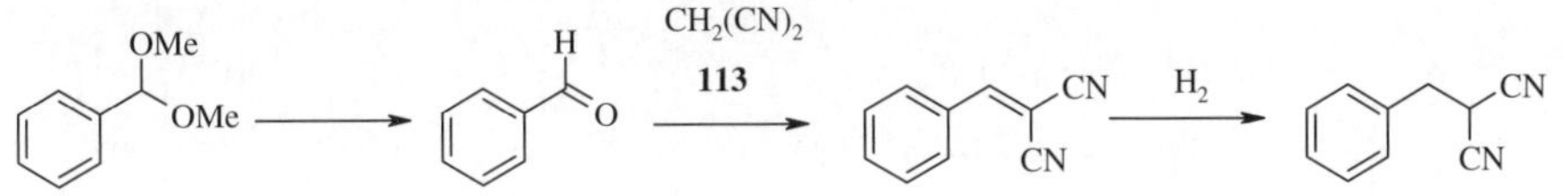

Scheme 3.32

stationary in the vessel by the magnet. Each catalyst was recovered in essentially pure form. To evaluate possible mutual catalyst contamination, the amount of sulfur, an elemental tag for the resin catalyst, was detected in the magnetic nanoparticle catalyst before and after reaction and it was essentially identical (0.07% S in the magnetic base catalysts before reaction and 0.05% S after reaction). The same is true of the amount of iron, an elemental tag for the magnetic catalyst, in the resin catalyst before and after reaction (0.01% Fe in the acid resin catalyst before reaction and 0.03% Fe after reaction). Reuse of the separated catalysts in the same reaction gave the same result with the same kinetic profile, indicating that no noticeable deactivation of the catalyst occurred upon combination of the two opposing catalysts. Notably, each catalyst on its own was unable to promote the conversion of **112** into **114**, indicating that the tandem action of two catalysts was required.

To show the versatility of this method, the recovered magnetic nanoparticle catalyst was then effectively used in a second multi-step, one-pot reaction, namely the conversion of benzaldehyde **5** into product **115** (Scheme 3.32) in conjunction with a new solid catalyst, Pt/Al_2O_3. Also, this tandem Knoevenagel–hydrogenation reaction was carried out in one-pot fashion with all reagents and catalysts added at time zero. In this case, benzaldehyde **5** was converted into **114** using the basic catalyst and this product was then hydrogenated to **115** by the supported platinum, with both reactions going to completion.

The reaction sequence was then extended to three steps, again with catalyst recovery, with conversion of **112** into **115** by adding the base catalyst supported on magnetic nanoparticles and the polymeric acid catalyst into the vessel along with the platinum catalyst enclosed in a membrane. The overall yield of the final product **115** was 78%.

A similar strategy starting from 4-nitrobenzaldehyde dimethyl acetal (**116**) involves the subsequent aldol reaction of the obtained 4-nitrobenzaldehyde (**109**) with acetone for the production of β-hydroxyketone **117** (Scheme 3.33).[57]

Commercially available Amberlite IR-120 (H^+-form) was used as an acid catalyst; this resin is a divinylbenzene-crosslinked partially sulfonated gel-type polystyrene. As a base catalyst, PEG-PS resin-supported proline was employed. The reaction was performed in water–acetone–tetrahydrofuran (1:1:1 v/v/v) at room temperature in the presence of 20 mol.% of resin-supported proline and Amberlite. After 20 h, the reaction mixture contained the starting 4-nitrobenzaldehyde dimethyl acetal, 4-nitrobenzaldehyde and the corresponding aldol product with acetone in a ratio of 4:9:87. This means that both the

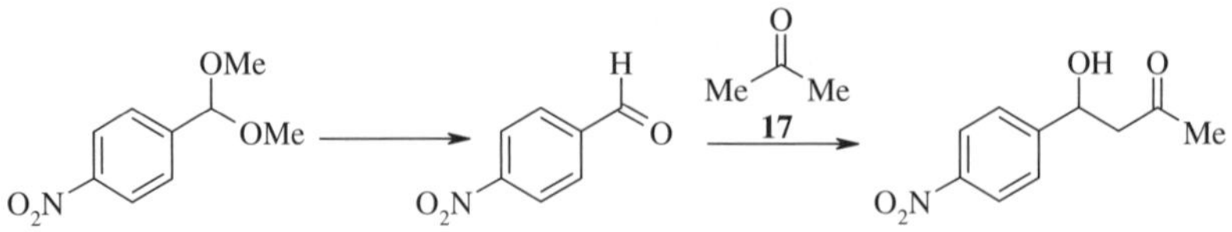

Scheme 3.33

Table 3.9 Reusability of the catalysts Amberlite IR-120/PEG-PS-proline in the two-step one-pot reaction for the production of (*R*)-4-(2-nitrophenyl)-4-hydroxybutanone.

Cycle	Yield (%)	ee (%)
1	89	73
2	88	77
3	89	77
4	87	77
5	84	79
6	86	76

hydrolysis of the acetal and the subsequent aldol reaction proceeded smoothly. Without Amberlite, the starting material was not consumed at all. Conversely, in the absence of the prolyl catalyst, 4-nitrobenzaldehyde became the only product; a similar result was obtained when *para*-toluenesulfonic acid was used instead of Amberlite, presumably due to neutralization of the weakly basic prolyl moiety.

Next, an enantioselective version of this aldol reaction was attempted using a peptide having the sequence D-Pro-Tyr-Phe supported on PEG-PS resin as a catalyst and 2-nitrobenzaldehyde dimethyl acetal as a starting material. The reaction occurred at a similar rate and the corresponding aldol product **117** was obtained with 73% ee. The reusability of the resin-supported catalysts was then examined: the resins were easily separated from the reaction mixture by simple decantation and were thoroughly dried (nitrogen flow and vacuum drying) before the next use. Although the rate of the aldol reaction somewhat decreased, almost the same level of catalytic activity was shown, even after the sixth use, with the same level of enantioselectivity (Table 3.9).

Mesoporous SBA-15 silica that contains both benzenesulfonic acid and propylamine was prepared by the sol–gel technique from tetraethyl orthosilicate (TEOS, **120**), 2-(4-chlorosulfonylphenyl)ethyl-trimethoxysilane (**119**) and (3-aminopropyl)trimethoxysilane (**118**) in the presence of poly(ethylene glycol)-*block*-poly(propylene glycol)-*block*-poly(ethylene glycol) [P123] (Scheme 3.34).[58]

The presence of intact organic functional groups was confirmed by ^{13}C and ^{29}Si NMR spectra. The activity of the catalyst was studied in the aldol condensation between 4-nitrobenzaldehyde and acetone (Scheme 3.35).

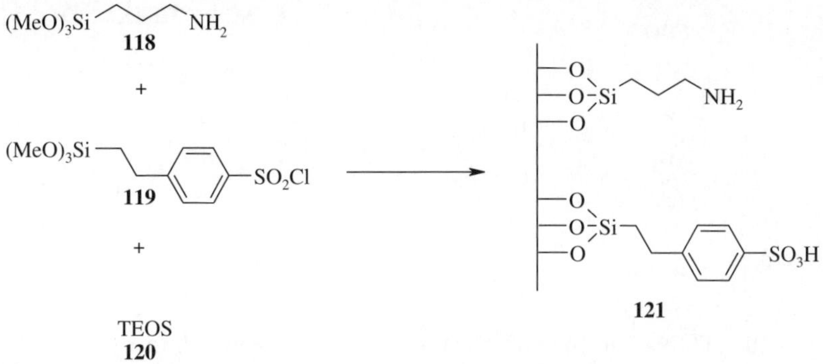

Scheme 3.34

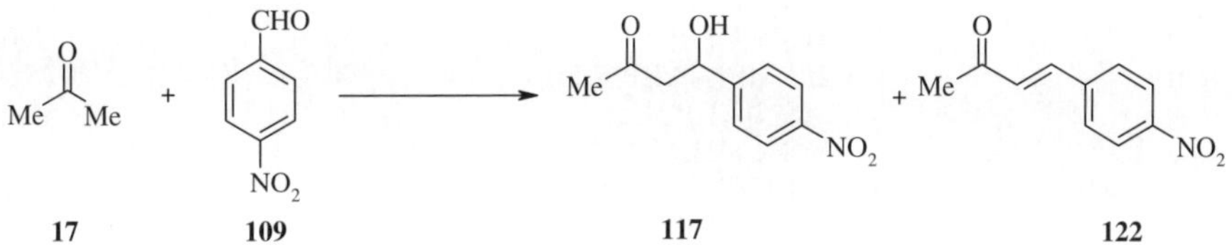

Scheme 3.35

The use of SBA-15 functionalized only with a sulfonic acid or amine gave significantly lower levels of conversion (SBA-15-SO_3H 16% yield and SBA-15-NH_2 33% yield *versus* SBA-15-SO_3H/NH_2 62% yield). This enhancement of catalytic activity is likely due to the cooperative interaction between neighbouring acid and base sites. The acid and base groups supported onto SBA-15 are in equilibrium between the free acid and base and the ion pair that results from neutralization. The equilibrium lies towards the poorly active ion pair in polar protic solvents as the proton exchange is rapid and the polar solvent stabilizes the ion pair. In fact, the conversion of 4-nitrobenzaldehyde in the model reaction was only 34% in water and raised to 88% upon moving to nonpolar solvents such as hexane.

A further example of a cooperative catalytic system is a bifunctionalized mesoporous silica nanosphere material (MSN) containing the ureidopropyl group (UDP) and 3-[2-(2-aminoethylamino)ethylamino]propyl group (AEP).[59] All catalysts were synthesized by a sol–gel procedure similar to the previously mentioned one by using different AEP/UDP molar ratios. The MSN-AEP/UDP catalyst was employed for the aldol, Henry and cyanosilylation reactions (Scheme 3.8 $R^1 = NO_2$, $R^2 = H$, Schemes 3.35 and 3.36, respectively) and TON values were compared with those observed by using MSN-AEP and MSN–UDP catalysts (Table 3.10).

The TON values of 2/8 MSN-AEP/UDP catalysts were higher than those of the monofunctionalized ones. These unusual catalytic enhancements are strong

Scheme 3.36

Table 3.10 TONs for the MSN-catalysed aldol, Henry and cyanosilylation reactions (see Schemes 3.8, 3.35 and 3.36, respectively).

Reaction	Catalyst	T (°C)	TON
Aldol	2/8 MSN-AEP/UDP	50	22.6
	MSN-AEP	50	5.4
	MSN-UDP	50	0.0
Henry	2/8 MSN-AEP/UDP	90	125.0
	MSN-AEP	90	55.9
	MSN-UDP	90	5.8
Cyanosilylation	2/8 MSN-AEP/UDP	50	276.1
	MSN-AEP	50	111.4
	MSN-UDP	50	45.9

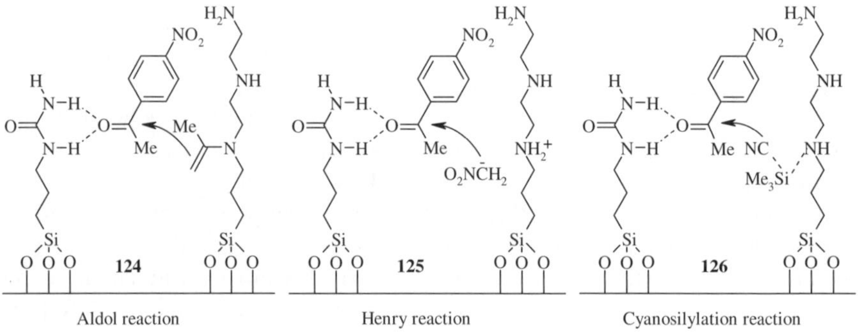

Figure 3.5

indications of the existence of cooperation between the general acid (UDP) and base (AEP) groups (Figure 3.5).

The secondary amine of the AEP group is responsible for the supported enamine formation with acetone (aldol reaction), the deprotonation of nitromethane (Henry reaction) and the generation of a potential nucleophile from trimethylsilyl cyanide through hypervalent silicate formation (cyanosilylation reaction). Therefore, the presence of both AEP and UDP groups in close proximity can cooperatively activate the electrophile (through hydrogen bond) and the nucleophile by enamine formation, thus enhancing the reaction rate.

MSN materials functionalized with AEP and a hydrophobic secondary group such as allyl (AL) and mercaptopropyl (MP) showed a great polarity effect in the nitroaldol condensation of 4-hydroxy- and 4-octyloxybenzaldehyde with nitromethane (Scheme 3.37).[60]

The MSN-AEP/AL material displayed a great selectivity towards **127b** with respect to **127a**, giving a molar product ratio **128b/128a** of 2.58 when **127a** was used in competition with **127b**. The AL and MP groups play a significant role in preferentially allowing the more hydrophobic reactant to penetrate into the mesopores and react with the AEP functionality.

Nanostructured heterogeneous catalysts were prepared by supporting amines on silica-alumina materials (SA-NR$_2$) *via* post-modification methodology. The promising adjacent position of acid and base sites on the SA-NR$_2$ allowed high catalytic activity for various organic transformations such as cyanoethoxycarbonylation (Scheme 3.38 route a) and the nitroaldol reaction (Scheme 3.38 route b).[61]

The cooperative effect of supported base and framework acidity was confirmed by comparison of the efficiency of SA-NR$_2$ with the silica-supported amines (SiO$_2$-NR$_2$) (Table 3.11).

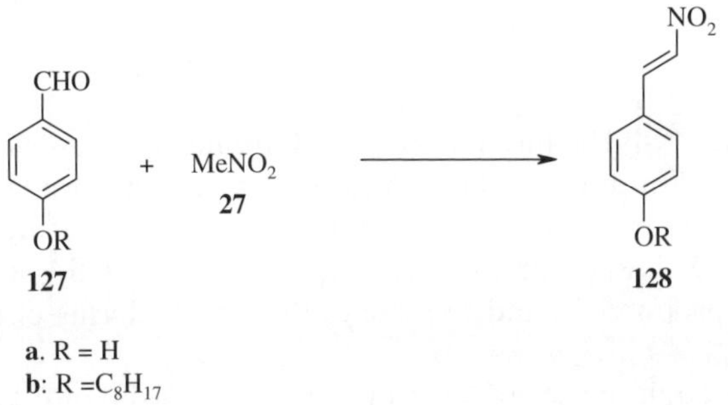

Scheme 3.37

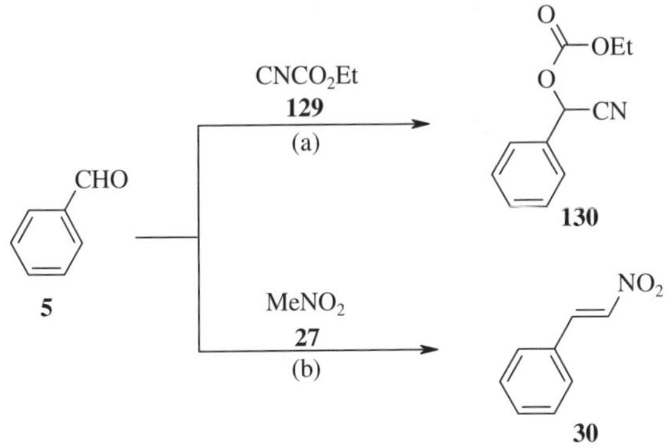

Scheme 3.38

Table 3.11   Activity of SA-NR$_2$ compared with that of SiO$_2$-NR$_2$ in nitroaldol and cyanoethoxycarbonylation reactions (see Scheme 3.38).

Catalyst	Reaction	Yield (%)
SA-NH$_2$	Nitroaldol	99
SiO$_2$-NH$_2$	Nitroaldol	37
SA-NEt$_2$	Cyanoethoxycarbonylation	95
SiO$_2$-NEt$_2$	Cyanoethoxycarbonylation	17

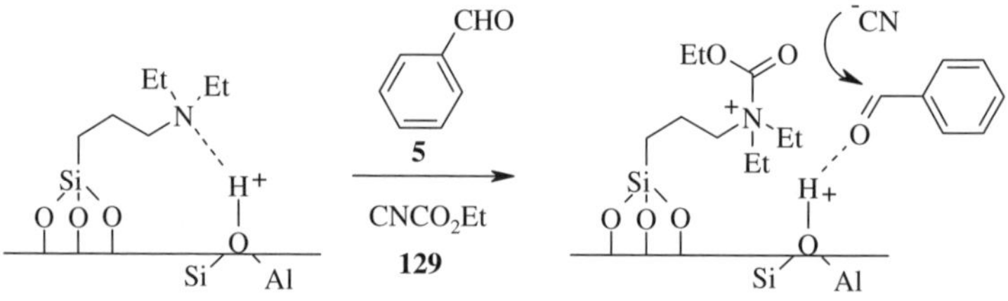

Scheme 3.39

Remarkably, SA-NH$_2$ and SA-NEt$_2$ were found to be excellent catalysts in both reactions, affording product **30** in 99% yield and product **130** in 95% yield, respectively, while the reactions scarcely proceeded using the silica supported base SiO$_2$-NR$_2$. Interestingly, a combination of SA and amine gave lower yield of products **30** and **130** due to the acid buffering effect to the basic amine.

The catalyst could be separated from the reaction mixture and reused with retention of high catalytic activity and selectivity in both reactions.

Concerning the mechanism, the authors suggest a reaction involving the dual activation of donor and acceptor substrate at the amine base site and the neighbouring Brønsted acid site on the SA surface (Scheme 3.39).

3.3.2   Michael Reactions

Complex multi-ring heterocyclic molecules like **136** have been prepared in one pot and under mild conditions by combining a base-catalyzed intermolecular Michael addition reaction of an α,β-unsaturated carbonyl compound and a suitable β-ketoamide pronucleophile with an acid-catalyzed intramolecular *N*-acyl iminium ion cyclization of the resulting adduct (Scheme 3.40).[62]

When the β-ketoamide **133**, containing an acidic methine group and pendant nucleophilic pyrrole substituent, was dissolved with methyl vinyl ketone (**53**) in dichloromethane and treated at room temperature with PS-BEMP (10%), the Michael adduct **134** was formed as the sole reaction product in 100% yield.

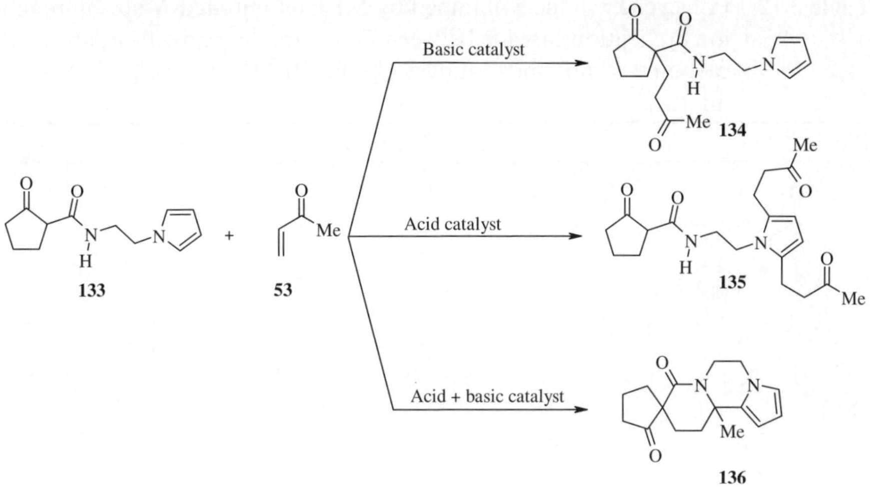

Scheme 3.40

Conversely, when the reaction was carried out by using Amberlyst A15 (200%) in the presence or absence of liquid BEMP (10%), the only observed reaction product was substituted pyrrole **135**, which was isolated in 68% yield; product **134** was not present at all in the reaction mixture. However, when the reaction was repeated by using a combination of PS-BEMP (10%) and Amberlyst A15 (200%), the sole reaction product was the desired tetracyclic product **136** – obtained in 83% yield as a 1:1 mixture of diastereoisomers. By using substoichiometric quantities of PS-BEMP (10%) and Amberlyst A15 (50%), an 85% conversion into **136** after 5 days was produced. These results demonstrate that PS-BEMP and Amberlyst A15 can operate as mutually compatible strongly basic and strongly acidic reagents, respectively, in the same vessel to facilitate the Michael-initiated N-acyl iminium ion cyclization cascade.

The methodology was then applied to various cyclic and acyclic β-ketoamides and α,β-unsaturated carbonyl compounds, producing polycyclic products in which a new 6,6-bicyclic ring system containing at least two new stereogenic centres was created (Table 3.12).

Both stereogenic centres may be fully substituted, and when an indole was used as the nucleophilic trap diastereoselectivities up to 3:1 were observed.

A multicatalyst system for the production of synthetically useful compounds **139** (Scheme 3.41) has been developed.[63]

The reaction involves the amine-catalyzed conversion of an aldehyde into a nitroalkene by reaction with nitromethane followed by a transition-metal-catalyzed Michael addition of β-dicarbonyl compounds in the same reaction vessel. Typically, amine catalysts and nickel complexes are incompatible due to their tendency to chelate and to render each other inactive.[64] However, microencapsulation of poly(ethyleneimine) (PEI) forms catalyst **140**, which can successfully be used in tandem with the nickel-based catalyst **141** (Figure 3.6).

Table 3.12 Polycyclic products obtained by Michael-initiated *N*-acyl iminium ion cyclization cascade between β-ketoamides and α,β-unsaturated carbonyl compounds catalysed by PS-BEMP and Amberlyst A–15 mixture.

Product	*Yield (%)*	*dr (%)*
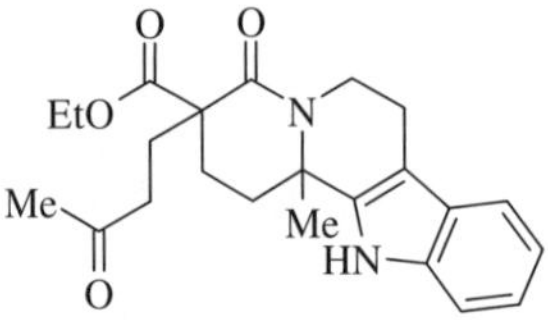	57	3 : 1
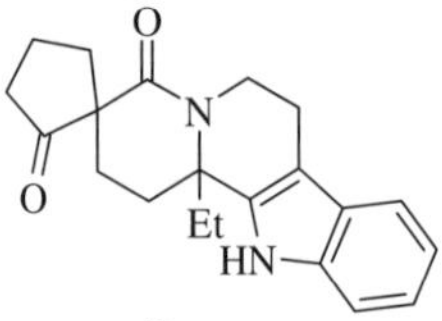	97	3 : 1
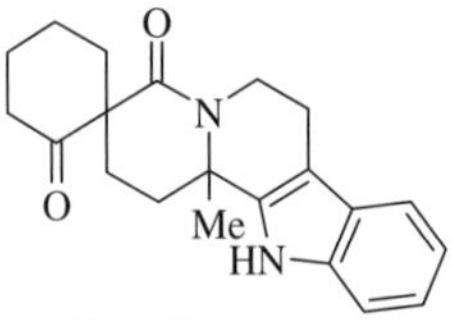	100	3 : 1
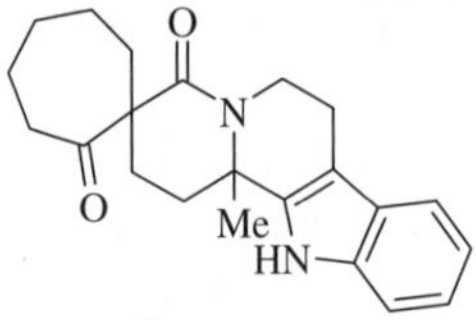	91	3 : 1
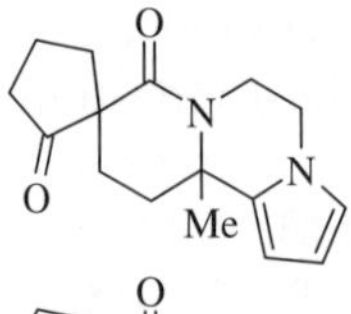	83	1 : 1
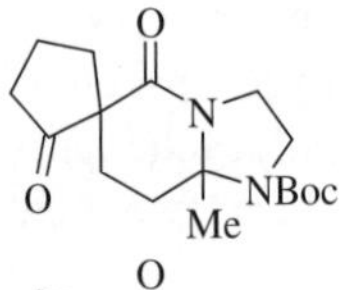	78	1 : 1
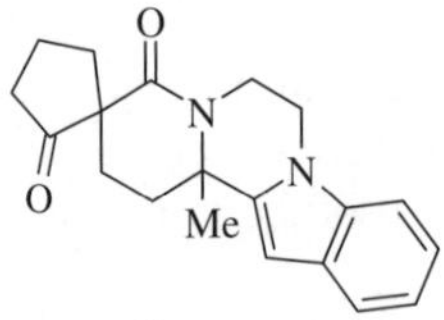	97	3 : 1
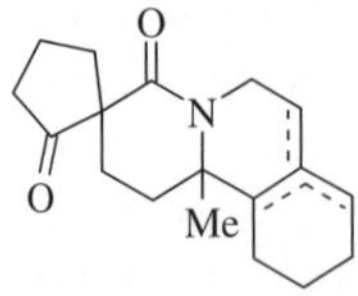	85	Mixture of isomers

Table 3.12   (*Continued*).

Product	Yield (%)	dr (%)
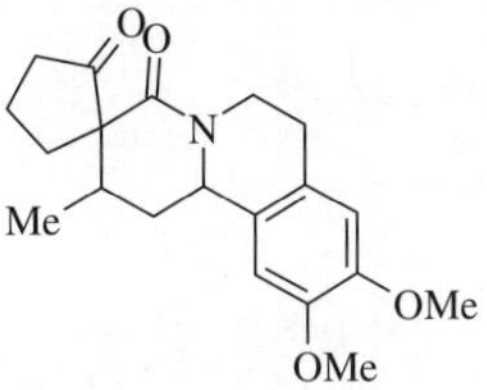	85	7 : 5 : 5 : 3
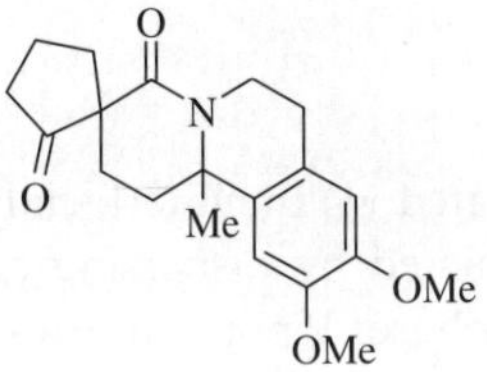	100	2 : 1
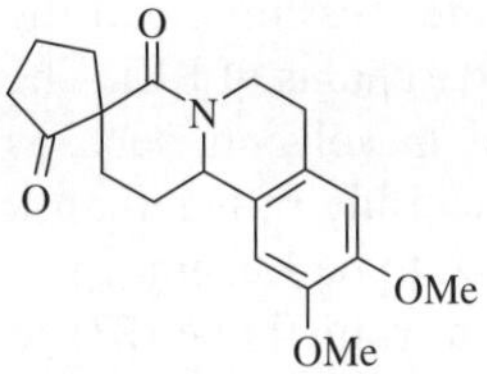	94	1 : 1
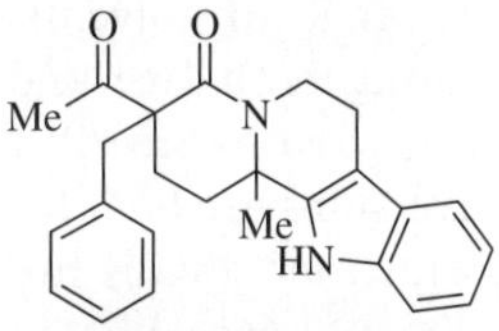	99	2 : 1
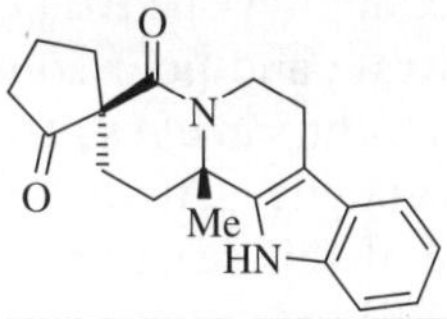	90	3 : 1

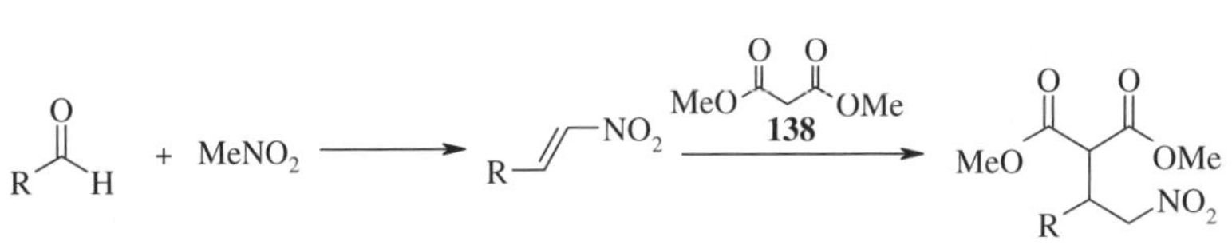

R = Ph, 4-MeC$_6$H$_4$, 4-MeOC$_6$H$_4$, 4-BrC$_6$H$_4$, 4-ClC$_6$H$_4$, 4-CNC$_6$H$_4$, 4-NO$_2$C$_6$H$_4$, Pr^i, Bu^i

Scheme 3.41

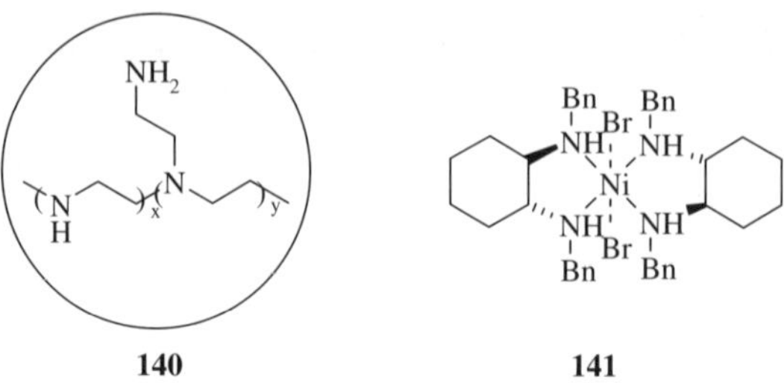

140   **141**

Figure 3.6

The amine-based Henry reaction catalyst was encapsulated *via* the interfacial polymerization of oil-in-oil emulsions. PEI was encapsulated by dispersing a methanolic PEI solution into a continuous cyclohexane phase. Upon emulsification, 2,4-tolylene diisocyanate (TDI) was added to initiate crosslinking at the emulsion interface, forming polyurea shells that contain free chains of PEI. The microcapsules crenate when dry and swell when placed in solvents such as methanol and dimethylformamide, suggesting a hollow capsule rather than a solid sphere formation. The catalyst loading was determined to be 1.6 mmol g^{-1}.

The reaction between benzaldehyde (**42**, R = Ph) and nitromethane (**27**) to produce nitrostyrene (**137**, R = Ph) was performed in a range of solvents. Swelling effects were separated from solvent effects by using both free and encapsulated PEI catalysts. The results for the reactions catalyzed by free PEI are not affected by swelling solvents; for each catalyst, conversions are high for ethanol, moderate for chloroform and low for acetone. However, this is not true for nonswelling solvents: the free PEI-catalyzed reactions revealed that toluene, diethyl ether and THF were relatively good, moderate and poor solvents, respectively, whereas encapsulated PEI did not produce the same results. Indeed, encapsulation resulted in an 80% decrease in catalytic activity in toluene. These results suggest that the solvent dependence of this reaction is two-fold: not only must the solvent be favourable for the PEI-catalyzed reaction but it also must be able to swell the microcapsules. Acetone swells the capsules but is a poor solvent for the reaction, while toluene is a good solvent for the reaction but is unable to swell the capsules. Only ethanol, a good solvent for the reaction that is also able to swell the capsules, is able to produce high conversions with both free and encapsulated PEI.

Concerning the reaction mechanism, the transformation of an aldehyde into a nitroalkene can occur *via* two different pathways: the first involves nitroalcohol formation through a traditional Henry reaction, which is then followed by water elimination to form the double bond; the second proceeds through an imine intermediate rather than the nitroalcohol. The authors predicted that nitroalkene formation goes through an imine intermediate (Scheme 3.9) rather than the nitroalcohol as they did not observe nitroalcohol formation at any point in the reaction. In addition, when the nitroalcohol is placed in the presence of swollen capsules, no nitroalkene formation is observed; this inability to

Table 3.13 Reaction between dimethyl malonate and various aromatic and aliphatic aldehydes catalysed by PEI microcapsules and nickel-based catalyst **141** mixture (see Scheme 3.41).

R	Yield (%)
Ph	80
4-MeC$_6$H$_4$	94
4-MeOC$_6$H$_4$	89
4-BrC$_6$H$_4$	43
4-ClC$_6$H$_4$	48
4-CNC$_6$H$_4$	<5
4-NO$_2$C$_6$H$_4$	<5
CH(Me)$_2$	71
CH$_2$CH(Me)$_2$	65

convert this potential intermediate into the final product provides further evidence against the nitroalcohol-elimination pathway.

The nitroalkene intermediate can either form the dinitro product or go through a Michael-type addition with the encapsulated PEI. Furthermore, because capsules swollen in methanol retain their catalytic activity when placed in toluene, the reaction can be run in a mixture of two different solvents. This allows both the encapsulated PEI and the nickel catalyst to operate in their respective ideal solvents, namely methanol and toluene. To demonstrate the scope of this one-pot process, the reaction was performed with dimethyl malonate (**138**) and various aromatic and aliphatic aldehydes. Table 3.13 shows the results.

Evidently, although the system tolerates both aromatic and aliphatic aldehydes, the introduction of an electron-withdrawing substituent on the aromatic substrate results in a decreased yield. To gain information about the mechanism of the overall tandem reaction, kinetic studies were carried out to identify the rate-determining step. Changing the catalyst concentration in the reaction between 3-methylbutyraldehyde, nitromethane and dimethyl malonate revealed that the reaction is first order in nickel catalyst, indicating that the Michael addition of dimethyl malonate to the nitroalkene is the rate-determining step.

An attractive feature of this two-step one-pot reaction is that it not only incorporates an innovative technique for site-isolation but also produces enantiomerically enriched useful intermediates. Indeed the Michael adducts that are produced represent precursors to γ-amino acids, allowing access to γ-amino butyric acid analogues. Pregabalin (**145**) is one such analogue that is approved for the treatment of both epilepsy and neuropathic pain. Following this one-pot synthetic strategy there is no need for nitroalkene isolation and by-product formation can be avoided. Scheme 3.42 depicts the total synthesis of pregabalin.

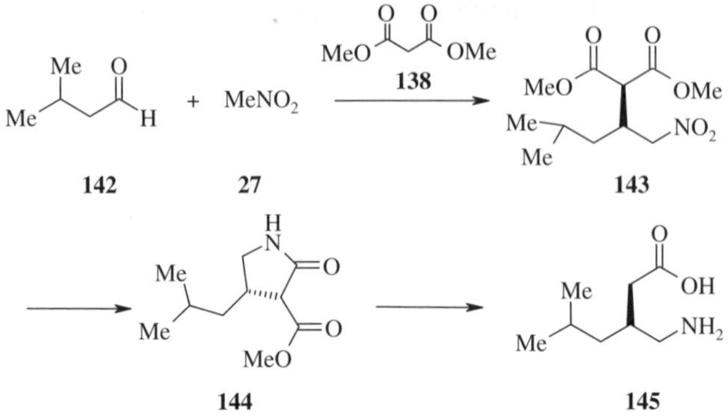

Scheme 3.42

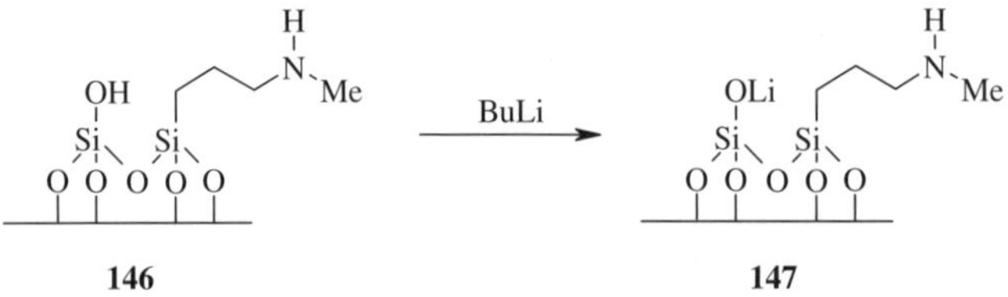

Scheme 3.43

Using 3-methylbutyraldehyde (**142**) as starting material and the chiral catalyst **141**, the tandem reaction afforded the corresponding Michael adduct (*S*)-**143** in 94% yield and 72% ee. Notably, this tandem catalysis system efficiently suppressed the yield of the undesired dinitro by-product to less than 5%. Overnight hydrogenation of (*S*)-**143** with Raney nickel gave nearly quantitative conversion into the ring-closed product **144**. Subsequent acid hydrolysis and decarboxylation produces the HCl salt of pregabalin in 95% yield, which retains an ee of 72%. Treatment of the 72% ee HCl salt with base followed by a single recrystallization from isopropyl alcohol–water afforded the final pregabalin **145** with 91.5% ee.

The catalytic domino Michael reaction of oxophorone and its derivatives **148** with various electron-poor olefins was promoted by *N*-methylaminopropylated silica (NMAP) combined with butyllithium that converts the surface silanols into the basic SiOLi groups (NMAP-Li) **147** (Scheme 3.43).[65]

The catalyst was prepared by treating *in situ* NMAP pellets with butyllithium in tetrahydrofuran. The oxophorone derivative **148** was added at − 70 °C followed by the Michael acceptor and some hexamethylphosphoramide; the reaction requires 10–18 h at room temperature. The domino Michael reaction of the kinetic enolate of **148** with an electron-poor carbon–carbon double bond furnished stereoselectively the *endo*-bicyclo[2.2.2]octane derivative **149** (Scheme 3.44).

The process can be applied to different α,β-unsaturated carbonyl compounds with moderate to excellent yields (Table 3.14).

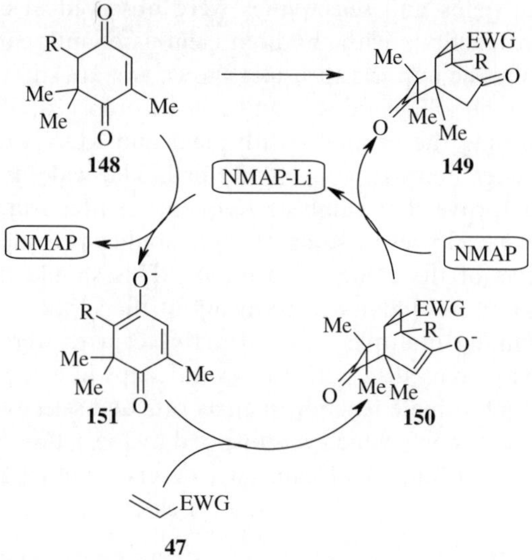

Scheme 3.44

Table 3.14   Domino conjugate reaction of oxophorones with various Michael acceptors catalysed by 4-NMAP (see Scheme 3.44).

R	EWG	HMPA (eqs)	Yield (%)
H	$CO_2C_6H_{11}$	0.4	90
H	CN	–	100
H	SO_2Ph	–	36
Me	$CO_2C_6H_{11}$	0.5	54
Allyl	$CO_2C_6H_{11}$	0.4	75
Allyl	$CO_2C_6H_{11}$	–	34

The catalyst was easily recovered, washed with diethyl ether and reused five times in the model reaction of oxophorone (**148**, R = H) with cyclohexyl acrylate, giving the corresponding product with unchanged high yield (75–94%).

3.4   Conclusions

We have shown the research findings related to the most important aspects of the preparation of supported organic bases and their use as heterogeneous catalysts in organic synthesis.

In particular, we have described selected examples of the application of monofunctional supported catalysts in reactions of considerable synthetic interest such as aldol, Michael and Knoevenagel condensations. In general, good to excellent yields and selectivities were observed in comparison with those obtained in reactions with the homogenous counterparts. However, in this context, despite the important results shown, we are still far from an ideal condition, wherein the solid catalysts can be easily prepared, fully characterized and utilized, affording the products with yield and TON values higher than those achieved under homogeneous conditions. The wide number of studies carried out so far prove that catalysts frequently suffer from poisoning and leaching phenomena. However, some of these studies provide insight into the mechanistic aspects of the catalytic reactions. This should help us to better understand the structure–activity relationship of the supported catalysts.

Catalysts containing multiple types of active centres were also described. These catalysts may be used to perform several steps in a sequence or to alter the characteristics of a single reaction such as rate and selectivity by a synergic cooperation. This approach, which was inspired by the action of enzymes as the best examples of multifunctional catalysts, deserves future intensive studies since in the last few years it has brought a series of very exciting results.

References

1. S. V. Ley, I. R. Baxendale, R. N. Bream, P. S. Jackson, A. G. Leach, D. A. Longbottom, M. Nesi, J. S. Scott, R. I. Storer and S. J. Taylor, *J. Chem. Soc. Perkin Trans. 1*, 2000, 3815.
2. R. A. Sheldon, *Chemtech*, 1994, 38.
3. J. M. Thomas, R. Raja, G. Sankar, R. G. Bell and D.W. Lewis, *Pure Appl. Chem.*, 2001, **73**, 1087.
4. P. M. P. M. J. H. Clark and D. J. Macquarrie, *J. Chem. Soc., Dalton Trans.*, 2000, 101.
5. *Chem. Rev.*, 2002, **102**, 3215-3892: entire volume devoted to recoverable catalysts and reagents, J. A. Gladysz guest Ed.
6. G. Gelbard and F. Vielfaure-Joly, *Reactive Funct. Polym.*, 2001, **48**, 65.
7. C. C. Leznoff, *Chem. Soc. Rev.*, 1974, **3**, 65.
8. R. A. Flowers II, X. Xu, C. Timmons and G. Li, *Eur. J. Org. Chem.*, 2004, 2988.
9. J. P. Mazaleyrat and D. J. Cram, *J. Am. Chem. Soc.*, 1981, **103**, 4584.
10. S. E. Denmark and R. A. Stravenger, *Acc. Chem. Res.*, 2000, **33**, 432.
11. S. E. Denmark, X. Su and Y. Nishigaichi, *J. Am. Chem. Soc.*, 1998, **120**, 12990.
12. K.-i. Shimizu, E. Hayashi, T. Inokuchi, T. Kodama, H. Hagiwara and Y. Kitayama, *Tetrahedron Lett.*, 2002, **43**, 9073.
13. A. Cauvel, G. Renard and D. Brunel, *J. Org. Chem.*, 1997, **62**, 749.
14. H. Hagiwara, H. Ono, N. Komatsubara, T. Hoshi, T. Suzuki and M. Ando, *Tetrahedron Lett.*, 1999, **40**, 6627.
15. J. Hamaya, T. Suzuki, T. Hoshi, K.-i. Shimizu, Y. Kitayama and H. Hagiwara, *Synlett*, 2003, 873.
16. A. Horváth, *Tetrahedron Lett.*, 1996, **37**, 4423.

17. N. Herron, *Inorg. Chem.*, 1986, **25**, 4714.
18. Y. V. Subba Rao, D. E. De Vos and P. A. Jacobs, *Angew. Chem., Int. Ed. Engl.*, 1997, **36**, 2661.
19. R. Sercheli, A. L. B. Ferreira, M. C. Guerreiro, R. M. Vargas, R. A. Sheldon and U. Schuchardt, *Tetrahedron Lett.*, 1997, **38**, 1325.
20. S. Jaenicke, G. K. Chuah, X. H. Lin and X. C. Hu, *Microporous Mesoporous Mat.*, 2000, **35–36**, 143.
21. G. Demicheli, R. Maggi, A. Mazzacani, P. Righi, G. Sartori and F. Bigi, *Tetrahedron Lett.*, 2001, **42**, 2401.
22. J. Hamaya, T. Suzuki, T. Hoshi, K.-i. Shimizu, Y. Kitayama and H. Hagiwara, *Synlett*, 2003, 873.
23. D. J. Macquarrie, J. H. Clark, A. Lambert, J. E. G. Mdoe and A. Priest, *React. Funct. Polym.*, 1997, **35**, 153.
24. D. Brunel, *Microporous Mesoporous Mat.*, 1999, **27**, 329.
25. D. Simoni, R. Rondanin, M. Morini, R. Baruchello and F. P. Invidiata, *Tetrahedron Lett.*, 2000, **41**, 1607.
26. T. D. Machajewski and C.-H. Wong, *Angew. Chem. Int. Ed.*, 2000, **39**, 1352.
27. A. S. Kucherenko, M. I. Struchkova and S. G. Zlotin, *Eur. J. Org. Chem.*, 2006, 2000.
28. D. Font, C. Jimeno and M. A. Pericàs, *Org. Lett.*, 2006, **8**, 4653.
29. F. Brackmann, H. Schill and A. de Meijere, *Chem. Eur. J.*, 2005, **11**, 6593.
30. R. Ballini, G. Bosica, D. Livi, A. Palmieri, R. Maggi and G. Sartori, *Tetrahedron Lett.*, 2003, **44**, 2271.
31. K.-i. Shimizu, H. Suzuki, E. Hayashi, T. Kodama, Y. Tsuchiya, H. Hagiwara and Y. Kitayama, *Chem. Commun.*, 2002, 1068.
32. H. Hagiwara, S. Inotsume, M. Fukushima, T. Hoshi and T. Suzuki, *Chem. Lett.*, 2006, **35**, 926.
33. D. Bensa, T. Constantieux and J. Rodriguez, *Synthesis*, 2004, 923.
34. Y. Mori, K. Kakumoto, K. Manabe and S. Kobayashi, *Tetrahedron Lett.*, 2000, **41**, 3107.
35. Y. V. Subba Rao, D. E. De Vos and P. A. Jacobs, *Angew. Chem., Int. Ed. Engl.*, 1997, **36**, 2661.
36. C. E. Song, J. W. Yang and H.-J. Ha, *Tetrahedron: Asymm.*, 1997, **8**, 841.
37. P. Hodge, E. Khoshdel and J. Waterjouse, *J. Chem. Soc., Perkin Trans. 1*, 1985, 2327.
38. M. Inagaki, J. Hiratake, Y. Yamamoto and J. Oda, *Bull. Chem. Soc. Jpn.*, 1987, **60**, 4121.
39. E. Angeletti, C. Canepa, G. Martinetti and P. Venturello, *Tetrahedron Lett.*, 1988, **29**, 2261.
40. R. K. Iler, *The Chemistry of Silica,* Wiley-Interscience, New York, 1979, Ch. 6, p 649.
41. D. J. Macquarrie and D.B. Jackson, *Chem. Commun.*, 1997, 1781.
42. D. J. Macquarrie, *Green. Chem.*, 1999, **1**, 195.
43. H. Hagiwara, A. Numamae, K. Isobe, T. Hoshi and T. Suzuki, *Heterocycles*, 2006, **68**, 889.

44. H. Hagiwara, K. Isobe, A. Numamae, T. Hoshi and T. Suzuki, *Synlett*, 2006, 1601.
45. D. A. Evans and G. C. Andrews, *Acc. Chem. Res.*, 1974, **7**, 4489.
46. B. M. Choudary, M. L. Kantam, P. Sreekanth, T. Bandopadhyay, F. Figueras and A. Tuel, *J. Mol. Catal. A: Chem.*, 1999, **142**, 361.
47. A. C. Blanc, D. J. Macquarrie, S. Valle, G. Renard, C. R. Quinn and D. Brunel, *Green Chem.*, 2000, **2**, 283.
48. A. Coelho, A. El-Maatougui, E. Raviña, J. A. S. Cavaleiro and A. M. S. Silva, *Synlett*, 2006, 3324.
49. M. L. Kantam, P. Sreekanth and P. Lakshmi Santhi, *Green Chem.*, 2000, **2**, 47.
50. C. G. Kruse, *Chirality in Industry*, ed. A. N. Collins, G. N. Sheldrake and J. Crosby, Wiley, Chichester, 1992, Ch. 14.
51. C. Ogawa, M. Sugiura and S. Kobayashi, *Chem. Commun.*, 2003, 192.
52. M. Tomoi, Y. Akada and H. Kakiuchi, *Makromolekulare Chemie., Rapid Commun.*, 1982, **3**, 537.
53. A. Corma, H. García and A. Leyva, *Chem. Commun.*, 2003, 2806.
54. L. F. Tietze, *Chem. Rev.*, 1996, **96**, 115.
55. N. T. S. Phan, C. S. Gill, J. V. Nguyen, Z. J. Zhang and C. W. Jones, *Angew. Chem., Int. Ed.*, 2006, **45**, 2209.
56. A. J. Rondinone, A. C. S. Samia and Z. J. Zhang, *J. Phys. Chem. B.*, 1999, **103**, 6876.
57. K. Akagawa, S. Sakamoto and K. Kudo, *Tetrahedron Lett.*, 2007, **48**, 985.
58. R. K. Zeidan, S.-J. Hwang and M. E. Davis, *Angew. Chem., Int. Ed.*, 2006, **45**, 6332.
59. S. Huh, H.-T. Chen, J. W. Wiench, M. Pruski and V. S.-Y. Lin, *Angew. Chem., Int. Ed.*, 2005, **44**, 1826.
60. S. Huh, H.-T. Chen, J. W. Wiench, M. Pruski and V. S.-Y. Lin, *J. Am. Chem. Soc.*, 2004, **126**, 1010.
61. K. Motokura, M. Tada and Y. Iwasawa, *J. Am. Chem. Soc.*, 2007, **129**, 9540.
62. A. W. Pilling, J. Boehmer and D. J. Dixon, *Angew. Chem., Int. Ed.*, 2007, **46**, 5428.
63. S. L. Poe, M. Kobašlija and D. T. McQuade, *J. Am. Chem. Soc.*, 2007, **129**, 9216.
64. S. L. Poe, M. Kobašlija and D. T. McQuade, *J. Am. Chem. Soc.*, 2006, **128**, 15586.
65. M. Fukushima, S. Endou, T. Hoshi, T. Suzuki and H. Hagiwara, *Tetrahedron Lett.*, 2005, **46**, 3287.

Task-specific Ionic Liquids for Fine Chemicals

CINZIA CHIAPPE

Dipartimento di Chimica e Chimica Industriale, Università di Pisa, via Risorgimento 35, 56126 Pisa, Italy

4.1 Introduction

Over the past decade, ionic liquids (ILs) have become one of the fastest growing research fields, mainly due to the unique physicochemical properties of these compounds. The application of these remarkable salts as solvents in reactions and extraction processes has been extensively investigated and reviewed.[1] Furthermore, several studies have evidenced the possibility to apply ILs as versatile "green" engineering liquids in various industrial fields: they can be used as heat transfer fluids, azeotrope-breaking liquids, lubricants, electrolytes, liquid crystals, supported IL membranes, plasticizers, and more.[2] Furthermore, ILs show improved properties when applied as electrolytes in electronic devices, as well as in processing metals and polymers – especially biopolymers such as cellulose.[3] The term "ionic liquid" identifies today a class of compounds liquid at/or near room temperature that are composed solely of ions, in contrast to conventional solvents that consist of covalent molecules. The ionic nature determines the lack of a measurable vapor pressure that, associated with a high thermal and chemical stability, makes these compounds intrinsically excellent candidates for industrial applications compared to common organic solvents. Many of the chemical reactions and separation processes used in the pharmaceutical and fine chemicals industry require the use of large amounts organic

RSC Green Chemistry Book Series
Eco-Friendly Synthesis of Fine Chemicals
Edited by Roberto Ballini

Published by the Royal Society of Chemistry, www.rsc.org

solvents. These solvents are often volatile and sufficiently water-soluble to contaminate air emissions and aqueous discharge streams, adding to the environment burden and the cost of downstream processing and recovery operations. For ecological and economic reasons using a solvent has an obvious downside and the question of whether to use a solvent at all and, if yes, which one is obviously very important. On the other hand, it has often been evidenced that ionic liquids can nicely complement, and even sometimes work better than, organic solvents in several industrial processes.[1,2] Currently, the most successful example of an industrial process using ionic liquid technology is the BASIL™ (biphasic acid scavenging utilizing ionic liquids) process; replacing triethylamine with 1-methylimidazole results in the formation of 1-methylimidazolium chloride, an ionic liquid that separates out of the reaction mixture as a discrete phase and can give again 1-methylimidazole, *via* base decomposition. The same group (BASF) has also developed processes for breaking azeotropes, dissolving and processing cellulose, replacing phosgene as a chlorinating agent with hydrochloric acid, and aluminium plating. Moreover, processes either operating or at pilot have been developed by other companies. This topic has been overviewed and extensively discussed in two recent reviews by Seddon and Plechkova and by Maase.[4]

Notably, however, despite the enthusiastic results evidenced by several research groups working in ILs chemistry, and the fact that the growing awareness of the pressing need for greener, more sustainable technologies has focused attention on the manufacture of fine chemicals and pharmaceutics, industrial applications are generally related to the production of "bulk chemicals."

Surely, the concepts and paradigms of ionic liquids are new and still not fully accepted in the wider community; many chemists consider with some skepticism the possibility of applying ILs as substitutes of common solvents in large-scale processes. Moreover, as has been recently evidenced,[5] there are several prerequisites that determine the choices of development chemists working in the fine chemical industry:

- The solvent must be commercially available at acceptable prices and with guaranteed quality. Its toxicity and other ecological factors must be well known.[5a]

As pharma manufacturing is highly regulated by government agencies (at present and before the European REACH regulation!), process modifications involving material substitution require complete approval. Process variations are therefore expensive and time-consuming, and are made only if necessary. Only solutions to problems that can be implemented in the short or intermediate terms and involve minimal alteration to the current pharmaceutical processes are taken into account.

- Fine chemicals are generally polar complex molecules (isomers, stereochemistry, functional groups) with limited thermal stability.[5]

The production of fine chemicals requires efficient synthetic protocols characterized by high regio-, chemo- and stereoselectivity, the activation of poorly reactive substrates and the minimization of side products formation. A drastic reduction of catalyst loading is also mandatory to avoid the risk of metal or ligand contamination in the final product. Chemists must be convinced not only that it is possible to perform reactions in ILs but also that it is possible to isolate the products from ILs and that it is possible to recycle the solvent and catalysts. The possibility of recycling ILs and catalysts reduces environmental impact and costs.

- Finally, in the fine chemical industry the time for the development of the production process for a new chemical entity is limited: between a few months in the pharmaceutical industry to 1–2 years for agrochemicals.[5a]

The consequences of these constraints on the choice of process technology are obvious: the first choice is always well-proven technology – new solutions with too many unknown parameters will be avoided. The extremely high number of ILs and the possibility to modulate their physicochemical properties through an appropriate structure modification, *i.e.*, the possibility to tailor the best ionic liquid for each specific application, may represent a limit instead of an advantage if time is a determining factor in the choice of reaction conditions.

In reality, many ILs can match at least some of the above-mentioned requisites. The price of ionic liquids depends on the constituent ions. At present, the price of imidazolium-based ionic liquids is 5–20 times higher than those of molecular solvents (on laboratory scale) but it follows the marketing rules (application on large scale should decrease the price), and recently it has been evidenced that ionic liquids with suitable properties can be obtained also using relatively cheaper materials. Related to the problem of the lack of toxicity information, research in this field is growing rapidly.[6] Although some toxicological works have evidenced that the ionic nature does not necessary render these compounds as "absolutely not-toxic materials," important data for a more rational design of environmental friendly salts are being collected.[6]

In addition, even if most applications reported so far are mainly related to the use of ionic liquids as solvents in the synthesis of simple model compounds, papers related to the preparation of more complex compounds have been published. The two-step synthesis of pravadoline has been described[7] in 2000 and subsequently complex multifunctional compounds have been synthesized in ILs and easily recovered from these media. Moreover, convenient recycling of the system has been evidenced.[1,2] Finally, related to the third requisite, although no exhaustive response has been given to date to the request for identification of a limited number of ionic liquids with broad applicability, the development of ILs bearing specific functional groups on cation and/or anion has allowed us to obtain new classes of salt liquids at room temperature having additional properties. These ILs are generally defined as task-specific ionic liquids (TSILs). The presence of selected functional groups on the cation and anion can enhance the capacity of ILs to interact with specific dissolved species (reagents, catalysts). When this interaction involves reagents or catalysts and is particularly strong, ILs are able to act as liquid supports and synthesis occurs in a homogeneous solution phase. Alternatively, when the interaction involves solid matrices, ILs functionalization can be used to obtain supported ionic solvents, which combine the advantages of ILs with those of heterogeneous materials.[8]

4.2   Task-specific Ionic Liquids (TSILs) in Batch Processing

Continuous processing has been employed for many years by petrochemical and bulk chemical producers to increase efficiency and overcome challenges inherent to specific reactions. In contrast, batch processing predominates in the pharmaceutical and fine chemical industry, in particular when (1) product life-cycles are relatively short; (2) overall product demand is limited; (3) production volumes are relatively low; (4) capital investments required for a continuous processing are high.[5b] Traditionally, most fine chemicals were synthesized without the use of catalysts because reagents with a sufficiently high reactivity were used and, if a catalyst was necessary at all, in most cases either a proton or a simple base was sufficient. More recently, however, the interest in greener and more sustainable technologies has favored the development of atom efficient catalytic methodologies also for the manufacture of fine chemicals and pharmaceuticals. Independent of the reaction conditions, the use of solvents is, however, much more common in fine chemicals production than in bulk chemicals because many starting materials and products are solids and do not tolerate high temperatures. The possibility of using alternative reaction media that circumvent the problems associated with many of the traditional volatile organic solvents may therefore be important in fine chemicals synthesis. Furthermore, the use of non-conventional reaction media may provide opportunities for facilitating the recovery and recycling of the catalyst, when catalytic methodologies are applied.

4.2.1   TSILs in Acid- and Base-catalyzed Reactions

Most ions constituting ionic liquids can be categorized according to their Lewis acid/base properties (*i.e.*, their capability to accept or to donate an electron pair); nevertheless, some ions may be considered according to the Brønsted definition, *i.e.*, on the basis of their ability to accept or donate a proton. Typical ionic liquids are those based on "neutral" or "very weakly basic" anions (BF_4^-, PF_6^-, NO_3^-, $CH_3SO_3^-$, Tf_2N^-) and neutral (tetraalkylammonium, dialkyl-pyrrolidium, trialkylsulfonium) or weakly acidic cations (1,3-dialkylimidazolium and 1,2,3-trialkylimidazolium) (Figure 4.1).

Imidazolium-based ionic liquids are the most investigated and used ILs to date. Related to their solvent properties, they are characterized by π^* values similar to water, by moderate α values, which depend primarily on the cation, and by moderate β values determined mainly by the anion. In particular, imidazolium ILs bearing a CH group between nitrogen atoms have larger α values than those in which a methyl group is present on the connecting carbon atom.

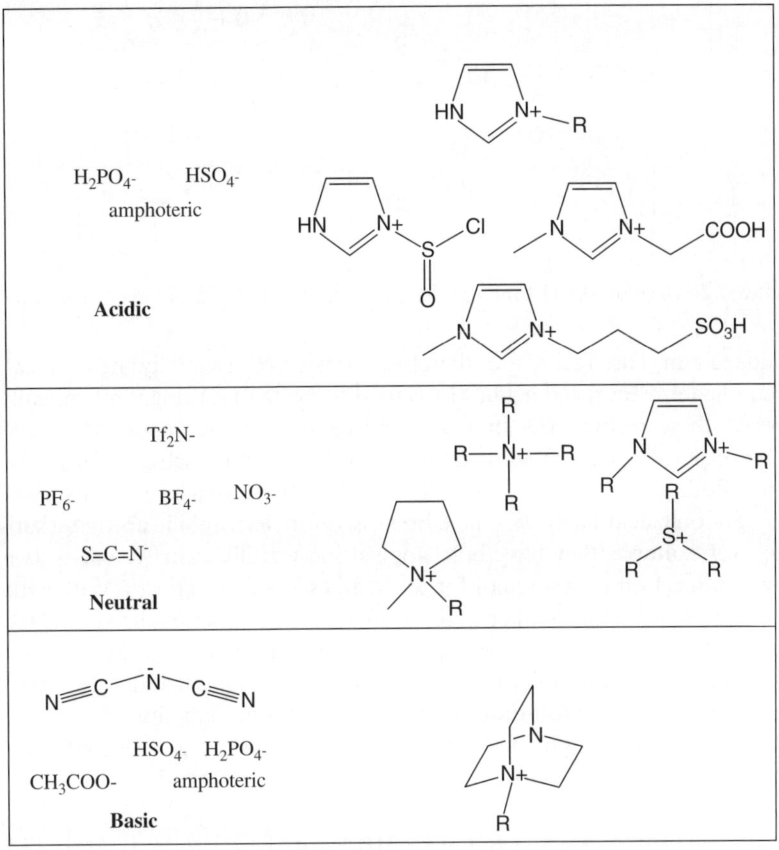

Figure 4.1 Examples of some common ionic liquids.

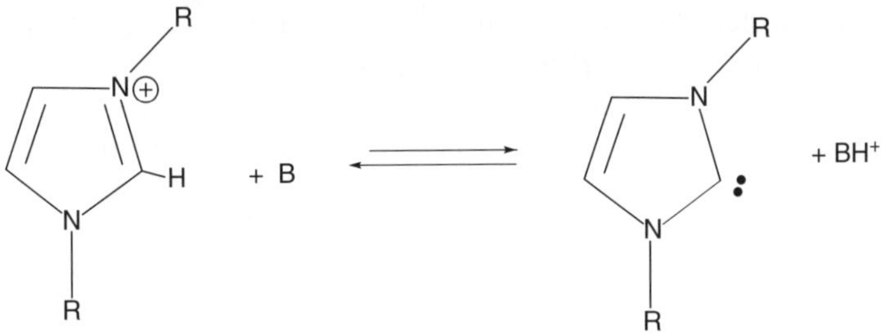

Scheme 4.1

Stereochemical and kinetic studies have confirmed[9] the enhancement of the hydrogen bond ability of the imidazolium cation on going from 1,2-dimethyl-3-alkylimidazolium salts to 1,3-dialkylimidazolium cations. However, deprotonation of the site between nitrogens is not particularly simple, it requires strong bases and depends on the ionic liquid counter-anion (Scheme 4.1).[10]

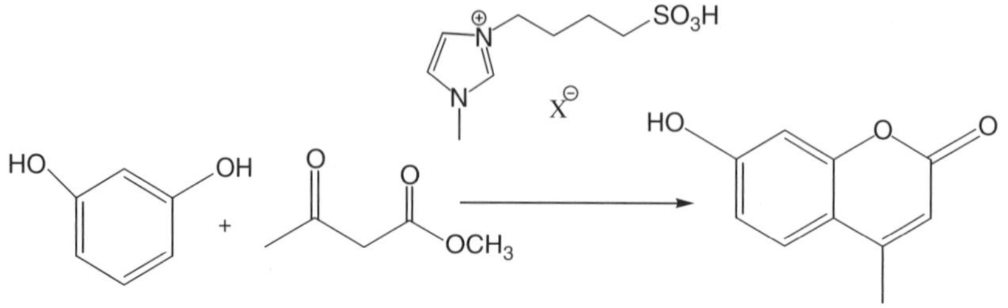

Scheme 4.2   Use of SO_3H-functionalized imidazolium ILs in the Pechmann reaction.

Imidazolium based ILs are therefore considered generally neutral cations. Indeed, the term Brønsted acidic ILs is used for salts that present intrinsically this property; these include ILs in which the cation contains appendages such as COOH or SO_3H, or ILs having mono or diprotic anions such as HSO_4^-, $H_2PO_4^-$, citrates. Both these classes of ionic liquids have been used as solvents and catalysts.[11] Sulfonic acid-based ILs have been used for electrophilic aromatic substitutions;[12] for example, they have been applied successfully as active catalysts in the Pechmann reaction of resorcinol for the synthesis of coumarins.[13] With activated phenols, at an oil bath temperature of 80 °C, SO_3H-functionalized imidazolium ionic liquids (5 mol.%) gave the expected product in high yield (Scheme 4.2).

Nevertheless, in the category of acidic ILs can be also included ILs based on protic ammonium, pyrrolidinium, pyridinium and imidazolium cations. These salts are formed through the transfer of a proton from a Brønsted acid to a Brønsted base:

$$B + A \rightarrow HB^+ + A^-$$

The process can be improved using a stronger acid and/or stronger base, hence leading to a greater driving force for the proton transfer. Ideally, in these ILs the proton transfer should be complete in order to have only cations and anions. In reality, the proton transfer is always less than complete and, consequently, neutral, acid and base species are present and aggregation and association of neutral and charged species can occur. Recently, MacFarlane and Seddon have proposed[14] to set a limit of 1% of neutral species in the IL for it to be classed as a "pure ionic liquid," with those with higher levels of neutral species perhaps better thought of as mixtures of ILs and neutral species. Several techniques have been proposed to provide information about the ionicity of such ILs (NMR, IR, changes of thermal properties, and ionic conductivity in the form of Walden plot); however, all these measurements generally give only a qualitative evaluation of the ionic nature of these liquids.[15] NMR and IR have also been used to determine the acidity of this class ILs; strong correlations between acidity and catalytic activity were evidenced.[15]

Related to their application in synthesis, protic ILs have been used in C–C bond-forming reactions (Knoevenagel condensations, Diels–Alder reactions, alkylation and Henry reactions, aldol condensations and Mannich reactions) and in C–O bond-forming reactions (Fisher esterifications, acetylation reactions, protection of alcohol groups).[15] Moreover, protic ILs have been used for isomerizations,[16] ether cleavage,[17] chlorination[18] and nitration[19] of aromatic compounds and, more recently, as substituted catalysts of oil of vitriol in acylation of salicylic acid for the synthesis of aspirin.[20]

In contrast, anions such as lactate, formate, carboxylates and dicyanamide form ionic liquids that can be defined as basic. The basic nature of these anions can impart important properties to the corresponding ILs and interest in these compounds is expected to grow in the coming years.[21] For example, using dicyanamide-based ionic liquids, the possibility of performing the acetylation of alcohols and carbohydrates with no additional base catalyst has been evidenced,[22] at least in some cases. Furthermore, the reactions are significantly faster and higher yielding in dicyanamide-based ionic liquids than in conventional solvents.[22] Finally, the anomeric ratio is shifted towards the β anomer in the presence of dicyanamide anion.[22]

However, the use of basic anions is not the sole approach followed to obtaining basic ILs. Liquid salts bearing this additional property can be prepared also by incorporation of a basic center into the cation. This approach generally affords more thermally stable ILs than those based on basic anions, which frequently present relatively low decomposition temperatures. Basic ionic liquids bearing aliphatic or aromatic amines on the side chain(s) have been synthesized[23] and, recently, some of these have been used as both the solvent and base for Heck, "copper free" Sonogashira, and for homocoupling reactions of terminal alkynes (Figure 4.2).

Ionic liquids with tertiary amines on the side chain are effective solvents for these reactions; furthermore, tertiary aliphatic amines may act as a reducing

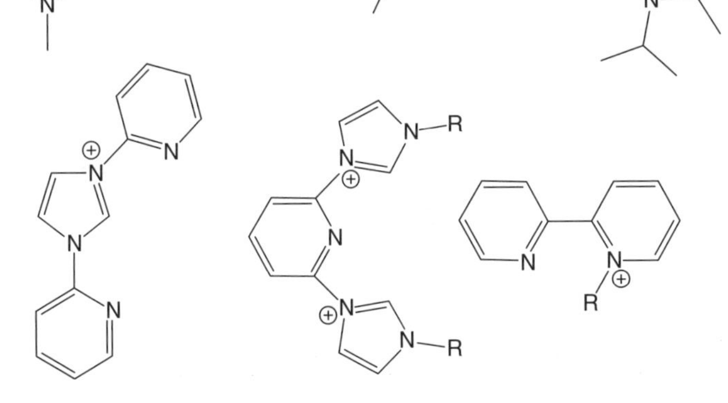

Figure 4.2 Basic IL cations.

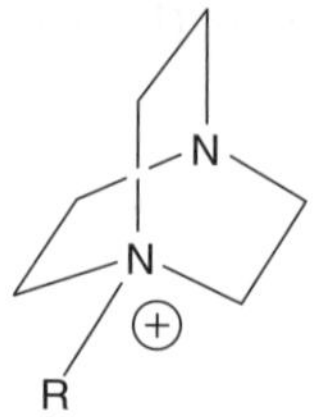

Figure 4.3 Dabco-based IL cations.

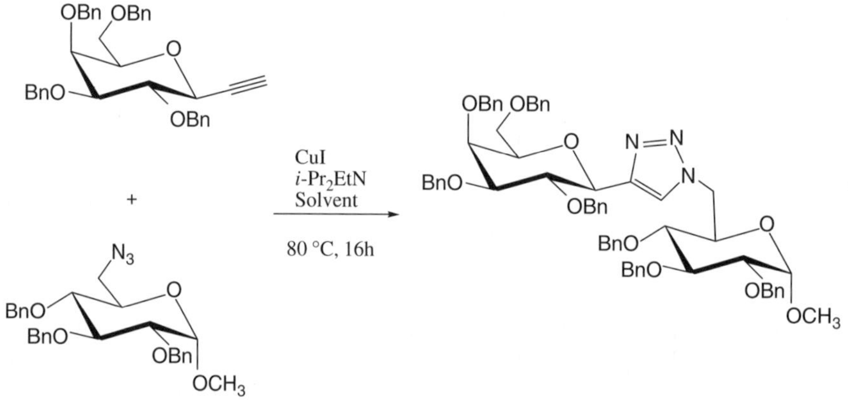

Scheme 4.3 Copper(i)-catalyzed azide-alkyne click reaction in ILs.

agent, and five recycles have been performed without loss in catalytic activity.[23] Nevertheless, basic ionic liquids can be also obtained by monoalkylation of diaza-bicyclo[2.2.2]octane (Dabco, Figure 4.3).[24]

Dabco-based ionic liquids have been recently used,[25] together other basic and neutral ILs, in the copper(i)-catalyzed azide-alkyne click reaction (Scheme 4.3). The model cycloaddition of a sugar azide with a sugar acetylene has been carried out in ten ILs. With one exception, in all investigated ILs the reaction afforded exclusively the 1,4-disubstituted triazole, namely a triazole-linked *C*-disaccharide, in high yields.

[C_8dabco][N(CN)$_2$] gave the highest yield (91%) of the pure isolated cycloadduct and the highest regioselectivity also when the reaction was carried out in the absence of the added Hünig's base. An active role of the IL anion in the Cu(i)-catalyzed reaction was evidenced on comparing the regiochemical behavior in the different investigated solvents.

4.2.2 TSILs in Metal-catalyzed Reactions

The use of ionic liquids to immobilize homogeneous catalysts and favor subsequent reuse has been one of the most fruitful areas of ionic liquids research to date. In general, ILs are good solvents for catalyst immobilization, and many

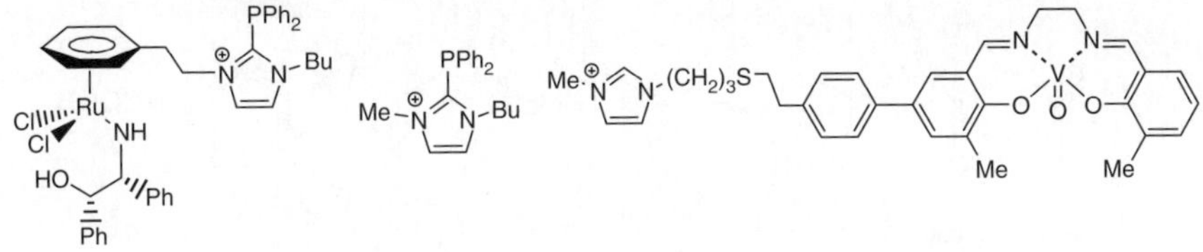

Figure 4.4 Task-specific catalysts or ligands.

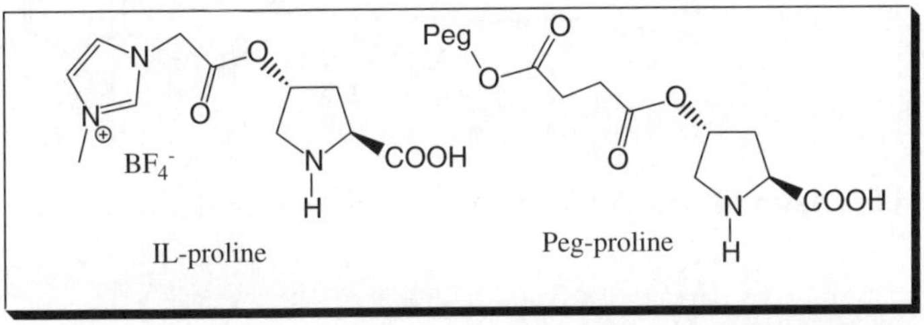

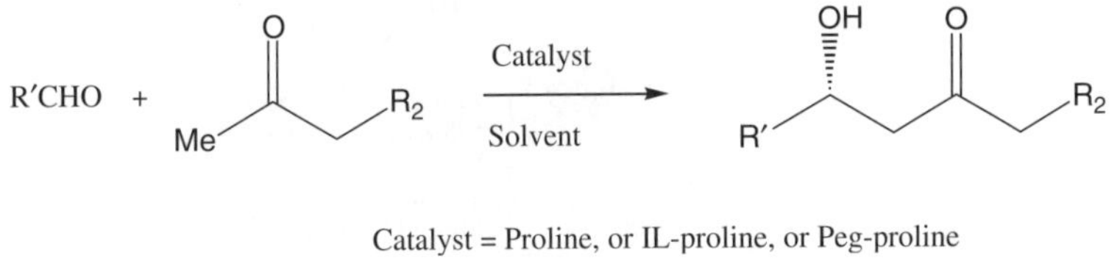

IL-proline

Peg-proline

Scheme 4.4 Proline catalyzed aldol reaction in molecular solvents.

catalysts can be separated at the end of the reaction by simple phase separation and recycled. However, common ILs still have a tendency to leach dissolved catalysts into the co-solvent used to extract the product(s). To overcome or reduce the catalyst leaching several approaches have been followed. The imidazolium salt motif was incorporated into the ligands, catalyst-precursors or support materials, obtaining functionalized catalytic systems with an increased affinity for the ionic medium, which can enhance catalyst reusability as well as, in some cases, activity in conventional ionic liquids (Figure 4.4).

For example, onium ion-tagged prolines have been synthesized and their catalytic activity in direct asymmetric aldol condensation was initially studied in molecular solvents.[26] The observed superior performance of IL-proline relative to Peg-proline or free proline, in terms of yields and enantiomeric excess, evidenced that the ionic moiety in IL-proline plays more than a silent or simply supporting role in the reaction (Scheme 4.4).[26]

More recently, greater chemical yields, higher enantioselectivities and more efficient recycles were obtained on performing the same reaction in ILs.[27] Two onium-tagged proline catalysts were employed (Scheme 4.5), evidencing the superior catalytic efficiency of ionic media for this kind of reaction.

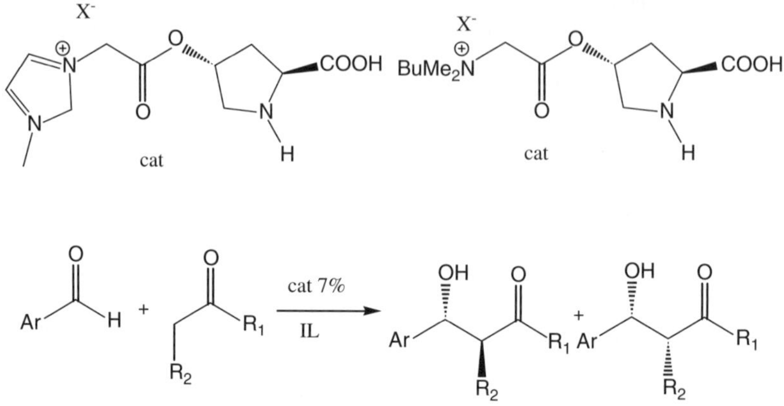

Scheme 4.5   Asymmetric aldol reaction using onium-tagged proline as catalyst.

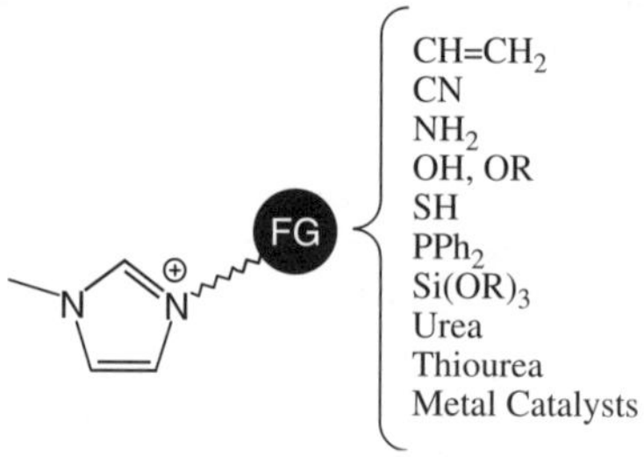

Figure 4.5   Functionalized IL cations.

Contemporaneously, the strategy to modify the IL to increase catalyst immobilization has been followed by some research groups. A range of different functional groups providing properties similar or identical to those of the major classes of organic solvents has been incorporated into IL cations, including vinyl and allyl, amine, amide, ether and alcohol, acid, urea and thiourea, fluorous chains, glycidyl chains, alkyne, phosphoryl, nitrile, thiols and ferrocenyl groups (Figure 4.5).[28]

The anion has been also functionalized, and ILs based on metal carbonyls, alkylselenites and functionalized borates have been synthesized.[29] Some of these ILs, bearing relatively simple functional groups, have been used as solvents in selected metal-catalyzed reactions,[30] evidencing that task-specific ILs can favor the activation of the catalyst, generate new catalytic species, and improve the catalyst stability. Moreover, they are able to optimize immobilization and recyclability, facilitate product isolation, and influence the selectivity of the reaction.

Nitrile-functionalized ILs have been used in Stille, Suzuki and Heck reactions[31] and their usefulness became apparent upon catalyst recycling; while in conventional ILs the activity rapidly decreased to zero, in nitrile-functionalized

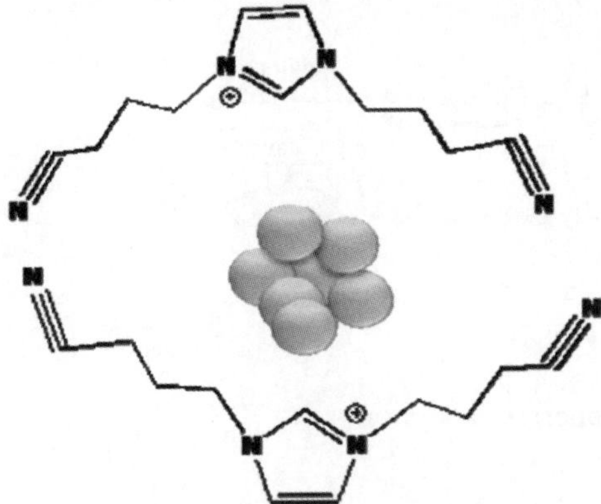

Figure 4.6 Pd nanoparticles in nitrile-imidazolium based ILs.

ILs little change was observed after four recycles. TEM analyses have revealed that while the nanoparticles immobilized in conventional ILs aggregated, those dissolved in [C$_3$CN][Tf$_2$N] remained evenly distributed. The role of the nitrile group on the IL cation is probably twofold: first it helps the formation of a protective film around the nanoparticles, preventing aggregation, and, second, it favors the stabilization of the active Pd(II) catalyst (Figure 4.6).

Nevertheless, nitrile functionalized ILs have been used for cobalt-catalyzed cyclotrimerization reactions of monosubstituted aromatic alkynes: also in this case, the nitrile functionality was able to stabilize the transient cobalt(I) catalytic species, ensuring good conversions.[32]

The ability of functionalized ILs to stabilize metal nanoparticles has been evidenced even in the case of other functionalized ILs; TSILs bearing thiol group(s) on the alkyl chains of the cationic moiety can stabilize gold nanoparticles.[33]

Despite the fact that promising results have been obtained using TSILs in metal-catalyzed reactions, the number of investigated ILs is still limited, and extremely limited is the number of explored reactions catalyzed by metals other than palladium.[34] The potentiality of these TSILs in the synthesis of fine-chemicals through metal-catalyzed reactions should therefore be explored more extensively.

4.2.3 TSILs in Supported Synthesis

An attractive feature of ionic liquids is their solubility. Depending on the choice of cation and anion, solubility can be tuned readily so that they can phase separate from organic as well as aqueous media. This behavior has suggested the possibility of using ionic liquids as soluble supports for organic synthesis. This novel liquid-phase strategy can embrace several possibilities: (1) supported catalysts, (2) supported reagents and (3) supported substrates.[35] In each case, the IL-supported species can be dissolved in a molecular solvent (usually more polar), and the reaction can be conducted in a homogeneous solution. After the

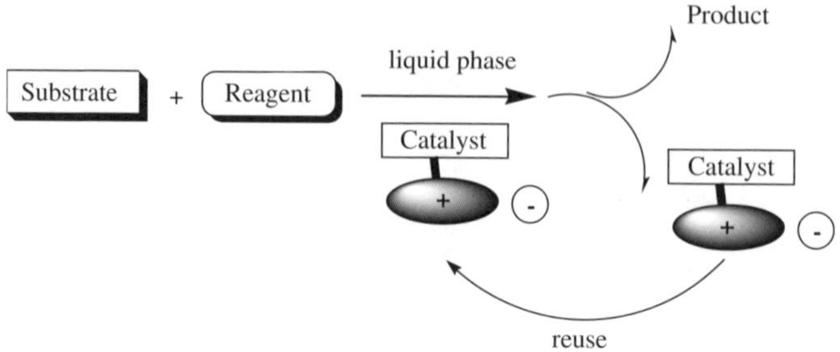

Scheme 4.6 IL-supported catalyst.

reaction, a less polar organic solvent is added in which the IL-supported species is not soluble, and the IL phase separates from the organic phase. The recovered IL-supported species can be regenerated (for reagent) or reused (for catalyst) or further reacted to give the final product, which would then be detached from the ionic liquid support.

4.2.3.1 IL-supported Catalysis

Davis was the first to recognize that functionalized ionic liquids can serve not just as reaction media but as catalysts as well.[36] One advantage of using an IL-supported catalyst is that the catalyst can be recovered simply by solubility difference (Scheme 4.6).

This is especially important in those cases where the catalyst has to be used in quite substantial amounts, as in the above-discussed (*S*)-proline-catalyzed asymmetric aldol reactions, where the catalyst is usually used at 30 mol.%.[37]

In addition, efficient recycling of IL-supported catalysts has suggested the possibility of applying the same approach in metal-catalyzed reactions, where the reuse of expensive ligands, metal, or both may be critical.

In 2003, Guillemine and Yao reported[38] in two independent papers the synthesis of IL-supported catalysts for ring-closing metathesis, whereas a IL-supported palladium complex was found to catalyze the Heck reactions with good recyclability of up to ten cycles.[39]

4.2.3.2 IL-supported Reagents

Recently, some examples of synthetic reagents anchored onto ionic liquids have been reported.[40] These new systems present the advantage of being readily separated from the reaction mixture by simple phase separation after the desired chemical transformation and can then be regenerated and reused. For example, IL-supported hypervalent iodine compounds have been used for the oxidation of alcohols to ketones (Scheme 4.7).

ILs can also be used as soluble supports for combinatorial synthesis, which initially was applied for the preparation of a small library of 4-thiazolidinones[41]

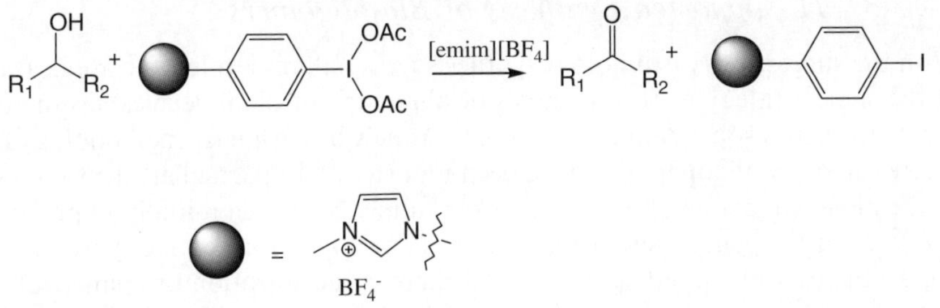

Scheme 4.7 IL-supported reagents.

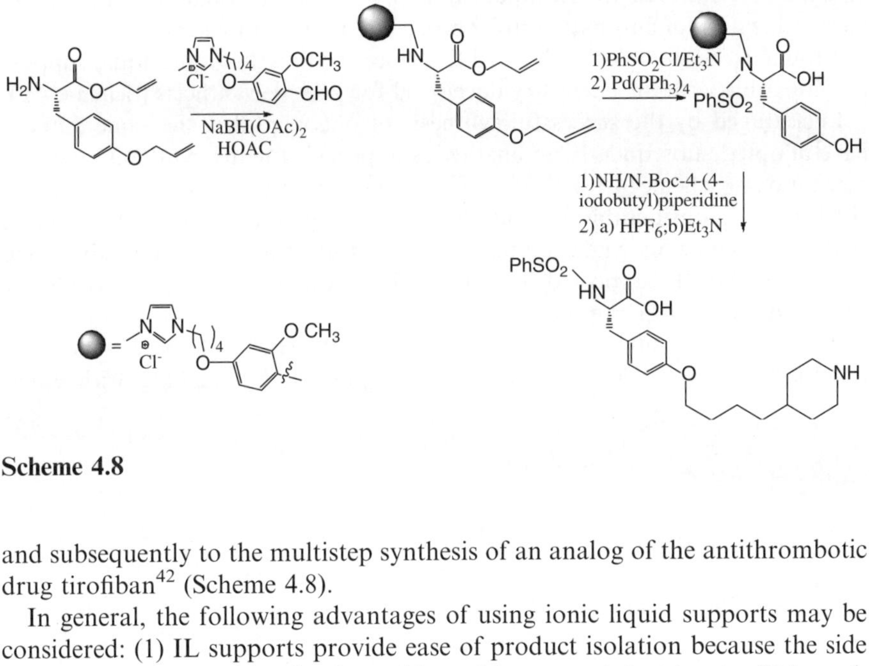

Scheme 4.8

and subsequently to the multistep synthesis of an analog of the antithrombotic drug tirofiban[42] (Scheme 4.8).

In general, the following advantages of using ionic liquid supports may be considered: (1) IL supports provide ease of product isolation because the side products are removed by simple washing with appropriate solvents; (2) in each step, the reaction can be followed by standard analytical techniques; (3) due to the high polarity of the ionic liquid supports, microwave dielectric heating can be easily applied to enhance the reaction. In the specific case of the synthesis of tirofiban, the authors evidence some other features as worthy of note. First, all five steps of the synthesis can be conducted in one pot using [bmim][PF$_6$] as reaction media. Second, the work up of every reaction step was done simply by consecutive extractions with diethyl ether and water to remove excess reagents. Third, in the cleavage of the desired product from the ionic liquid support it is necessary to use HPF$_6$ instead of the more commonly used trifluoroacetic acid (TFA). This was because the excess TFA underwent anion exchange with [bmim][PF$_6$] to give [bmim][TFA] as contaminant in the product.

4.2.3.3 *IL-supported Synthesis of Bio-oligomers*

With the success of IL-supported synthesis for small molecules, the application of the same strategy to the syntheses of oligomers of biomolecules has invited exploration and some data on the use of this new liquid-phase methodology for the synthesis of oligopeptides have been reported.[43] In particular, the synthesis of the bioactive pentapeptide Leu5-enkephalin was chosen initially as probe to verify: (1) if the general synthetic protocols developed for peptide synthesis are applicable to ionic liquid support; (2) if there is racemization or epimerization of the peptide units during attachment of the first amino acid to the ionic liquid moiety (loading), the coupling steps and the detachment step; (3) if the solubility of the IL-supported oligopeptide molecule is modified by the growing peptide chain such that purification by simple washings with organic and aqueous solvents is no longer practical. Leu5-enkephalin, obtained without any recrystallization or chromatography procedure in the entire synthetic sequence, was found to be >90% pure by HPLC, demonstrating that ionic liquid support is compatible with the chemistry developed for peptide synthesis (Scheme 4.9).

Encouraged by the successful synthesis of oligopeptides, the same authors have adopted subsequently an analogous approach for the synthesis of oligo-saccharides (Scheme 4.10).[44] The efficient synthesis of oligosaccharides is a challenge of considerable magnitude because the traditional solution-phase synthesis is laborious and requires purification by chromatography after each step. For the IL-supported synthesis, the β-thioglycoside was covalently anchored onto the IL support.

As in the case of IL-supported peptide synthesis,[43] all the IL-anchored intermediates in Scheme 4.10 were simply purified by washing with ether,

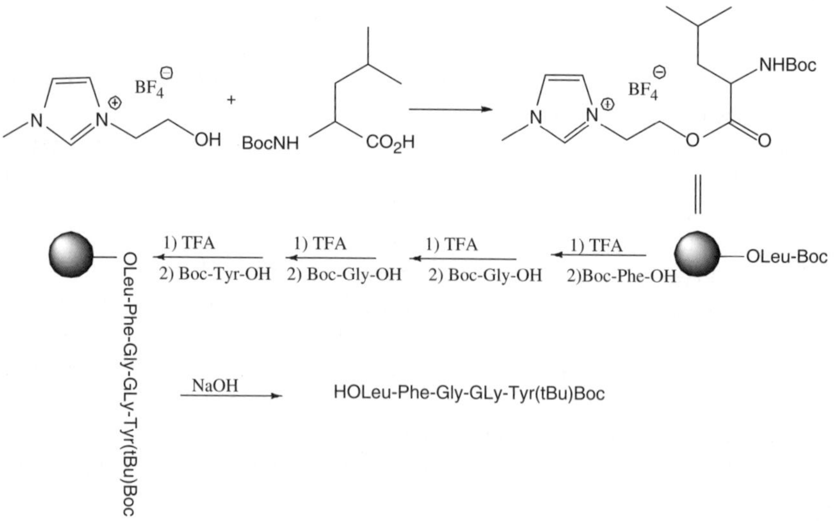

Scheme 4.9

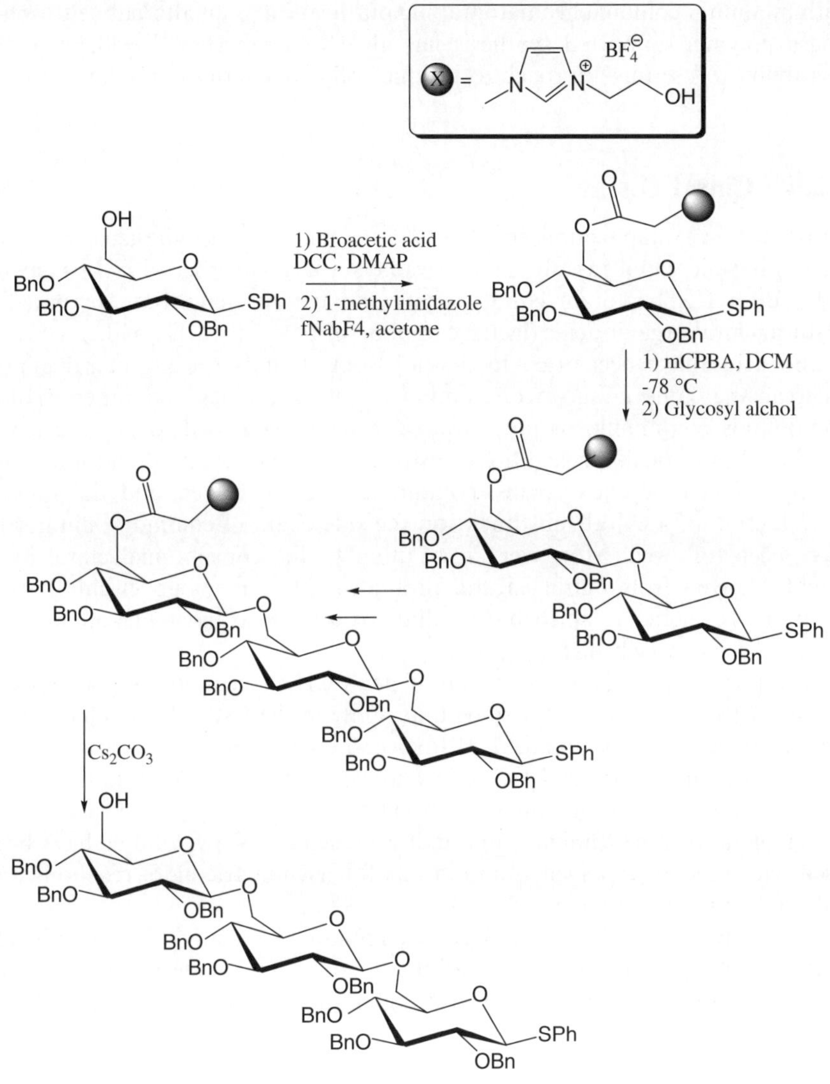

Scheme 4.10

aqueous media, or both, and their structures were characterized by conventional analytical techniques, including NMR and MS. Chromatographic purification was, however, required in each reaction step. At the end, the trisaccharide obtained by the IL-supported synthesis was found to be NMR and TLC pure and identical to a sample prepared *via* unsupported synthesis. As evidenced by the authors, this example suggests that the coupling conditions and stereoselectivity developed for classical solution-phase synthesis in the carbohydrate field can be translated to IL-supported synthesis. More recently, the efficient and cost-effective synthesis of a tetrasaccharide on an IL support with minimum column chromatographic purification as an alternative to solid-phase polymer-supported synthesis has also been reported,[45] evidencing the possibility to use this approach to produce oligosaccharides on a large scale.

4.2.4 Chiral ILs

Since the first example published[46] in 1997 the number of synthesized chiral ILs (CILs) has increased rapidly, and nowadays a large number of CILs is available. Both ILs based on chiral anions and cations have been prepared.[47] Natural chiral organic acids (lactic, camphorsulfonic, mandelic, malic, *etc.*) and amino acids have been used to develop ionic liquids bearing chiral anions whereas the same amino acids, amino alcohols, amines and other natural compounds containing, or not, a nitrogen have been used to prepare chiral cations. Both the approach to transform polyfunctionalized natural compounds in heterocyclic systems, containing a chiral center, and the simpler introduction of a chiral substituent on the side chain of common onium salts have been followed. Moreover, in addition to the conventional chiral ionic liquids derived from chiral natural products, a library of novel chiral spiro compounds, including spiro bis(pyridinium) and spiro bis(imidazolium) salt, has also been described (Figure 4.7).[48]

However, despite the rapid design of news CILs, successful applications in synthesis remained elusive for some time. Only in the last 2–3 years have some significant results been obtained. Leitner and co-workers in 2006 reported[49] a high enantiomeric excess (84% ee) by using a chiral anion containing ionic liquid for an aza-Baylis–Hillman reaction (Scheme 4.11), whereas CILs with an imidazolium or a benzimidazolium unit attached to (*S*)-pyrrolidine have been used with success as solvents or catalysts for asymmetric aldol reactions and Michael additions to nitroolefins (ee up to 99%).[50]

Unfortunately, modest results have been obtained in many other investigated reactions. A recent review reports all these data.[51] The modest or poor results found in many reactions performed using chiral ILs are really not surprising.

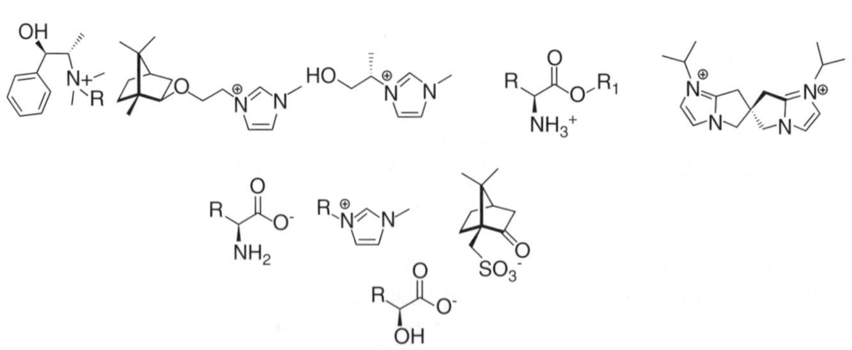

Figure 4.7 Some examples of novel chiral ILs.

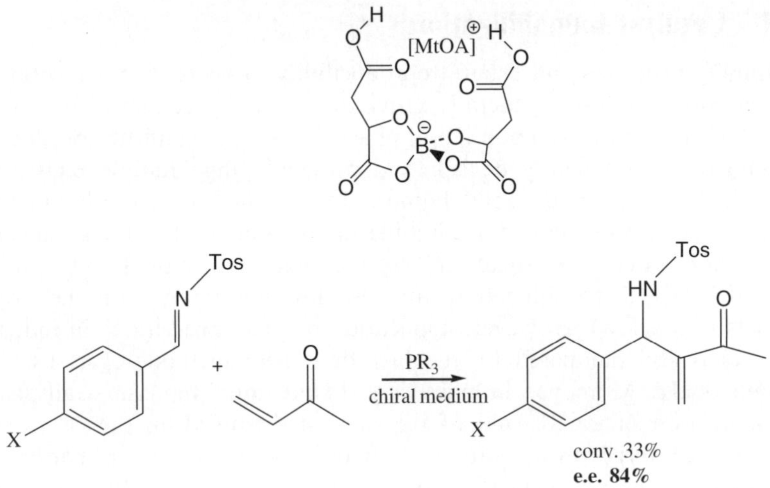

Scheme 4.11

Generally, the use of chiral solvents in organic synthesis has always given modest results;[52] the ability of ionic liquids to induce asymmetry at least in some kind of reactions is therefore a remarkable feature of these media. This positive property may be related to some peculiarities arising from the ionic nature of these solvents, such as the polymer like behavior,[53] the high degree of organization[54] or the ability to interact strongly with a selected transition state, through ion pairing phenomena.[55]

Considering that CILs are generally not expensive and therefore they might be used as solvents for large-scale synthesis, a deeper understanding of the ability of these media to induce asymmetry, at least in some processes, may be important for a rational design of new CILs to be applied with success in the industrial synthesis of chiral fine chemicals.

4.3 Task-specific Ionic Liquids in Continuous Processing

Although batch processing predominates in the pharmaceutical and fine chemical industry, recent advances in technologies supporting chemical process R&D seem to favor the introduction of continuous processing also in these areas. The application of continuous processing can give new opportunities in the case of (1) fast reactions (*i.e.*, complete within seconds to minutes) with instable intermediates or products that would otherwise degrade over extended reaction times of batch processing, (2) reactions involving hazardous reactants prepared and consumed *in situ*, (3) reactions possessing potential for runaway exothermic hazard, (4) batch reactions requiring intimate mixing of reactants in multiple phases and (5) reactions occurring rapidly upon close contact with specific energy sources (*e.g.*, microwave, UV irradiation and sonication).

4.3.1 Catalyst Immobilization

To simplify processes and ensure reproducibility, process chemists within the pharmaceutical industry generally gravitate towards reaction chemistry that can be run in a single common liquid phase. However, significant productivity advantages can frequently be achieved by employing multiple phases (*i.e.*, liquid–liquid, gas–liquid, solid–liquid). In the last decade, liquid–liquid biphasic processes involving ionic liquids and organic product phase have been the subject of intense investigation since the ionic catalyst phase often induces excellent catalyst immobilization and also low miscibility with the organic products (see below). However, application of these procedures in industrial processes is still hampered by the fact that these methodologies use large amounts of ILs. Moreover, in many cases liquid–liquid biphasic catalysis uses only a fraction of the IL and of the catalyst dissolved; in the case of fast reactions, due to the low mass transfer often the process occurs primarily at the inter-phase of the diffusion layer of the IL catalyst phase. In this context, advances may be found in the use of immobilized catalysts and solvents for which monolithic columns can be used as process reactors.

The concept of immobilized ionic liquids entrapped, for instance, on the surface and pores of various porous solid materials (supported ionic liquid phase, SILP) is rapidly become an attractive alternative. In addition, the SILPs can also answer other important issues, such as the difficult procedures for product purification or IL recycling, some toxicity concerns and the problems for application in fixed-bed reactors, which should be addressed for future industrial scale-up. This new class of advanced materials shares the properties of true ILs and the advantages of a solid support, in some cases with an enhanced performance for the solid material. Nevertheless, a central question for the further development of this class of materials is to understand how much the microenvironment provided by the functional surfaces is similar or not to that imparted by ILs. Recent studies carried out[56] using the fluorescence of pyrene to evaluate the polarities of a series of SILPs based on polymeric polystyrene networks reveal an increase in polarity of polymers, whereas the polymer functional surfaces essentially maintain the same polarity as the bulk ILs. However, this is surely not a simple task, in particular if we consider that the basic knowledge of pure ILs is still in its infancy, and we are just starting to understand the fundamentals of pure ILs when used as solvents.

SILPs as catalytic systems have been proposed for olefin hydroformylation and hydrogenation (RH-catalyzed), Heck reactions (Pd-catalyzed) and hydroamination (Rh-, Pd- and Zn-catalyzed).[57] In these systems the catalysts are composed of a transition-metal complex dissolved in a thin film of the ionic liquid, which is held on a porous solid with high surface area by physisorption, tethering or covalent anchoring. An important question related to the use of SILPs is the immobilization of the ionic liquid on the carrier. Covalent anchoring can be performed either *via* the anion or cation involving the hydroxyl groups of the surface.[58] This is conveniently carried out, for instance, *via* a reaction with the appended alkyl–silyl side branch tethered into the anion or cation, respectively. This method, however, principally allows only anchoring of relatively isolated cation–anion molecule pairs and does not give a true liquid layer. The first supported Lewis acidic ionic liquid systems were prepared and explored for selective low-temperature alkylations of olefins to generate hydrocarbons fuels, and, subsequently, the related systems were developed[59] for Friedel–Crafts reactions. Here, the catalysts were prepared by confining pre-formed catalytically active ILs on the support by impregnation. Reaction between the chloroaluminate anions and the support surface OH groups results in a covalent anchoring of anions to the support and formation of HCl (Scheme 4.12). Depending on the characteristic of the support materials, different amounts of IL can be maintained on the support surface (SiO_2 > Al_2O_3 > Ti_2O > Zr_2O).

To avoid the formation of HCl in the synthesis of supported Lewis acidic ILs, ionic liquids bearing an alkoxylsilyl group on the alkyl chain, which can be covalently bound (grafted) to the surface, were used. In this procedure the aluminium halide was introduced subsequently, giving a highly acidic ionic complex on the surface of the support (Scheme 4.13).

An alternative sol–gel strategy has been reported for the preparation of a closely related material. Using the approach of the initial grafting of the imidazolium cation onto pre-dried silica, followed by treatment with $PdCl_2$, non-acidic supported palladium-based ILs were also prepared and used in Suzuki–Miyaura coupling reactions.[60]

Analogously, oxidation catalysts were prepared[61] by using an interesting integrated approach that combines sol–gel entrapped perruthenate as aerobic catalyst, an encapsulated ionic liquid as solubility promoter, and $scCO_2$ as

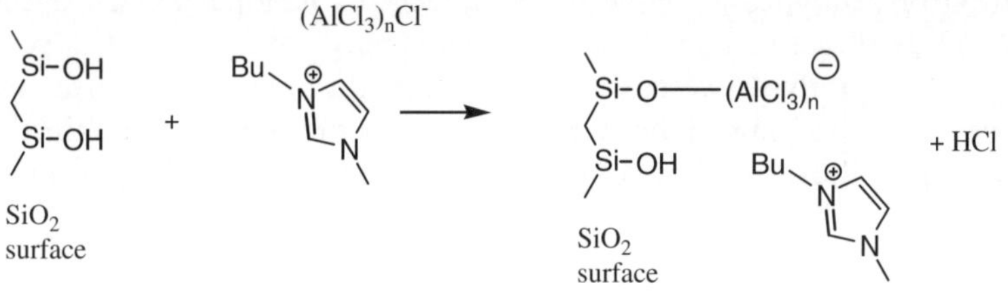

Scheme 4.12

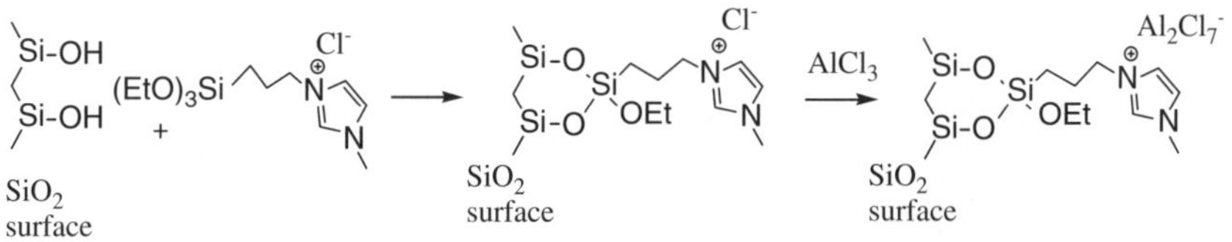

Scheme 4.13

reaction solvent. In this case, a methodology that synergically combines the advantages of immobilized metal catalyst and ionic liquid with that of the use of dense-phase carbon dioxide as reaction and extraction solvent was developed.

In the simplest methodology, however, the immobilization approach involves the treatment of a solid with a substantial amount of ionic phase. The IL, or the IL-metal complex, is sucked into the support by capillary forces, resulting into a seemingly "dry" texture. In contrast to the previously described situations, this approach is based on non-bonding attachment of the IL to the support and results in the formation of multiple layers of free IL on the carrier, which then acts as an inert phase to dissolve catalysts. The resulting ionic phase behaves both as reaction medium and catalyst.

For example, various ionic liquids (ILs), together with Pd-metal species, were immobilized on a high-surface area, of an active carbon support.[62] The resulting SILCA catalysts were studied[63] in the production of fine chemicals, and in particular in the hydrogenation of unsaturated aldehydes, citral and cinnamaldehyde (Figure 4.8).

These model molecules are challenging, due to the possibility of several parallel and consecutive reactions that can occur, depending on the experimental conditions and the nature of the catalyst. Using different ILs, the authors evidence the feasibility of altering the selectivity profiles, not only by means of altered reaction conditions but also by variation of the nature of the IL present in the catalytic layer. XPS, FESEM and EFTEM analyses show that Pd derived from a $Pd(acac)_2$-precursor, initially dissolved into the ionic liquid, undergoes a change of oxidation state from Pd^{2+} to Pd^{4+}, regardless of the ionic liquid in question. However, upon decomposition of the precursor, at 373 K under H_2-flow, a transition to either Pd^+ or Pd^0 occurred, indicating the formation of catalytically active Pd complexes or nanoparticles, respectively. The amounts of active metal palladium entrapped into the IL was very low; however, the catalysts were very active and durable against deactivation. Moreover, the leaching of both Pd and ionic liquids was determined to be negligible and, thus, the slow deactivation was attributed to the accumulation

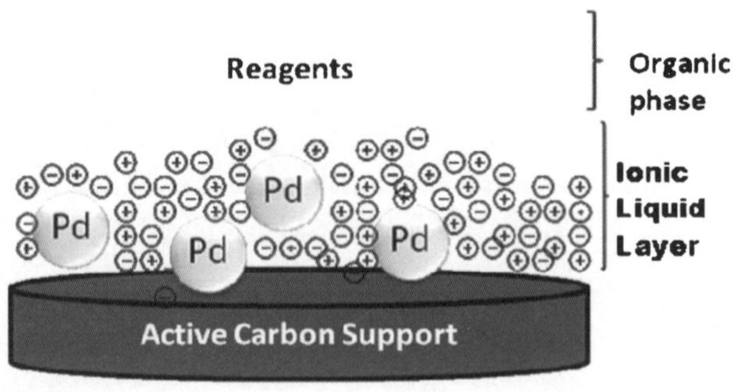

Figure 4.8   A SILCA catalyst.

of various organic impurities and water as well as slow agglomeration of Pd particles into larger clusters.

Recently, a solid-state NMR investigation of a series of supported catalysts consisting of a Lewis acidic function ($[Pd(DPPPF)(CF_3CO_2)_2]$) and a Brønsted acid function immobilized in a thin film of imidazolium salts on silica support (SILFs) has evidenced[64] that the imidazolium cations form a solvent cage around the palladium complexes, thereby establishing a long-range ordered system, which determines a reduced mobility (Figure 4.9).

The possibility of using the ordering effect to induce unusual properties in the supported complexes was suggested by the authors, who included among the peculiar features the enhanced metal–substrate interactions, the reorientational of substrate molecules within the solvent cage and the possibility of directing the approach of molecules to catalytically active centers.

In agreement with these data, more recently an unexpected surface-mediated selectivity modulation was observed[65] when SILFs were used for the cyclopropanation reactions using 5,5′-isopropylidenebis[(4R)-4-phenyl-4,5-dihydro-1,3-oxazole] as chiral ligand for CuI. When the thin film of ionic liquid was supported on a clay, the system behaves as a nearly two-dimensional nanoreactor in which the restrictions in rotational mobility and the close proximity to the surface support produce variations in the stereo- and enantioselectivities, leading to a complete reversal of the overall selectivity with respect the bulk solvent. Notably, only layered solids with negative charges on the layers (clays) give rise to this behavior, suggesting that the formation of ion pairs between the catalyst and the surface may be a decisive factor (Figure 4.10).

Although selectivity, both in bulk and SILFs, is not impressive these data evidence differences between bulk and supported ILs and suggest that a deeper understanding of the surface-complex–substrate interactions in the supported ILs may be important to design new, more efficient systems that may be able to give products that are stereochemically different to those obtained in solution.

Supported ILs were used also for no-metal-catalyzed reactions. Recently, Hünig's base tethered ammonium ILs have been used[66] to catalyze the

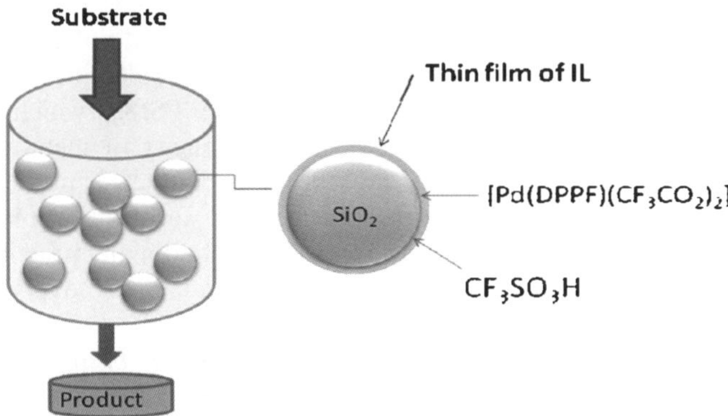

Figure 4.9 Silica-supported catalyst.

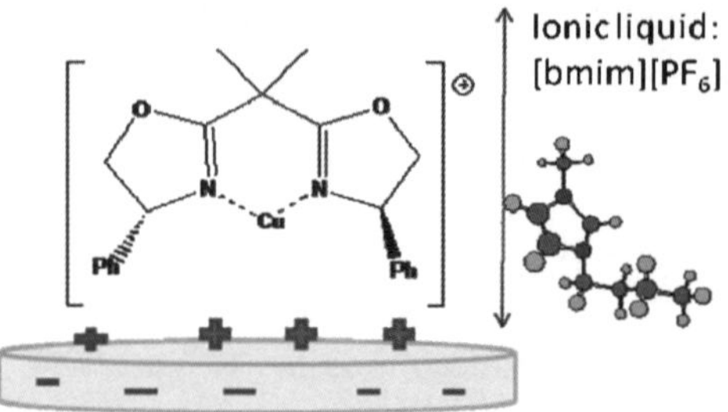

Figure 4.10   A supported thin film of ionic liquid employed for cyclopropanation reactions.

Knoevenagel condensation of aldehydes/ketones with malonitrile and ethyl cyanoacetate. The reactions were carried out under homogeneous and biphasic conditions, including the use of liquid-silica supported IL, with the biphasic system employing cyclohexene as the second phase. Although supported ILs showed a reduced initial activity, in general an excellent recyclability was observed, with the reaction repeated over five times without leaching of the IL into the extractant phase or reduction in activity.

A further approach that has been used to immobilize ILs is the incorporation of the IL and eventual catalysts into the porous framework of a membrane. Supported IL membranes containing suitable catalysts have been used for olefin hydrogenation,[67] in oligomerization reactions[68] and for separation applications.[69] These latter applications ranged from isomeric amine separation[70] to the enzyme-facilitated transport of (*S*)-ibuprofen through a supported liquid membrane.[71] In this case the selective separation of the (*S*)-isomer from the racemic mixture was obtained by utilizing the lipase-facilitated transformation of this latter isomer; the formed (*S*)-ibuprofen ester dissolved in the supported liquid membrane and diffused across to the receiving phase where it was hydrolyzed. *Candida rugosa* (CRL) facilitated selective esterification in the feed phase while lipase from *Porcine pancreas* (PPL) was responsible for hydrolysis of ester in receiving phase (Figure 4.11).

On the other hand, polymeric beads of supported ILs ("polymer-supported imidazolium salt," PSIS) were also prepared *via* the covalent anchoring of an imidazolium salt to a polystyrene resin.[72] These PSISs, which have the advantage of significantly enhancing the nucleophilicity of metal salts compared with conventional methods, have been used as efficient catalysts for nucleophilic fluorination and for other nucleophilic substitution reactions. In particular, the authors found that the applied PSIS had many practical merits: product recovery and purification was simple and catalyst recovery and reuse were easy (Figure 4.12).

Finally, several imidazolium-based polymers have been synthesized,[73] including poly[3-(4-vinylbenzyl)-1-methylimidazolium chloride and bis(trifluoromethylsulfonyl)imide], and recent studies have evidenced their ability to stabilize metal nanoparticles (Figure 4.13).[74]

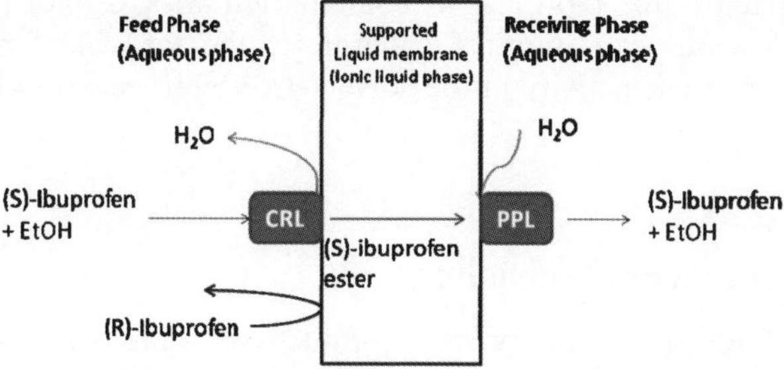

Figure 4.11 Enzyme-facilitated transport of (*S*)-ibuprofen through a supported liquid membrane.

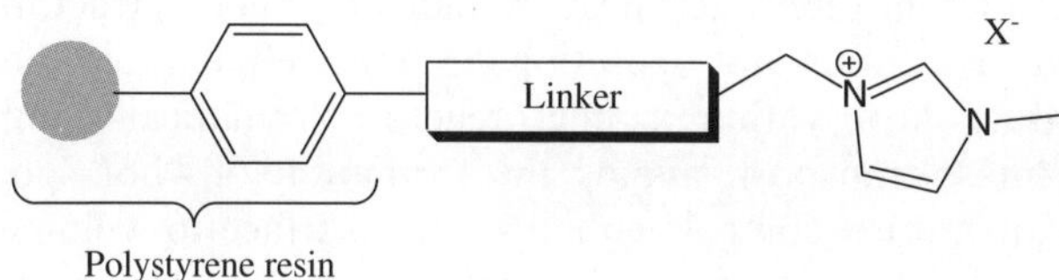

Figure 4.12 Example of a polymer-supported imidazolium salt (PSIS).

Figure 4.13 Poly[3-(4-vinylbenzyl)-1-methylimidazolium chloride].

In particular, highly stable Pd nanoparticles, protected by an imidazolium based ionic polymer in a functionalized IL, can be easily prepared. These Pd nanoparticles are excellent pre-catalysts for Suzuki, Heck and Stille coupling reactions and can be stored without undergoing degradation for at least two years.[75]

On the other hand, polymer-supported task-specific ILs in which the imidazolium cations couple L-proline *via* the ionic-pair interaction have also been synthesized and applied in metal scavenging and heterogeneous catalysis.[76] The novel materials displayed considerable ability for metal scavenging onto their surface [*e.g.*, CuI, Pd(OAc)$_2$, Pd^0 and IrCl$_3$] without the aid of a non-immobilized ionic liquid. Moreover, attempts to use these materials in the CuI-catalyzed N-arylation of nitrogen-containing heterocycles revealed that these systems are characterized by a much higher activity and recycling ability than those exhibited by free L-proline in combination with CuI in [bmim][BF4], under homogeneous conditions. Furthermore, the Pd-soaked material also shows higher catalytic activity in the solvent-free hydrogenation of styrene to ethylbenzene.

4.3.2 Microreactor Technology

The potential for significant process productivity gains has also generated considerable interest in microreactor technology.[77] Microreactors are continuous flow systems with internal channel widths in the 50–300 µm range and volumes most typically in the microlitre range. Thanks to modern advances in microfabrication technology, microreactors with precise and complex internal geometries and connections can be constructed, and specific devices, such as micropumps, micromixers, microheat exchangers, microextractors, *etc.*, have been designed for various unit operations. Extensively reduced dimensions for reactant mixture flow within a microreactor dramatically improves heat transfer and mass transport, avoids the formation of "hot spots," thereby increasing temperature control, enhances the surface-to volume ratio, and increases the operational safety. Certainly, these features may be used to advantage in (1) controlling reactions that are highly exothermic, (2) allowing reactions normally conducted at cryogenic temperatures to be run at substantially higher temperatures, and (3) minimizing undesired secondary reactions. Generally, the flow characteristics in a microchannel lie in the laminar regime, where mixing is dominated by molecular diffusion. Fast mixing can be achieved at scale, however, through multilamination and recombination, where shortened diffusion paths and increased fluid interfaces are made possible. Thus, in situations where the reaction rate is diffusion limited, the domain for reaction is essentially reduced to a series of parallel planes between the two reactant streams. With zoned control of localized temperature within the microreactor and short, controlled reaction times, secondary reactions that would normally be evident under batch conditions can frequently be minimized or eliminated. Moreover, the limited volume of active chemistry within a continuous flow microreactor provides greatly improved process control and safety. In principle, an additional benefit of microreactor technology is also scalability, since the technology is amenable to "numbering up" the effective microreactor channels rather than "scaling-up" vessel size.

With the aid of microfluidics engineering technology, biphasic reaction systems are now capable of being designed in which homogeneous catalysts are localized in ionic liquids in a manner that allows them to be cleanly separated from product streams and recycled. Although in miniaturized chemical reactions heterogeneous catalysts are generally used – probably because heterogeneous catalysis can take advantage of the high volume-to-surface ratio ensured by the microchannels – the possibility of using this technology associated with ILs has been evaluated with an homogeneous catalyzed reaction. A low-viscosity ionic liquid, [bmim][Tf$_2$N], was specifically screened[78] for effectiveness as a catalyst carrier with enhanced flow and mixing properties in a continuous microreactor application for the Pd-catalyzed Mizoroki–Heck reaction of iodobenzene and butyl acrylate. A continuous flow system was achieved with efficient catalyst recycling by using [bmim][Tf$_2$N] and an automated microreactor system, in conjunction with an efficient self-designed microextraction/catalyst recycling system.

4.4 Ionic Liquids in Biocatalysis for Fine Chemicals

Biocatalysis are increasingly being used to assist in synthetic routes to complex molecules of industrial interest.[79] In the last years, there has been a particular interest in the use of biotechnological processes to transform complex natural compounds (biopolymers) into more simple compounds. More in detail, fermentation processes have been performed to transform biomass in primary or secondary metabolites to produce energy and "building blocks."[80] However, the biggest role of biocatalysis performed both using whole cells and purified enzymes still remains in pharmaceutical sector,[81] which much more than others has particular demands.[82] Generally, the first requirement for a biocatalytic process to be economically viable is to achieve product concentrations comparable to chemical processes of at least 50–100 g L^{-1}. Since, enzymes in nature work at millimolar levels of substrate, enzymes and cells must operate away from their natural conditions and maintain under these conditions their activity. Furthermore, enzymes are often expensive, techniques such as immobilization are important to increase the ability to recycle the catalyst. Nevertheless, for biopharmaceutical products it is also necessary to avoid the presence of unwanted biological materials; biocatalytic processes cannot be employed in the final steps.

The applicability of biocatalysis to industrial processes, in particular in preparation of enantiopure pharmaceuticals is therefore increased when it has been noticed[83] that enzymes could also work in non-aqueous media such as organic solvents able to solubilize the mostly hydrophobic substrates not soluble in water and often not stable in this environment. Although organic solvents are extensively employed in biocatalysis, they suffer from severe drawbacks and their application in biocatalysis dramatically decreases the sustainability of the use of enzymes as catalysts. In this sense, ionic liquids have recently emerged as a promising alternative to perform biocatalyzed reactions in non-aqueous media.[84] Furthermore, due to their tunability ILs offer the possibility to apply the so-called "solvent engineering" to biocatalyzed reactions. Finally, enzymes of widely different types are catalytically active in ILs.[85] Lipases, and more in general hydrolases, maintain their activity in these media; the enantioselectivity and stability are often better than in traditional solvents.[86] Generally, enzyme compatible ILs do not interact strongly with enzymes or cause the latter to dissolve; therefore, to enhance enzymatic activity different strategies have been adopted, including the addition of small amounts of water to the ionic liquid or the enzyme immobilization on solid supports.[87]

Furthermore, to increase solubility and enzyme activities initially functionalized ILs with polyethylene glycol chains have been used,[88] whereas, more recently, new classes of "biocompatible" ILs able to solubilize and/or stabilize proteins has been synthesized.[89] These ILs consist of "biocompatible" anions such as dicyanamide, citrate, saccharinate and dihydrogen phosphate associated to different cations, ranging from 1-butyl-1-methylpyrrolidinium to choline. Cytochrome *c* from horse, used as model protein, can be dissolved in 1-butyl-1-methylpyrrolidinium dihydrogen phosphate containing 10–20 wt% of water up to 3 mM, whereas choline citrate–water has been used[90] with success in oxidation reactions catalyzed by chloroperoxidase from *Caldariomyces fumago*. In addition, the unconventional solvent properties of ILs have been used in biocatalyst recycling and product recovery systems that are not feasible in common traditional solvents. Bioreactors with covalently supported SILPs able to absorb enzymes (*Candida antarctica* lipase B, CALB) were prepared, leading to highly efficient and robust biocatalyst.[91]

However, despite the fact that data collected in recent years evidence the possibility of using ILs in biocatalyzed reactions, probably much more work is necessary before they can become extensively used as solvents in fine chemical biosynthesis. For example, it is necessary to develop suitable methods for product isolation, in particular if they are of limited or no volatility, and it is also necessary to increase recyclability of the biocatalytic systems over long time applications.

4.5 Conclusions

Nowadays, industry is facing a substantial challenge to modify its processes to cleaner, environmentally friendly ones. This concerns mainly pharmaceutical companies, whose processes consist of several steps, involving large amounts of volatile organic solvents (VOSs) and releasing undesirable waste. Processes based on ILs are emerging as a solution for this problem, mainly for the efficient replacement of VOSs. In particular, ILs can be considered effective reaction media when "task-specific ionic liquids" are combined with emerging technologies. The breadth of literature cited in this review illustrates the range of productivity enhancement opportunities available to the pharmaceutical and fine chemical industries using ILs and emerging technologies. Research developed in recent years evidences great promise for increasing process reliability while lowering manufacturing costs. However, only the will-power for innovation of the industrial world can determine the final application.

References

1. *Ionic Liquids IIIB: Fundamentals, Progress, Challenges, and Opportunities - Transformations and Processes*, ed. R. D. Rogers and K. R. Seddon, ACS Symp. Ser., vol. 902, American Chemical Society, Washington D.C., 2005; *Ionic Liquids IIIA: Fundamentals, Progress, Challenges, and Opportunities -*

Properties and Structure, ed. R. D. Rogers and K. R. Seddon, ACS Symp. Ser., vol. 901, American Chemical Society, Washington D. C., 2005; *Ionic Liquids in Synthesis*, ed. P. Wasserscheid and T. Welton, Wiley-VCH Verlag GmbH, Weinheim, 2nd edn., 2007; J. Scammells, J. L. Scott and R. D. Singer, *Aust. J. Chem.*, 2005, **58**, 155; J. L. Anderson, D. W. Armstrong and G.-T. Wei, *Anal. Chem.*, 2006, **78**, 2893; J. Earle and K. R. Seddon, *Pure Appl. Chem.*, 2000, **72**, 1391; T. Welton, *Chem. Rev.*, 1999, **99**, 2071; F. Endres and S. Z. El Abedin, *Phys. Chem. Chem. Phys.*, 2006, **8**, 2101; F. van Rantwijk, R. M. Lau and R. A. Sheldon, *Trends Biotechnol.*, 2003, **21**, 131.

2. *Ionic Liquids: Industrial Applications for Green Chemistry*, Ed. R. D. Rogers and K. R. Seddon, ACS Symp. Ser., vol. 818, American Chemical Society, Washington D.C., 2002; *Electrochemical Aspects of Ionic Liquids*, ed. H. Ohno, Wiley-Interscience, Hoboken, 2005; *Ionic Liquids: The Front and Future of Material Development*, ed. H. Ohno, CMC Press, Tokyo, 2003; H. Zhao, *Chem. Eng. Comm.*, 2006, **193**, 1660.

3. (a) A. Takada and Y. Takahashi, *Cellulose Comm.*, 2007, **14**, 142; (b) Y. Fukaya, K. Hayashi, M. Wada and H. Ohno, *Green Chem.*, 2008, **10**, 44.

4. (a) N. V. Plechkova and K. R. Seddon, *Chem. Soc. Rev.*, 2008, **37**, 123; (b) M. Maase, Industrial application of ionic liquids, in *Ionic Liquids in Synthesis, ed. P. Wasserscheid and T. Welton,* Wiley-VCH Verlag GmbH, Weinheim, 2nd edn., 2007.

5. (a) H. V. Blaser and M. Studer, *Green Chem.*, 2003, **5**, 112; (b) A. E. Rubin, S. Tummala, D. A. Both, C. Wang and E. J. Delaney, *Chem. Rev.*, 2006, **106**, 2794.

6. (a) C.-W. Cho, T. P. T. Pham, Y. C. Jeon and Y.-S. Yun, *Green Chem.*, 2008, **10**, 67; (b) D.J. Couling, R. J. Bernot, K. M. Docherty, J. K. Dixon and E. J. Maginn, *Green Chem.*, 2006, **8**, 82; (c) M. T. Garcia, N. Gathergood and P. J. Scammells, *Green Chem*, 2005, **7**, 9; (d) K. J. Kulacki and G. A. Lamberti, *Green Chem.*, 2008, **10**, 104; (e) A. Latala, P. Stepnowski, M. Nedzi and W. Mrozik, *Aquat. Toxicol.*, 2005, **73**, 91; (f) M. Matzke, S. Stolte, K. Thiele, T. Juffernholz, J. Arning, J. Ranke, U. Welz-Biermann and B. Jarstoff, *Green Chem.*, 2007, **9**, 1198; (g) J. Pernak, K. Sobaszkiewich and L. Mirska, *Green Chem.*, 2003, **5**, 52; (h) C. Pretti, C. Chiappe, D. Pieraccini, M. Gregori, F. Abramo, G. Monni and L. Intorre, *Green Chem.*, 2006, **8**, 238; (i) J. Ranke, S. Stolte, R. Stormann, J. Arning and B. Jastorff, *Chem. Rev.*, 2006, **107**, 2183.

7. M. J. Earle, K. R. Seddon and P. B. McCormac, *Green Chem.*, 2000, **2**, 261.

8. (a) Z. Fei, T. J. Geldbach, D. Zhao and P. J. Dyson, *Chem. Eur. J.*, 2006, **12**, 2122; (b) S. Lee, *Chem. Commun.*, 2006, 1049.

9. (a) C. Chiappe and D. Pieraccini, *J. Phys. Org. Chem.*, 2005, **18**, 275; (b) C. Chiappe and D. Pieraccini, *J. Org. Chem.*, 2004, **69**, 6059; (c) G. Ranieri, J. P. Hallet and T. Welton, *Ind. Eng. Chem. Res.*, 2008, **47**, 638; (d) L. Crowhurst, N. L. Lancaster, J. M. P. Arlandis and T. Welton, *J. Am. Chem. Soc.*, 2004, **126**, 1154.

10. (a) Y.-J. Kim and A. Streitwieser, *J. Am. Chem. Soc.*, 2002, **124**, 5757; (b) E. Elliot and H. T. Scott, *Curr. Org. Synth.*, 2007, **4**, 381.

11. K. E. Johnes, R. M. Pagni and J. Bartness, *Monatsch. Chem.*, 2007, **138**, 1077.

12. H. Xing, T. Wang, Z. Zhou and Y. Deng, *Synth. Commun.*, 2006, **36**, 2433.

13. Y. Gu, J. Zhang, Z. Duan and Y. Deng, *Adv. Synth. Catal.*, 2005, **347**, 512.

14. D. R. MacFarlane and K. R. Seddon, *Aust. J. Chem.*, 2007, **60**, 3.

15. T. L. Greaves and C. J. Drummond, *Chem. Rev.*, 2008, **108**, 206.

16. E. Janus, M. Lozinski and J. Pernak, *Chem. Lett.*, 2006, **35**, 210.

17. G. Driver and K. E. Johnson, *Green Chem.*, 2003, **5**, 163.

18. C. Chiappe, E. Leandri and M. Tebano, *Green Chem.*, 2006, **8**, 742.

19. K. K. Laali and V. J. Gettwrt, *J. Org. Chem.*, 2001, **66**, 35.

20. D. Jiang, W. Li, C.-D. Xu and L.-Y. Dai, *Yingyong Huaxue*, 2007, **24**, 1080.

21. D. R. Mac Farlane, J. P. Pringle, K. M. Johansson, S. A. Forsyth and M. Forsyth, *Chem. Commun.*, 2006, 1905.

22. S. A. Forsyth, D. R. Mac Farlane, R. J. Thomson and M. von Itzstein, *Chem. Commun.*, 2002, 714.

23. (a) R. Wang, J.C. Xiao, B. Twamley and J. M. Shreeve, *Org. Biomol. Chem.*, 2007, **5**, 671; (b) C. Ye, J. C. Xiao, B. Twamley, A. D. LaLonde, M. G. Norton and J. M. Shreeve, *Eur. J. Org. Chem.*, 2007, 5095.

24. A. Wykes and S. L. MacNeil, *Synlett*, 2007, 107.

25. A. Marra, A. Vecchi, C. Chiappe, B. Melai and A. Dondoni, *J. Org. Chem.*, 2008, **73**, 2458.

26. (a) W. Miao and T. H. Chan, *Acc. Chem Res.*, 2006, **39**, 897; (b) W. Miao and T. H. Chan, *Adv. Synth. Catal.*, 2006, **348**, 1711.

27. M. Lombardo, F. Pasi, S. Easwar and C. Trombini, *Adv. Synth. Catal.*, 2007, **349**, 2061.

28. (a) W. Chen and F. Liu, *J. Organomet. Chem.*, 2003, **673**, 5; (b) M. Sorm and S. Nespurek, *Acta Polym.*, 1985, **36**, 433; (c) D. Zhao, Z. Fei, T. J. Geldbach, R. Scopelliti, G. Laurenczy and P. J. Dyson, *Helv. Chim. Acta*, 2005, **88**, 665; (d) F. Mazille, Z. Fei, D. Kuang, D. Zhao, S. M. Zakeeruddin, M. Gratzel and P. J. Dyson, *Inorg. Chem.*, 2006, **45**, 1585; (e) K. M. Lee, Y. T. Lee and Y. J. B. Lin, *J. Mater. Chem.*, 2003, **13**, 1079; (f) L. C. Branco, J. N. Rosa, J. J. Moura Ramos and C. A. M. Afonso, *Chem. Eur. J.*, 2002, **8**, 3671; (g) A. P. Abbott, G. Capper, D. L. Davies and R. Rasheed, *Inorg. Chem.*, 2004, **43**, 3447; (h) A. C. Cole, J. L. Jensen, I. Ntai, K. L. T. Tran, K. J. Weaver, D. C. Forbes and J. H. Jr Davis, *J. Am. Chem. Soc.*, 2002, **124**, 5962; (i) J. D. Holbrey, W. M. Reichert, I. Tkatchenko, E. Bouajila, O. Walter and I. Tommasi, *Chem. Commun.*, 2003, 28; (j) N. Brausch, A. Metlen and P. Wasserscheid, *Chem. Commun.*, 2004, 1552; (k) Z. Fei, D. Zhao, T.J. Geldbach, R. Scopelliti and P. J. Dyson, *Chem. Eur. J.*, 2004, **10**, 4886; (l) D. Li, F. Shi, J. Peng, S. Guo and Y. Deng, *J. Org. Chem.*, 2004, **69**, 3582; (m) Z. Fei, W. H. Ang, T. J. Geldbach, R. Scopelliti and P. J. Dyson, *Chem. Eur. J.*, 2006, **12**, 4014; (n)

Y. R. Mirzaei, H. Xue and J. M. Shreeve, *Inorg. Chem.*, 2004, **43**, 361; (o) Z. Mu, W. Liu, S. Zhang and F. Zhou, *Chem. Lett.*, 2004, **33**, 524.

29. (a) R. J. C. Brown, P. J. Dyson, D. J. Ellis and T. Welton, *Chem. Commun.*, 2001, 1862; (b) P. J. Dyson, J. S. McIndoe and D. Zhao, *Chem. Commun.*, 2003, 508; (c) H. S. Kim, Y. J. Kim, J.Y. Bae, S. J. Kim, M. S. Lah and C. S. Chin, *Organometallics*, 2003, **22**, 2498; (d) D. Zhao, Z. Fei, C. A. Ohlin, G. Laurenczy and P. J. Dyson, *Chem. Commun.*, 2004, 2500.

30. (a) Z. Fei, T. J. Geldbach, D. Zhao and P. J. Dyson, *Chem. Eur. J.*, 2006, **12**, 2122; (b) S. Lee, *Chem. Commun*, 2006, 1049.

31. (a) D. Zhao, Z. Fei, T. J. Geldbach, R. Scopelliti and P. J. Dyson, *J. Am. Chem. Soc.*, 2004, **126**, 15876; (b) C. Chiappe, D. Pieraccini, D. Zhao, Z. Fei and P. J. Dyson, *Adv. Synth. Catal.*, 2006, **348**, 68; (c) Z. Fei, D. Zhao, D. Pieraccini, W. H. Ang, T. J. Geldbach, R. Scopelliti, C. Chiappe and P. J. Dyson, *Organometallics*, 2007, **26**, 1588.

32. M. Lombardo, F. Pasi, C. Trombini, K. R. Seddon and W. R. Pitner, *Green Chem.*, 2007, **9**, 321.

33. H. Itoh, K. Naka and Y. Chujo, *J. Am. Chem. Soc.*, 2004, **126**, 3026.

34. T. J. Geldbach and P. J. Dyson, *J. Am. Chem. Soc.*, 2004, **126**, 8114.

35. W. Miao and T. A. Chan, *Acc. Chem. Res.*, 2006, **39**, 897.

36. J. H. Davis and Jr. K. J. Forrester, *Tetrahedron Lett.*, 1999, **38**, 1621.

37. B. List, *Tetrahedron*, 2006, **62**, 556.

38. (a) N. Audic, H. Clavier, M. Mauduit and J. C. Guillemin, *J. Am. Chem. Soc.*, 2003, **125**, 9248; (b) Q. Yao and M. Sheets, *J. Organomet. Chem.*, 2005, **690**, 3585.

39. Q. J. C. Xiao, B. Twamley and J. M. Schreeve, *Org. Lett.*, 2004, **6**, 3845.

40. S. T. Handy and M. Okello, *J. Org. Chem.*, 2005, **70**, 2874.

41. J. Fraga-Dubruil and J. P Bazureau, *Tetrahedron*, 2003, **59**, 6121.

42. M. de Korb, A. W. Tuin, S. Kuiper, S. Everkleeft, G. A. van der Marel and R. C. Buijsman, *Tetrahedron Lett.*, 2004, **45**, 2171.

43. W. Miao and T. H. Chan, *J. Org. Chem.*, 2005, **70**, 3251.

44. X. He and T. H. Chan, *Synthesis*, 2006, 1645.

45. A. K. Pathak, C. K. Yerneni, Z. Young and V. Pathak, *Org. Lett.*, 2008, **10**, 145.

46. J. Horwarth, K. Hanlon, D. Fayne and P. McCormac, *Tetrahedron Lett.*, 1997, **38**, 3097.

47. (a) A. D. Headley and B. Ni, *Aldrichimica Acta*, 2007, **40**, 107; (b) X. Chen, X. Li, A. Hu and F. Wang, *Tetrahedron: Asymmetry*, 2008, **19**, 1.

48. M. L. Patil, C. V. L. Rao, K. Yonezawa, S. Takizawa, K. Onitsuka and H. Sasai, *Org. Lett.*, 2006, **8**, 227.

49. R. Gausepohl, P. Buskens, J. Kleinen, A. Bruckmann, C. W. Lehman, J. Klankermayer and W. Leitner, *Angew. Chem. Int. Ed.*, 2006, **45**, 3689.

50. (a) S. Z. Luo, L. Zhang, X. Mi, Y. Qiou and J. P. Cheng, *J. Org. Chem.*, 2007, **72**, 9350; (b) B. Ni, Q. Zhang and A. D. Headly, *Green Chem.*, 2007, **9**, 737; (c) B. Ni, Q. Zhang and A. D. Hadley, *Tetrahedron Lett.*, 2008, **49**, 1249.

51. K. Bica, P. Gaertner, *Eur. J. Org. Chem.*, 2008, **19**, 3235.

52. D. Seebach and H. A. Oei, *Angew. Chem., Int. Ed. Engl.*, 1975, **14**, 634.

53. D. Xiao, D. R. Rajian, S. Li, R. A. Bartsch and E. L. Quitevis, *J. Phys. Chem. B*, 2006, **110**, 16174.

54. C. Chiappe, *Monatsh Chem.*, 2007, **138**, 1035.

55. P. F. Schulz, N. Müller, A. Bösmann and P. Wasserscheid, *Angew. Chem., Int. Ed.*, 2007, **46**, 1293.

56. M. I. Barguete, F. Galindo, E. Garcia-Verdugo, N. Karbass and S. V. Luis, *Chem. Commun.*, 2007, 3086.

57. C. P. Mehnert, *Chem. Eur. J.*, 2005, **11**, 50.

58. C. P. Mehnert, R. A. Cook, N. C. Dispenziere and A. Afeworki, *J. Am. Chem. Soc.*, 2002, **124**, 12932.

59. (a) E. Benazzi, Y. Chauvin, A. Hirschauer, N. Ferrer, H. Olivier, J. Y. Bernard *US Patent*5 693 585 (1997); (b) C. de Castro, E. Sauage, M. H. Valkenberg and W. F. Hölderich, *J. Catal.*, 2000, **196**, 86.

60. H. Hagiwara, K. H. Ko, T. Hoshi and T. Suzuki, *Chem. Comm.*, 2007, 2838.

61. R. Ciriminna, P. Hasemann, J. J. E. Moreau, M. Carraro, S. Campestrini and M. Pagliaro, *Chem Eur. J.*, 2006, **12**, 5220.

62. (a) J.-P. Mikkla, P. P. Virtanen, K. Kordas, H. Karhu and T. O. Salmi, *Appl. Cat. A*, 2007, **328**, 68; (b) P. Virtanen, H. Karhu, K. Kordas and J.-P. Mikkla, *Chem. Eng. Science*, 2007, **62**, 3660.

63. P. Virtanen, H. Karhu, K. Kordas and J. P. Mikkola, *Chem. Eng. Sc.*, 2007, **62**, 3660.

64. C. Sievers; o. Jimenez, T. E. Müller, S. Steuernagel and J. A. Lercher, *J. Am. Chem. Soc.*, 2006, **128**, 13990.

65. M. R. Castillo, L. Fousse, J. M. Fraile, G. I. Gracìa and J. A. Mayoral, *Chem. Eur. J.*, 2007, **13**, 287.

66. C. Paun, J. Barklie, P. Goodrich, H. Q. N. Gunaratne, A. McKeown, V. I. Pârvulescu and C. Hardacre, *J. Mol. Cat. A*, 2007, **269**, 64.

67. T. H. Cho, J. Fuller and R. T. Carlin, *High Temp. Mater. Processes*, 1998, **2**, 543.

68. R. T. Carlin, T. H. Cho and J. Fuller, *Proc. Eletrochem. Soc.*, 1998, **98**, 180.

69. (a) G. P. Pez and R. T. Carlin, *J. Membr. Sci.*, 1992, **65**, 21; (b) T. Schaefer, C. A. Rodrgues, A. M. A. Carlos and J. G. Crespo, *Chem. Commun.*, 2001, 1622.

70. L. C. Branco, J. G. Crespo and C. A. M. Afonso, *Angew. Chem., Int. Ed.*, 2002, **41**, 2721.

71. (a) T. Schaefer, C. A. Rodrgues, A. M. A. Carlos and J. G. Crespo, *Chem. Commun.*, 2001, 1622; (b) E. Tiiyko, T. Maruyme, N. Komiya and M. Goto, *Chem. Commun.*, 2003, 2926.

72. D. W. Kim and D. Y. Chi, *Angew. Chem., Int. Ed.*, 2004, **43**, 483.

73. (a) X. D. Mu, J. Q. Meng, Z. C. Li and Y. Kou, *J. Am Chem. Soc*, 1995, **127**, 9694; (b) N. Yan, C. Zhao, C. Luo, P. J. Dyson, H. Liu and Y. Kou, *J. Am. Chem. Soc.*, 2006, **128**, 8714; (c) B. Altava, M. Barguete, E. Garcia-

Verdugo, N. Karbass, S. V. Luis, A. Puzary and V. Sans, *Tetrahedron Lett.*, 2006, **47**, 211.

74. D. Zhao, Z. Fei, W. H. Ang and P. J. Dyson, *Small*, 2006, **2**, 879.

75. X. Yang, Z. Fei, D. Zhao, W A Hang, Y. Li, P. J. Dyson, *Inorg. Chem.*, 2008, **47**, 3292.

76. W. Chen, Y. Zhang, L. Zhu, J. Lan, R. Xie and J. You, *J. Am. Chem. Soc.*, 2007, **129**, 13879.

77. T. Wirth, ed. *Microreactors in Organic Synthesis and Catalysis,* Wiley-VHC Verlag GmbH, Weinheim, 2008.

78. S. Liu, T. Fukuyama, M. Sato and I. Ryum, *Org. Proc. Res. Develop.*, 2004, **8**, 477.

79. (a) A. Liese, K. Seelbach and C. Wandrey, Eds. *Industrial Biotrasformations,* Wiley-VCH Verlag GmbH, Weinheim, 2000; (b) K. Faber 2004 Biotransformations in Organic Chemistry (4th edn) Springer.

80. A. Albin, J. Bader, E. Mast-Gerlach and U. Stahl, *J. Biotechnol.*, 2007, 160.

81. R. N. Paterl, *Curr. Opin. Drug. Discov. Devel.*, 2003, **6**, 902.

82. D. J. Polard and J. M. Woodley, *Trend in Biotechnol.*, 2006, **25**, 66.

83. A. M. Klibanov, *CHEMTECH*, 1986, **16**, 354.

84. S. Cantone, U. Hanefeld and A. Basso, *Green Chem.*, 2007, **9**, 954.

85. (a) S. V. Malhotra, V. Kumar and S. V. Parmar, *Curr. Org. Synth.*, 2007, **4**, 370; (b) J. R. Harjani, P. U. Naik, S. J. Nara and M. M. Salunkhe, *Curr. Org. Synth.*, 2007, **4**, 354; (c) F. Van Rantwijk and R. A. Sheldon, *Chem. Rev.*, 2007, **107**, 2757.

86. (a) C. Chiappe, E. Leandri, S. Lucchesi, B. D. Hammock and C. Morisseau, *J. Mol. Catal. B.*, 2004, **27**, 243; (b) C. Chiappe, E. Leandri, B. D. Hammock and C. Morisseau, *Green Chem.*, 2007, **9**, 162–168.

87. J. A. Laszlo and D. L. Compton, *Biotechnol. Bioeng.*, 2001, **75**, 181.

88. M. B. Turner, S. K. Spear, J. G. Huddlston, J. D. Holbrey and R. D. Rogers, *Green Chem.*, 2003, **5**, 443.

89. K. Fujita, D. R. Macfarlane and M. Forsyth, *Chem. Commun.*, 2005, 4804.

90. C. Chiappe, L. Neri and D. Pieraccini, *Tetrahedron Lett.*, 2006, **47**, 5089.

91. P. Lozano, E. Garcia-Verdugo, R. Piamtongkam, N. Karbass, T. De Diego, M. I. Burguete, S. V. Luis and J. L. Iborra, *Adv. Synth. Catal.*, 2007, **349**, 1077.

Green Procedures for the Synthesis of Useful Molecules Avoiding Hazardous Solvents and Toxic Catalysts

BRINDABAN C. RANU AND KALICHARAN CHATTOPADHYAY

Department of Organic Chemistry, Indian Association for the Cultivation of Science, Jadavpur, Kolkata 700032, India

5.1   Introduction

Chemical industries, in general, use a large volume of organic solvents and catalysts in the manufacturing processes. However, the toxic and volatile nature of many organic solvents, particularly chlorinated hydrocarbons, have posed a serious threat to the environment. Hence, industries are finding it very difficult to meet the standards set by pollution regulatory boards while using traditional methods. Thus, redesign of chemical processes avoiding hazardous organic solvents and toxic catalysts is highly desirable. This chapter deals with this issue and highlights the recent developments primarily on:

1. Reaction in aqueous medium
2. Reaction on the surface of benign solid inorganic supports under organic solvent-free condition
3. Neat reactions
4. Reaction in an ionic liquid.

RSC Green Chemistry Book Series
Eco-Friendly Synthesis of Fine Chemicals
Edited by Roberto Ballini
© Royal Society of Chemistry 2009
Published by the Royal Society of Chemistry, www.rsc.org

5.2 Reaction in Aqueous Medium

Water is a unique reaction medium. However, the reaction in water remained unexplored for a long period, probably because of two common beliefs:

1. The reaction cannot occur unless the reacting substances are soluble in a reaction medium. As organic molecules are non-polar and insoluble in water, organic reaction is not favored in water.
2. Functional groups may be affected in water.

However, it has been found that this assumption does not apply to organic reactions, in general. Many organic reactions are faster and more efficient in water than in organic solvents.[1] In an aqueous medium, non-polar (organic) molecules have a general tendency to form intermolecular aggregates so as to diminish the organic molecule–water interfacial area. This property of water, called the "hydrophobic effect,"[2] brings the non-polar reactants together and consequently rate of reaction increases due to close proximity of the reactants in the transition state. The hydrophobic effect is a principal contributor to acceleration of organic reactions in water. The H-bonding between H_2O molecule and reagents also plays a significant role in rate enhancement. The H-bonding properties,[3] high dielectric constant and the complex solvating properties of water make the ionic and asymmetric reaction faster and more selective in water compared to organic solvents.[4]

5.2.1 Indium-mediated Reactions

Allylation of a carbonyl group is an important reaction for C–C bond formation. Indium-mediated allylation (Scheme 5.1) has received wide acceptance because the reaction is carried out in water and is compatible with sensitive functionalities such as OH, NH_2, COOH, *etc.*[5]

Indium-mediated coupling of α-(bromomethyl)acrylic acid with carbonyl compound in aqueous medium furnished the corresponding γ-hydroxy-α-methylene carboxylic acid. The reaction has been applied to the synthesis of (+)-3-deoxy-D-glycero-D-galacto-nonulosonic acid and *N*-acetylneuraminic acid[6] (Scheme 5.2).

Li *et al.*[7] showed a very useful application of In-mediated propargylation for the synthesis of antitumor, antiviral compound (+)-goniofufurone (Scheme 5.3).

5.2.2 Aldol Reaction

The classical aldol reaction usually employs a basic catalyst in a protic solvent. However, this process is often associated with undesirable side products. The

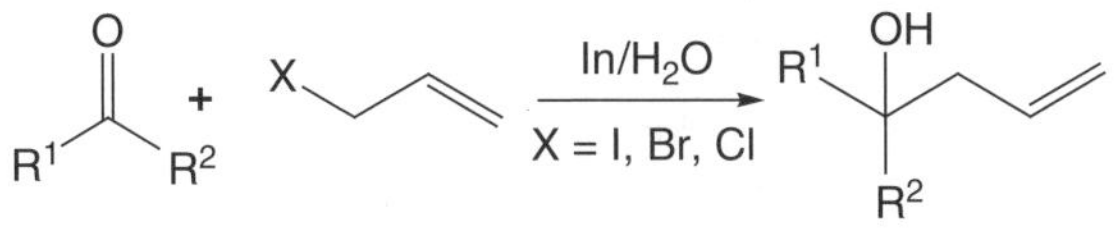

Scheme 5.1

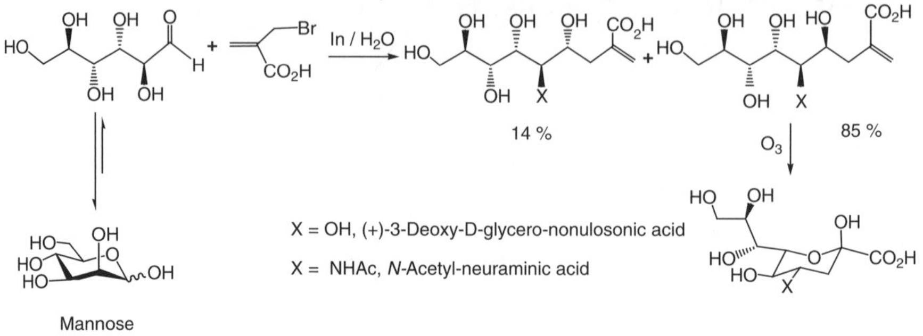

X = OH, (+)-3-Deoxy-D-glycero-nonulosonic acid

X = NHAc, N-Acetyl-neuraminic acid

14 %

85 %

Mannose

Scheme 5.2

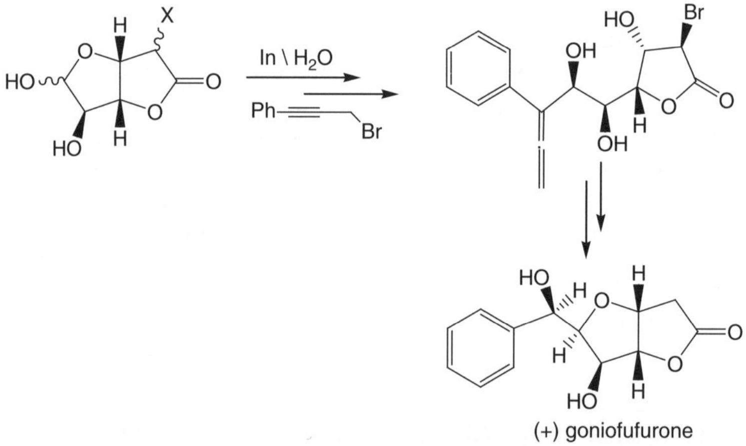

(+) goniofufurone

Scheme 5.3

Lewis acid catalyzed aldol reaction or aldol-type reaction in general provides better yields. Early work on the aldol reaction in H_2O medium focused primarily on lanthanide triflates (Scheme 5.4) as catalyst.[8]

In this respect, the use of water as a solvent for enantioselective reaction has been explored.[9] In 2006 Barbas and co-workers[10] developed an efficient proline-derived chiral catalyst (Scheme 5.5) for aldol condensation in water with high reactivity, diastereoselectivity and enantioselectivity.

This catalyst has also been successfully employed for enantioselective Michael reaction[11] (Scheme 5.6).

5.2.3   Diels–Alder Reaction

Breslow's pioneering work[12] demonstrated that the Diels–Alder reaction (Scheme 5.7) of cyclopentadiene with butenone in water was manifoldly faster and more stereoselective than reactions in either isooctane or methanol.

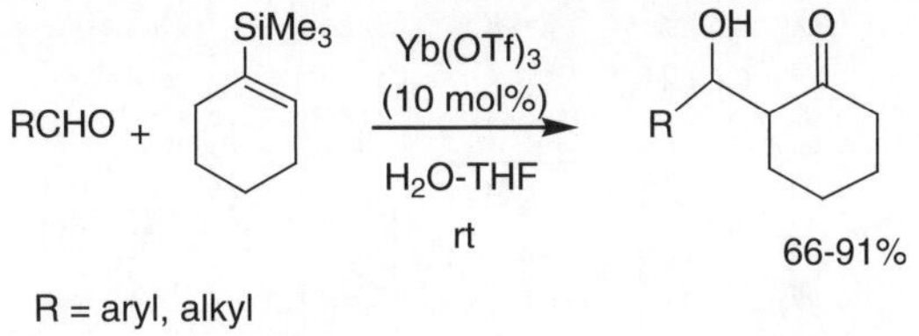

RCHO + (SiMe3-cyclohexenyl silane) → (R-CH(OH)-2-cyclohexanone)

Yb(OTf)$_3$ (10 mol%), H$_2$O-THF, rt

66-91%

R = aryl, alkyl

Scheme 5.4

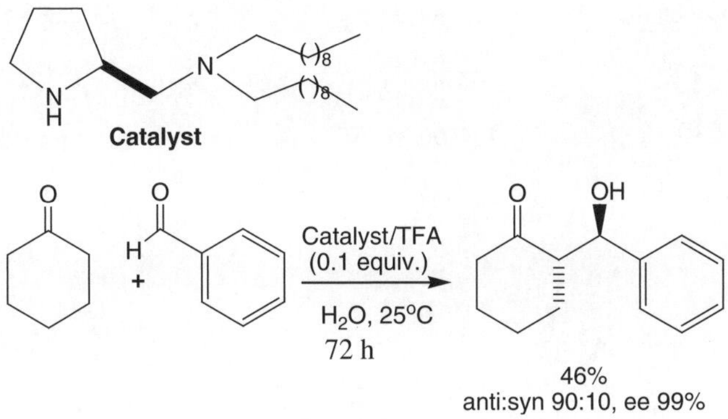

Catalyst

(cyclohexanone) + (benzaldehyde) → (product)

Catalyst/TFA (0.1 equiv.), H$_2$O, 25°C, 72 h

46%
anti:syn 90:10, ee 99%

Scheme 5.5

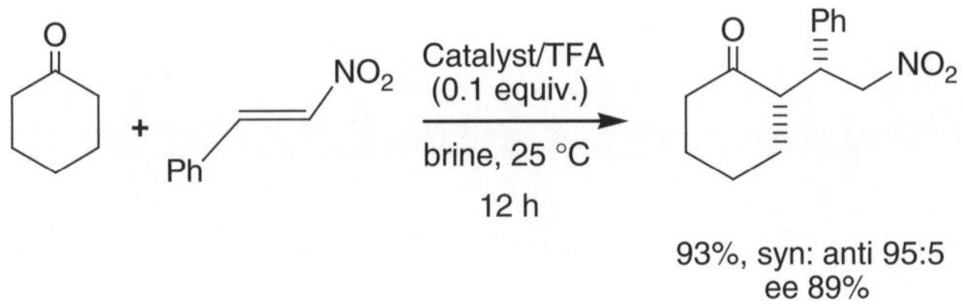

(cyclohexanone) + (Ph-CH=CH-NO$_2$) → (product)

Catalyst/TFA (0.1 equiv.), brine, 25 °C, 12 h

93%, syn: anti 95:5
ee 89%

Scheme 5.6

The Diels–Alder reaction in water has been used as a key step in the synthesis of ($\pm$)-11-ketotestosterone (Scheme 5.8) by De Clercq.[13] The rate enhancement in water medium is due to the enforced hydrophobic interaction between diene and dienophile. The H-bonding properties of H$_2$O molecules to the activated complex play a major role in selectivity.

5.2.4 Radical Reactions

Although the use of water in S_N1 reactions is well known, its use in radical reactions has been less explored. Oshima *et al.*[14a] showed that the radical cyclization reaction of a 2-iodoalkanoic acid to γ-lactone (Scheme 5.9) in water in the presence of a water-soluble radical initiator proceeds very efficiently.

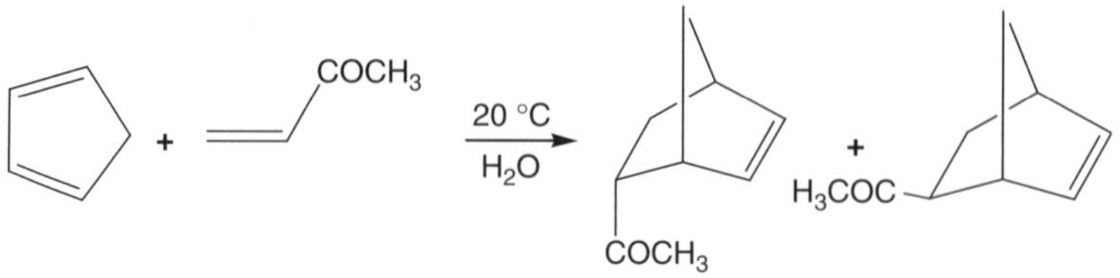

Solvent	Rate Constant ($k_2 \times 10^5$, $M^{-1} s^{-1}$)	endo:exo
Isooctane	5.94 ± 0.3	3.85 : 1
MeOH	75.5	8.5 : 1
H_2O	4400 ± 70	21.4 : 1

Scheme 5.7

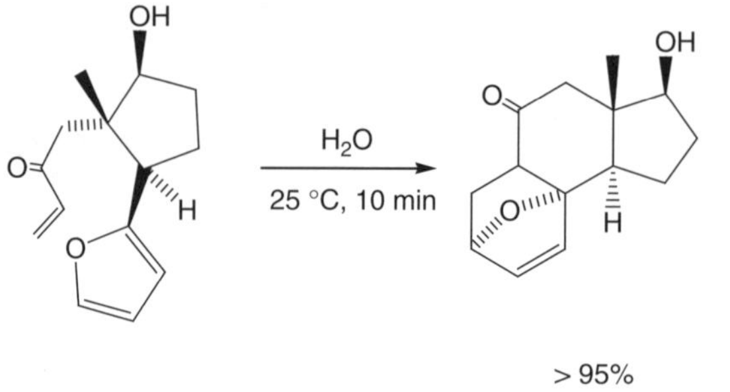

> 95%

Scheme 5.8

radical initiator

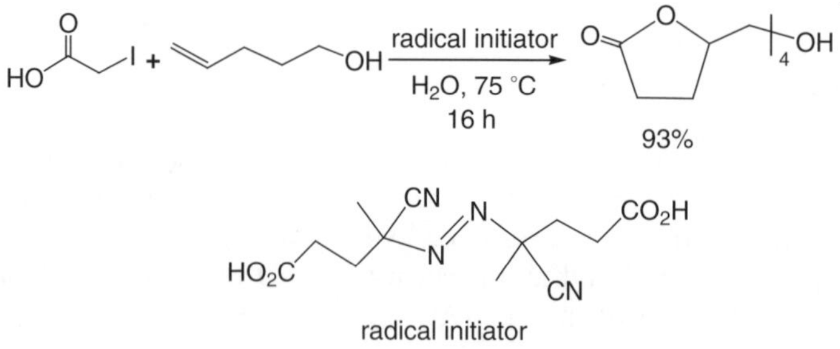

Scheme 5.9

The atom transfer radical cyclization of allyl iodoacetate is also more efficient
in water than in benzene or hexane. For instance, treatment of the iodoacetate
with triethylborane in benzene or hexane did not give any desired lactone while
in water the reaction proceeds with high yield.[14b] It is suggested that the large

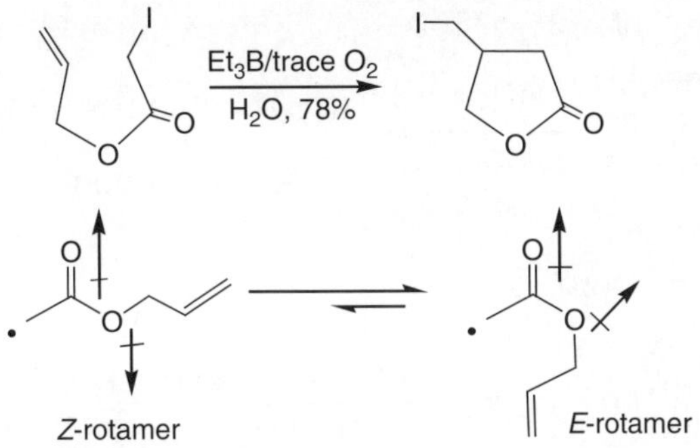

Scheme 5.10

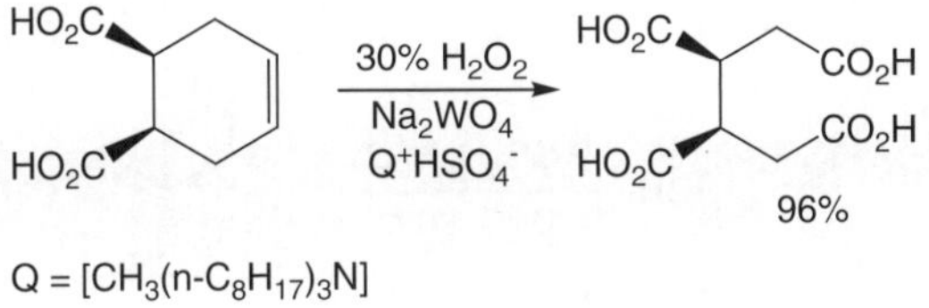

Scheme 5.11

dielectric constant of water promotes the rotation of the (Z)-rotamer to (E)-rotamer (Scheme 5.10) and makes the cyclization easier. Water also strongly forces the reactive species to reduce the volume and so enhances cyclization rate.

5.2.5 Oxidation Reactions

Oxidative cleavage of an alkene to carboxylic acid and ketone is one of the most useful reactions in organic synthesis. Commonly used reagents for this transformation include O_3 and metal oxides such as OsO_4 and RuO_4 in combination with a co-oxidant like $NaIO_4$ and oxone. However, these reagents are toxic and are not desirable in the context of green chemistry. More recently, a few green procedures involving relatively less toxic reagents have been reported;[15] the most elegant one by Noyori *et al.*[15c] uses H_2O_2 as outlined in Scheme 5.11.

An alternative green procedure has been developed by Ranu's group[16] for the oxidation of alkenes to the corresponding carboxylic acid or ketones using *tert*-butyl hydroperoxide in water in the presence of a catalytic amount of $InCl_3$ (Scheme 5.12). The oxidized products, adipic acid and suberic acid from cyclohexene and cyclooctene, respectively, are industrially very important. A few complex olefins containing peptide bonds underwent cleavage to the corresponding carboxylic acids without any damage to these sensitive functionalities by this procedure.

Sato *et al.* have reported[17] the oxidation of alkenes to 1,2-diols using H_2O_2 as oxidant in the presence of resin-supported sulfonic acid in aqueous medium

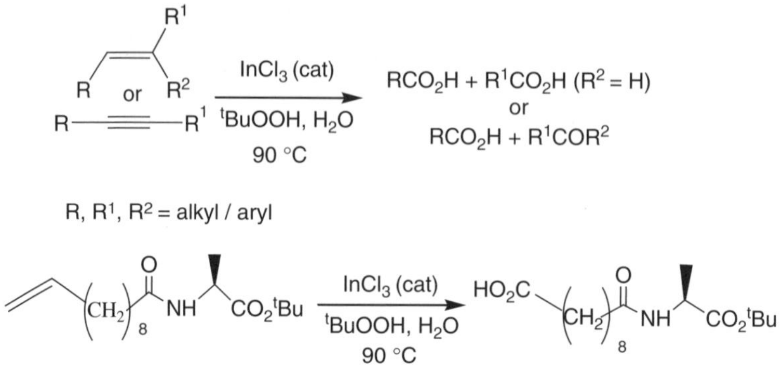

Scheme 5.12

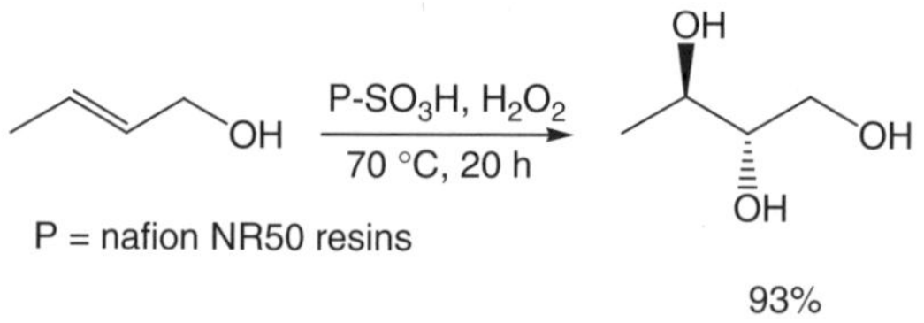

Scheme 5.13

(Scheme 5.13). The products (1,2-diols) are widely used as intermediates in fragrance, perfume and cosmetic industries. No over-oxidation product of the diol was observed.

5.2.6   Claisen Rearrangement

The accelerating influence of water as a solvent on the rate of Claisen rearrangement has also been demonstrated.[18] The use of glucose as chiral auxiliary by Auge *et al.*[18] led to moderate asymmetric induction in the Claisen rearrangement. After removal of the glucose moiety followed by separation of diastereomers, an enantiomerically pure product was obtained (Scheme 5.14).

5.2.7   Aza-Michael Reaction

β-Amino carbonyl compounds are versatile intermediates for the synthesis of biologically important natural products, antibodies and β-amino alcohols.[19]

The conjugate addition of an amine to an α,β-unsaturated carbonyl compound/nitrile has been achieved[20] in water without any catalyst at room temperature (Scheme 5.15). The rate acceleration of the reaction was explained by H-bonding of H_2O to the amine and carbonyl group of alkene.

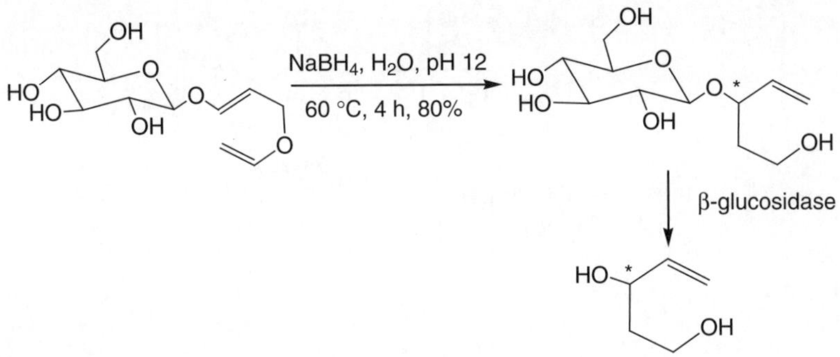

Scheme 5.14

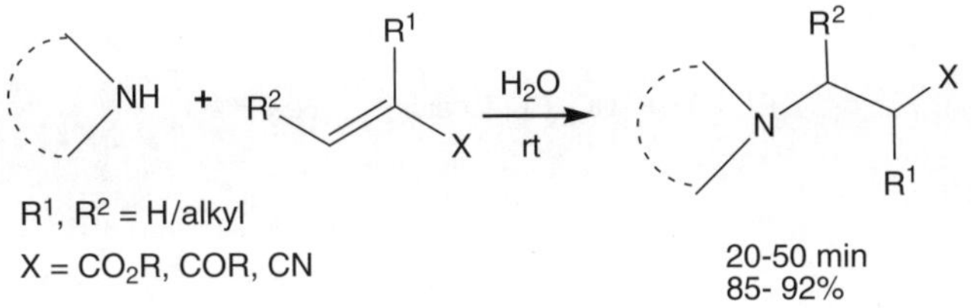

Scheme 5.15

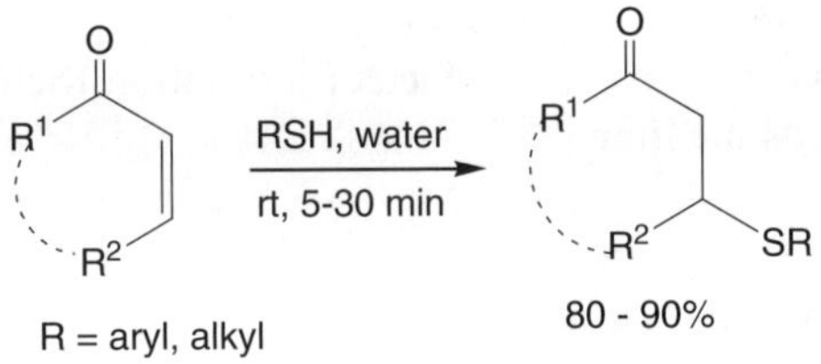

R^1, R^2 = H/alkyl
X = CO$_2$R, COR, CN

20-50 min
85- 92%

Scheme 5.16

R = aryl, alkyl

80 - 90%

5.2.8 Thia-Michael Addition

Chakraborti *et al.*[21] have described an efficient catalyst-free method for water-mediated conjugate addition of thiols to α,β-unsaturated carbonyl compounds, providing β-sulfido carbonyl compounds (Scheme 5.16) at room temperature. The reactions were fast and highly chemoselective. Water played a dual role in simultaneously activating the α,β-unsaturated carbonyl compound and the thiol.

Similarly, thiols undergo addition reaction with non-activated alkene[22] to give selectively the *anti*-Markovnikov adduct in water without any catalyst or additive (Scheme 5.17).

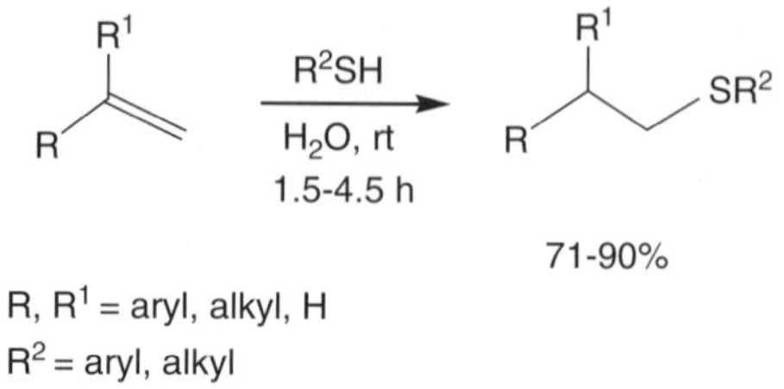

R, R^1 = aryl, alkyl, H

R^2 = aryl, alkyl

71-90%

Scheme 5.17

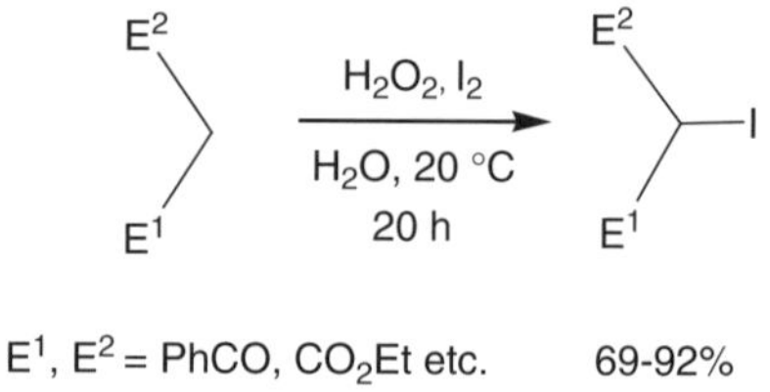

E^1, E^2 = PhCO, CO$_2$Et etc.   69-92%

Scheme 5.18

5.2.9 Iodination

Iodination is of much importance for access to iodinated organic molecules that are frequently used for C–C bond formation, iodine–metal exchange, *etc.*[23] Several iodo-substituted molecules are biologically active and have attracted considerable attention as therapeutic drugs.[24]

Jereb *et al.*[25] have demonstrated efficient iodination (Scheme 5.18) using I$_2$ and H$_2$O$_2$ in aqueous medium.

5.2.10 Mannich Reaction

Several procedures for a one-pot Mannich-type reaction in water to give β-amino carbonyl compounds catalyzed by either Lewis acid or Brønsted acid with or without addition of surfactants have been developed.[26] The reactions are high yielding; however, the diastereoselectivities are moderate. The HBF$_4$-catalyzed reaction[27] between aldimines and ketene silyl acetals in a water/SDS mixture provides high stereoselectivity with very good yields (Scheme 5.19).

5.2.11 Cyclization Reactions

Nitrogen-containing heterocycles are of great importance as they constitute subunits in many natural products and biologically active pharmaceuticals.[28]

The direct syntheses of *N*-aryl-azacycloalkanes, 4,5-dihydrophazine derivatives and pyrazoles by double alkylation of hydrazine derivatives with alkyl

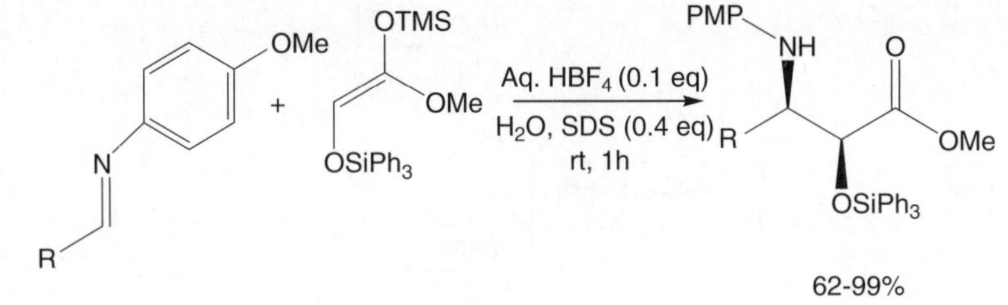

Scheme 5.19

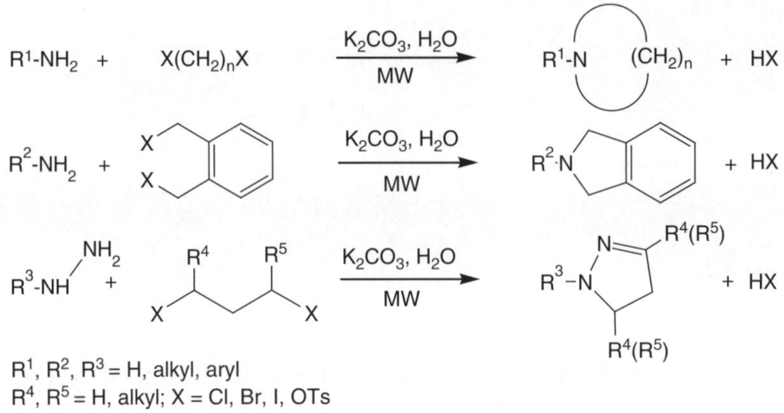

R^1, R^2, R^3 = H, alkyl, aryl
R^4, R^5 = H, alkyl; X = Cl, Br, I, OTs

Scheme 5.20

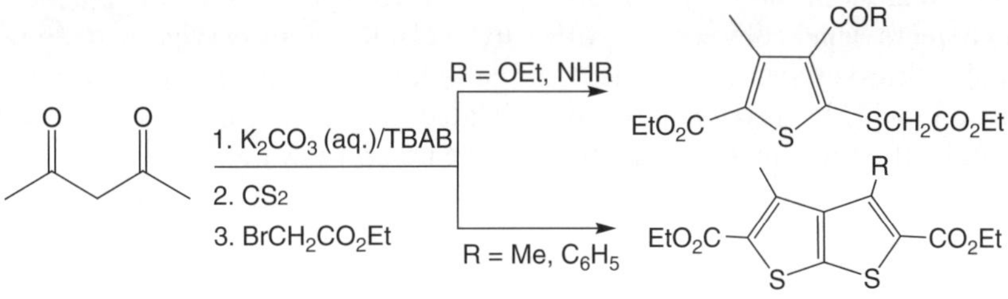

Scheme 5.21

halides or ditosylates in water under microwave irradiation (Scheme 5.20) have been demonstrated by Varma *et al.*[29]

Thiophene derivatives were also afforded in water by Dong and Liu.[30] The polysubstituted thiophene and polysubstituted thieno[2,3-b]thiophenes have been obtained in high yields. The reaction of 1,3-dicarbonyl compounds and bromoacetic ester was catalyzed by aqueous tetrabutylammonium bromide (TBAB) in the presence of CS_2 and K_2CO_3 (Scheme 5.21). After isolation of product the aqueous solution of TBAB was recycled for subsequent reactions.

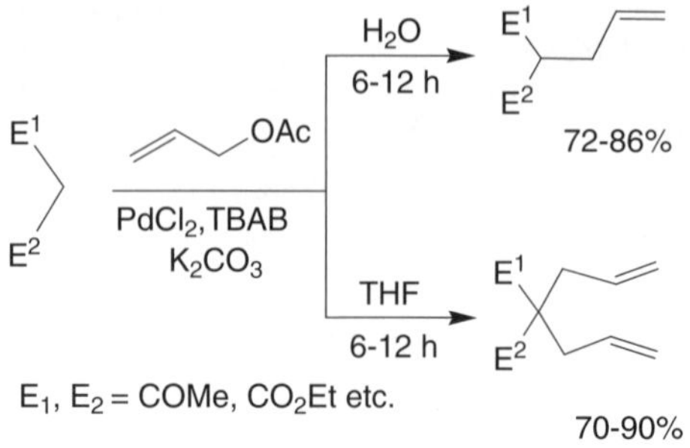

E$_1$, E$_2$ = COMe, CO$_2$Et etc.

72-86%

70-90%

Scheme 5.22

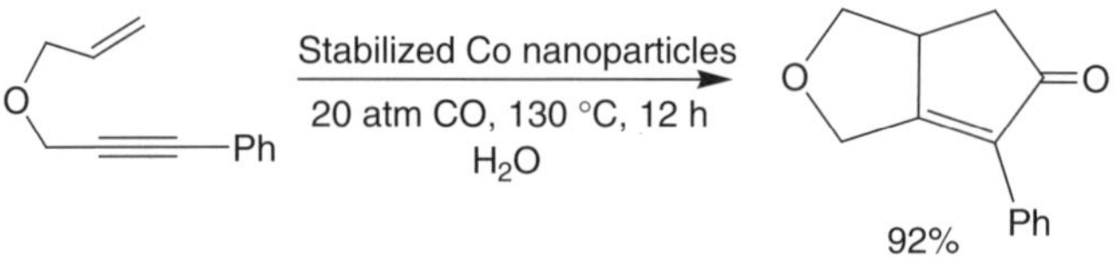

92%

Scheme 5.23

5.2.12 Solvent Controlled Mono- and Bis-allylation of Active Methylene Compounds

The allylation of active methylene compounds with allyl alcohols or their derivatives, called the Tsuji–Trost reaction, is a widely used process in academia as well as in industry.[31] Ranu *et al.*[32] have reported that the reaction of active methylene compounds with allyl acetate catalyzed by palladium(0) nanoparticles (Scheme 5.22) led to mono-allylation in water, whereas the reaction in THF provided the bis-allylated product. This is a remarkable example of controlling the direction of a reaction by water.

5.2.13 Pauson–Khand Reaction

The Pauson–Khand reaction is a very useful tool for access to cyclopentanone derivatives. The intramolecular Pauson–Khand reaction in water (Scheme 5.23) has successfully been carried out using colloidal Co-nanoparticles as catalyst in water.[33]

5.2.14 Heck Reaction

The Heck reaction has been widely used for the synthesis of many natural and unnatural products. The Heck reaction catalyzed by palladium acetate in water using a water-soluble ligand has been employed for the synthesis of clavicipitic acid (Scheme 5.24).

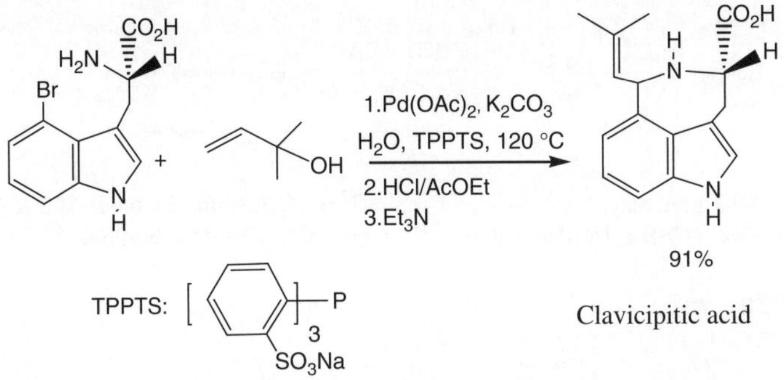

Scheme 5.24

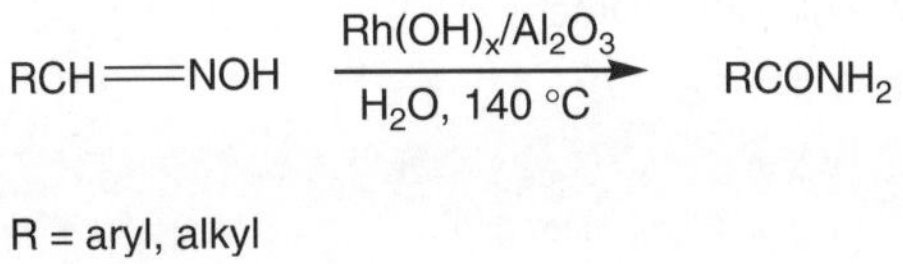

$$RCH{=}NOH \xrightarrow[\text{H}_2\text{O, 140 °C}]{\text{Rh(OH)}_x/\text{Al}_2\text{O}_3} RCONH_2$$

R = aryl, alkyl

Scheme 5.25

5.2.15　Beckmann Rearrangement

A heterogeneous rhodium catalyst, $Rh(OH)_x/Al_2O_3$,[35] acts as an efficient catalyst for a one-pot synthesis of primary amide from aldoximes (Scheme 5.25). The amides are used extensively in detergent and lubricant purification.

5.2.16　Synthesis of 1,3-En-yne Derivatives

The 1,3-en-yne moiety is of considerable interest in organic synthesis as many naturally occurring compounds with promising biological activity[36] contain this moiety, *e.g.*, terbinafine, a potent drug for suprafacial infection, and cali-chemicine γ_1, an effective antitumor antibiotic. A new efficient route[37] for the synthesis of enyne carboxylic ester and nitrile has been developed by Heck coupling of *vic*-di-iodoalkene with conjugated carboxylic ester and nitriles in aqueous medium (Scheme 5.26) using *in situ* generated palladium(0) nanoparticles. The high dielectric constant of H_2O is assumed to contribute this type of coupling reaction with higher efficiency.

5.2.17　Hiyama Coupling

Hiyama coupling is an efficient protocol for the synthesis of biaryl derivatives. A very fast (5 min) Hiyama coupling of organosilanes with aryl bromides or iodides in aqueous medium at 90 °C, using Pd nanoparticles, has been reported recently by Ranu and his co-workers[38] (Scheme 5.27).

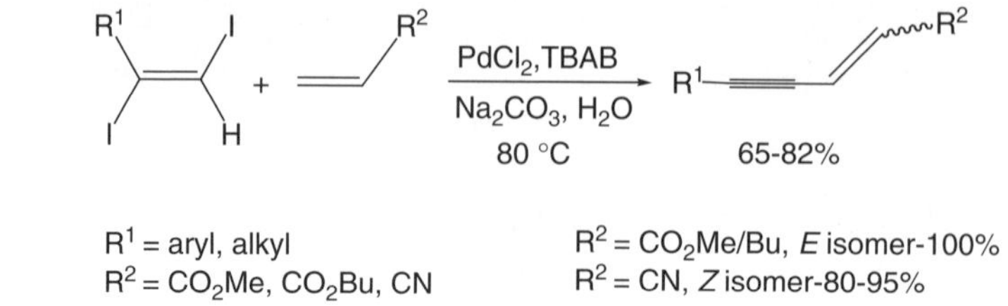

PdCl2,TBAB
Na2CO3, H2O
80 °C
65-82%
R1 = aryl, alkyl
R2 = CO2Me, CO2Bu, CN
R2 = CO2Me/Bu, E isomer-100%
R2 = CN, Z isomer-80-95%

Scheme 5.26

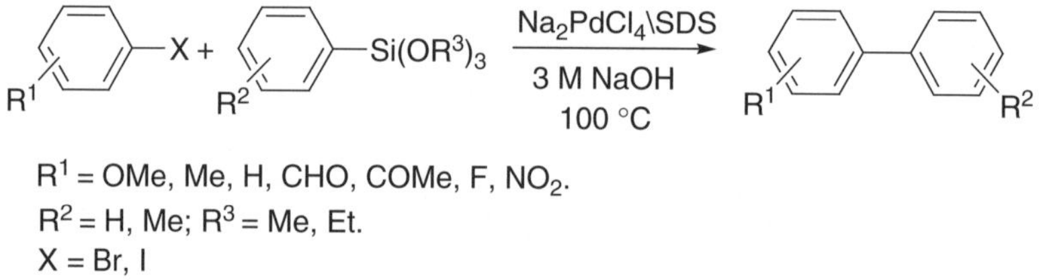

Na2PdCl4\SDS
3 M NaOH
100 °C
R1 = OMe, Me, H, CHO, COMe, F, NO2.
R2 = H, Me; R3 = Me, Et.
X = Br, I

Scheme 5.27

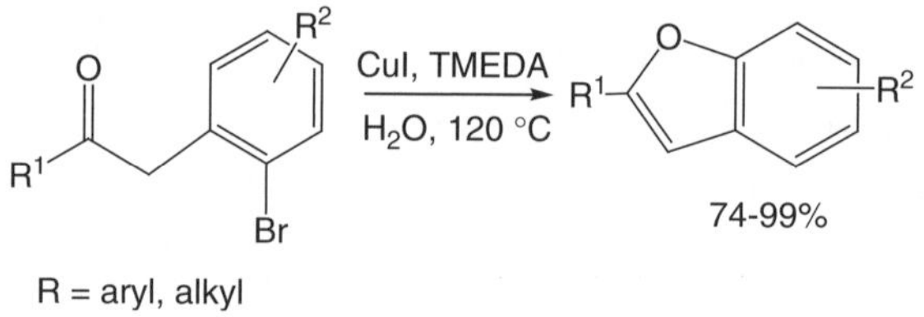

CuI, TMEDA
H2O, 120 °C
74-99%
R = aryl, alkyl

Scheme 5.28

5.2.18 Synthesis of Benzo[*b*]furan

Benzo[*b*]furans are of great interest because of their application in pharmacology and their wide occurrence in nature. For example, 2-aryl-benzofurans and their derivatives exhibit a broad range of biological activities as antineoplastic, antiviral and antioxidative agents.[39] Dominguez *et al.*[40] have reported a preparative method involving Cu-TMEDA complex for an efficient synthesis of benzofuran using water as a solvent (Scheme 5.28).

5.3 Surface-mediated Reaction

One of the most fundamental examples of a surface-mediated reaction is the heterogeneous catalytic hydrogenation of olefins and acetylenes on charcoal-supported transition metals like Pd, Pt, Ni, *etc.*[41]

It is assumed that the effectiveness of surface-mediated reactions under solvent-free condition is due to the following factors:

1. The pores present in the supports behave as enzyme pockets where both substrate and reactant are constrained, consequently lowering the entropy of activation.
2. The surface provides a microenvironment of different polarity with acidic and basic sites, causing appropriate activation of substrates and stabilization of the intermediates.
3. The products are, in general, obtained with higher purity compared to their homogeneous counterparts.

The combination of all these factors probably contributes to the selectivity and efficiency often achieved in surface-mediated reactions. Various heterogeneous surfaces such as clay, silica gel and alumina are widely used.

5.3.1 Reaction on an Al_2O_3 Surface

Al_2O_3 has been found to have significant applications both as a surface and as an adsorbent for various selective organic transformations. Generally, unactivated alumina contains a top layer of –OH groups and underlying Al^{3+} cations that cannot influence the organic reaction significantly. On heating, the peripheral H_2O molecules are driven from the surface, exposing Al^{3+} cation and O^{2-} anion, which then behave as an amphoteric catalyst.[42]

5.3.1.1 Diels–Alder Reaction

Remarkable *endo* selectivity has been observed in the Diels–Alder reaction of cyclopentadiene and *N*-phenylmaleimide (Scheme 5.29) using activated Al_2O_3 under solvent-free mild conditions.[43]

5.3.1.2 Amide Synthesis

Pagni and Kabalka[44] reported acid-catalyzed hydrolysis of nitrile selectively into amide on the surface of unactivated Al_2O_3. The Al_2O_3 surface is polar and contains a layer of OH groups, which could serve as the source of water in this reaction (Scheme 5.30).

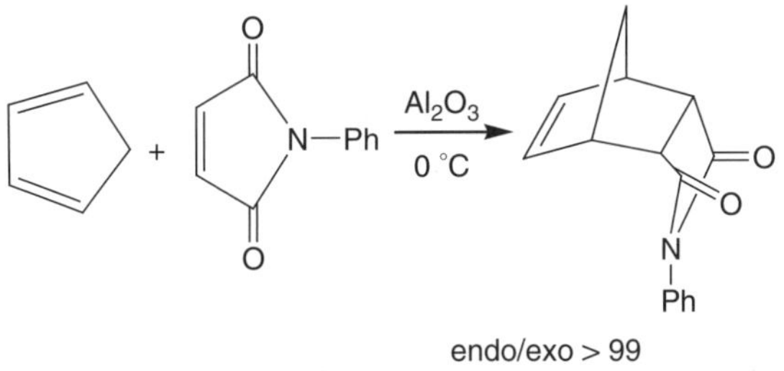

Scheme 5.29

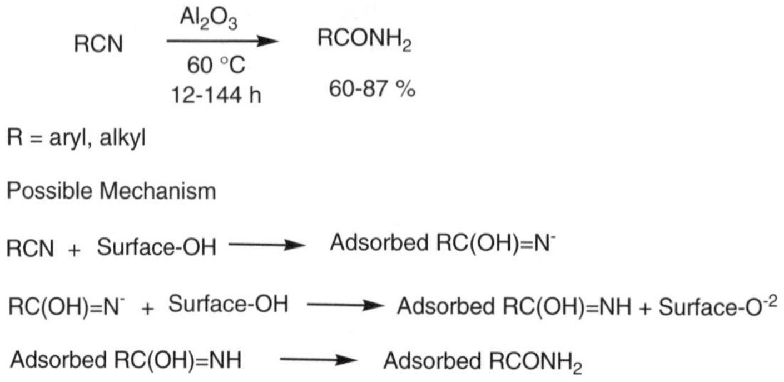

$$\text{RCN} \xrightarrow[\substack{60\ ^\circ\text{C} \\ 12\text{-}144\ \text{h}}]{\text{Al}_2\text{O}_3} \text{RCONH}_2$$

60-87 %

R = aryl, alkyl

Possible Mechanism

$$\text{RCN} + \text{Surface-OH} \longrightarrow \text{Adsorbed RC(OH)=N}^-$$

$$\text{RC(OH)=N}^- + \text{Surface-OH} \longrightarrow \text{Adsorbed RC(OH)=NH} + \text{Surface-O}^{-2}$$

$$\text{Adsorbed RC(OH)=NH} \longrightarrow \text{Adsorbed RCONH}_2$$

Scheme 5.30

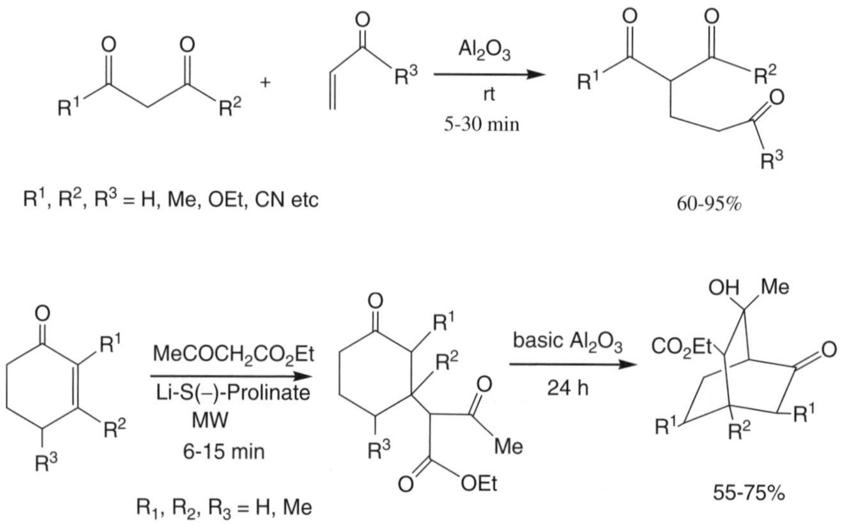

R^1, R^2, R^3 = H, Me, OEt, CN etc

60-95%

R_1, R_2, R_3 = H, Me

55-75%

Scheme 5.31

5.3.1.3 Michael Reaction and Synthesis of Bicyclo[2.2.2]octanone Derivatives

The Michael reaction of several β-dicarbonyl compounds and α,β-unsaturated aldehyde, ketones and esters can be carried out very efficiently on the surface of Al_2O_3 without any solvent and catalyst (Scheme 5.31).[45a]

A simple route to a bicyclo[2.2.2]octanone system was achieved by Michael reaction under microwave irradiation of a mixture of cyclohexenones and ethyl acetoacetate adsorbed on the surface of solid lithium (*S*)-(–)-prolinate followed by intramolecular aldolization on a column of basic Al_2O_3[45b] (Scheme 5.31). The bicyclo[2.2.2]octane system is the core unit of several biologically active natural products and a useful intermediate for synthetic manipulation.[46]

5.3.1.4 Synthesis of Ferrocene Derivatives

Ferrocene derivatives have been the subject of current interest due to their potential uses as chiral ligands in organic synthesis and as useful materials.[47]

Friedel–Crafts acylation using $AlCl_3$ and acid chloride has serious disadvantages in the context of green synthesis as these reagents are highly toxic. Moreover, a mixture of mono- and diacylated product was often formed. The monoacylated ferrocene was obtained in high yield by a reaction of ferrocene with carboxylic acid and $(CF_3COO)_2O$ on the solid surface of Al_2O_3[48] (Scheme 5.32). A wide range of structurally diverse monoacylated ferrocene derivatives have been prepared by this procedure.

5.3.1.5 Synthesis of Propargyl Amines

Propargyl amines are versatile intermediates in organic synthesis and an important structural motif in drug molecules. In addition, these β-amino-alkynes are also of importance as its two potential sites can be modified further to fit particular applications.

An efficient and environmentally benign method for the synthesis of β-amino-alkyne (Scheme 5.33) has been developed by Kabalka and co-workers[49] using CuI-doped alumina as a catalyst under solvent-free microwave irradiation.

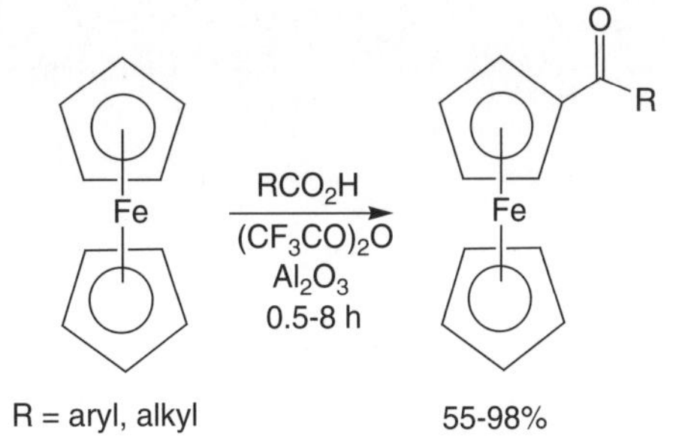

RCO₂H
(CF₃CO)₂O
Al₂O₃
0.5-8 h
Fe
Fe
O
R
R = aryl, alkyl
55-98%

Scheme 5.32

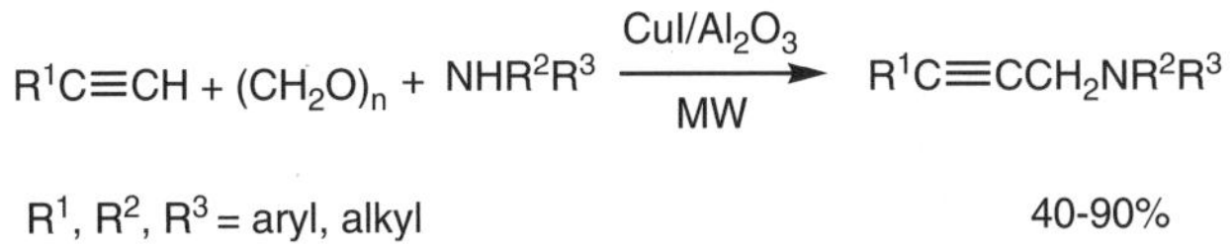

R¹C≡CH + (CH₂O)ₙ + NHR²R³
CuI/Al₂O₃
MW
R¹C≡CCH₂NR²R³
R¹, R², R³ = aryl, alkyl
40-90%

Scheme 5.33

5.3.2 Reaction on a SiO$_2$ Surface

5.3.2.1 Diels–Alder Reaction

Like alumina, SiO$_2$ also shows catalytic efficiency for organic reactions under solvent-free conditions. A good example is catalysis of the Diels–Alder reaction (Scheme 5.34) by silica gel with high regio- and stereoselectivity.[50]

5.3.2.2 Epoxide Ring Cleavage

Silica gel has been used as reagent and absorbent as well as catalyst for many nucleophilic processes. Kotsuki and Shimanouchi have demonstrated[51] an elegant epoxide ring opening using LiX supported silica gel (Scheme 5.35) in the absence of any solvent.

5.3.2.3 Synthesis of Spirooxindoles

The spirooxindole system is the core unit of many natural products such as spirotryprostatin A, isopeteropodine, *etc.*[52] Perumal and co-workers have developed a fast, clean and simple method[53] for the synthesis of spirooxindoles and spiroindenoquinoxaline derivatives catalyzed by silica gel impregnated indium(III) chloride under solvent-free microwave irradiation (Scheme 5.36).

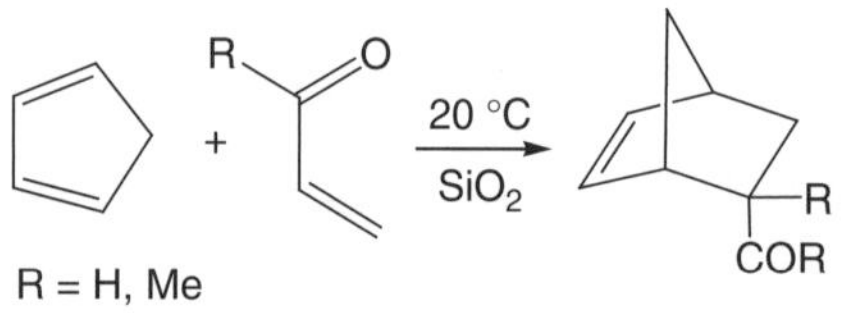

R = H, Me

Scheme 5.34

X = Cl, Br, I.

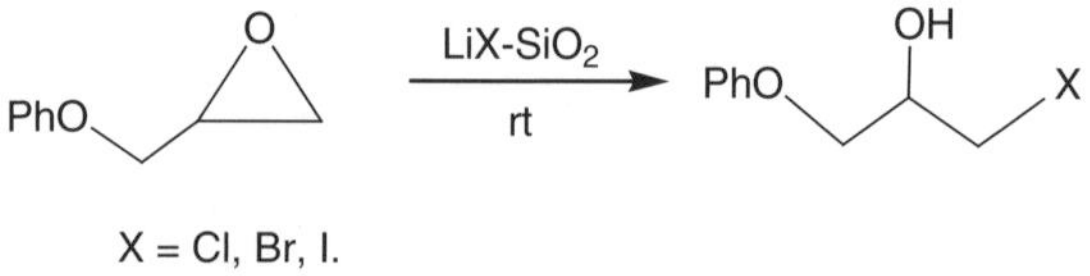

Reagent	Reaction time (h)	Yield (%)
LiBr-SiO$_2$	1	100
LiBr	48	44
LiBr in CH$_2$Cl$_2$	22	100

Scheme 5.35

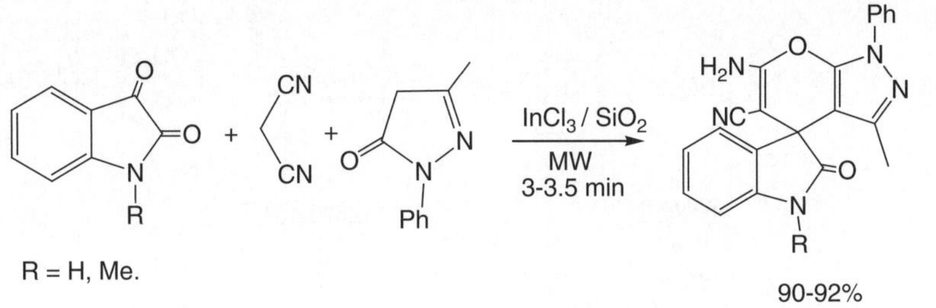

InCl3 / SiO2
MW
3-3.5 min
R = H, Me.
90-92%

Scheme 5.36

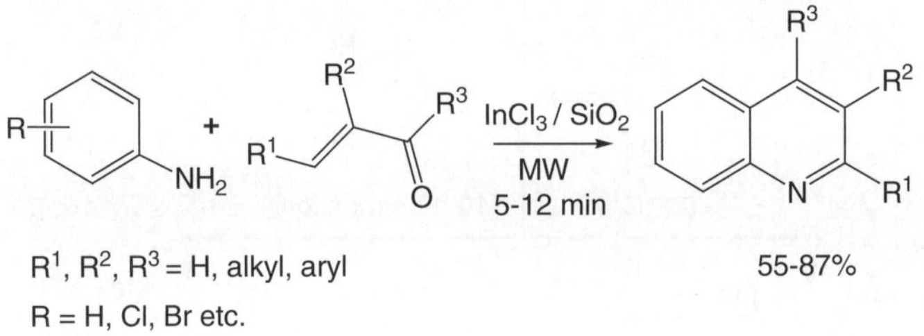

InCl3 / SiO2
MW
5-12 min
R1, R2, R3 = H, alkyl, aryl
R = H, Cl, Br etc.
55-87%

Scheme 5.37

5.3.2.4 Quinoline Synthesis

The classical method of quinoline synthesis involves Skraup's procedure.[54] However, it requires a large amount of H_2SO_4 at elevated temperature and the reaction often becomes violent.

A simple and efficient method for the synthesis of 4-alkyquinolines by one-pot reaction of aniline with alkyl vinyl ketone on the surface of SiO_2-impregnated $InCl_3$ under microwave irradiation has been demonstrated by Ranu and his group[55] (Scheme 5.37).

5.3.3 Reaction on the Surface of a Clay

Clay, because of its inherent acidity, has been found to catalyze several acid-catalyzed reactions. The high acidity of clay is attributed to the presence of exchangeable cations, which polarize coordinated water molecules and induce their dissociation.

5.3.3.1 Michael Reaction

The Michael reaction of 1,3-dicarbonyl compounds with an enone on the surface of Cu^{2+} montmorillonite provides 1,5-dioxo synthons (Scheme 5.38) that are useful intermediates in organic synthesis.[56] The catalyst was recycled for several runs; the presence of solvent retards the reactions.[57]

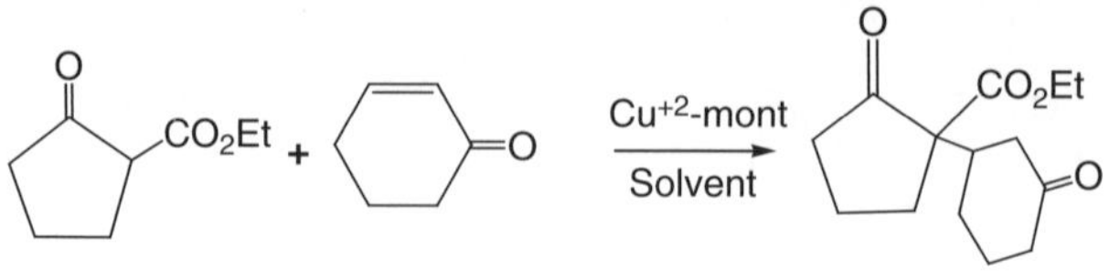

Solvent	Yield (%)
neat	96
EtOH	76
DMSO	0
Toluene	46

Scheme 5.38

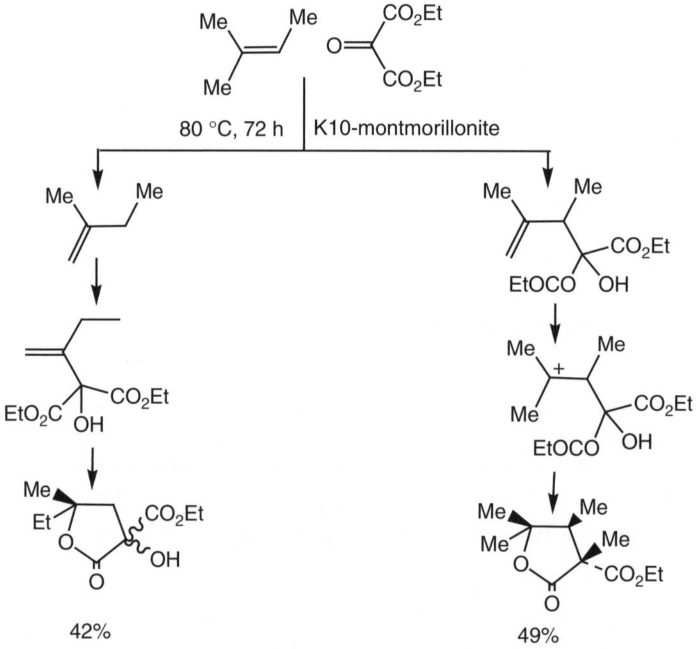

Scheme 5.39

5.3.3.2   *Ene Reaction*

Acid-catalyzed condensation of 2-methyl-2-butene with diethyl 2-oxopropane-1,3-dioate on montmorillonite followed by cyclization led to the formation of isomeric lactones, as reported by Foucaud *et al.*[58] (Scheme 5.39).

5.3.4   Reaction on the Surface of a Zeolite

Zeolite as a catalyst is very useful for organic reactions, providing high selectivity. As they are available in different cavity sizes, they are very effective in allowing certain reactants to fit in their cavities, and at the same time preventing others, and thus reactions are highly selective and undesired side reactions are restricted.

5.3.4.1 Selective Methylation of Toluene

The methylation of toluene with CH_3OH over ZSM-5 zeolite impregnated H_3PO_4 led to *p*-xylene selectively[59] (Scheme 5.40).

The other isomers (*o,m*) of xylene isomerize rapidly inside the pores. The *p*-xylene diffuses out reasonably fast, while other isomers are rather slow and thus these isomers undergo isomerization to *p*-xylene before escaping from the channel system.

5.3.4.2 Selective Meerwein–Ponndorf–Verley Reduction

Zeolite BEA is a stereoselective and regenerable heterogeneous catalyst for the Meerwein–Ponndorf–Verley (MPV) reduction of 4-*tert*-butylcyclohexanone to *cis*-4-*tert*-butylcyclohexanol (Scheme 5.41), as demonstrated by Bekkum *et al.*[60] The *cis*-isomer is very important in the fragrance industry and is very difficult to obtain by other methods.

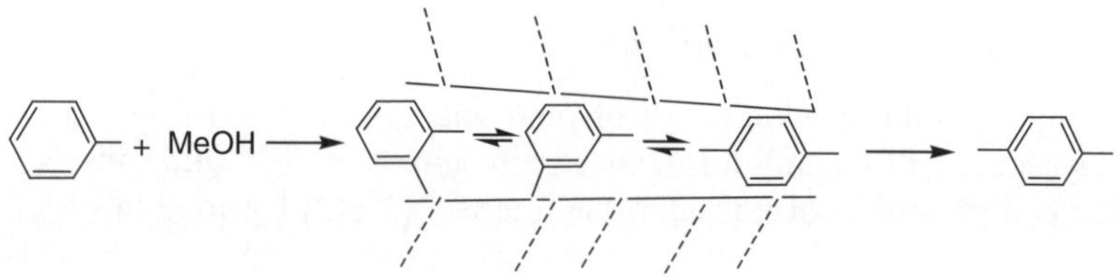

Scheme 5.40

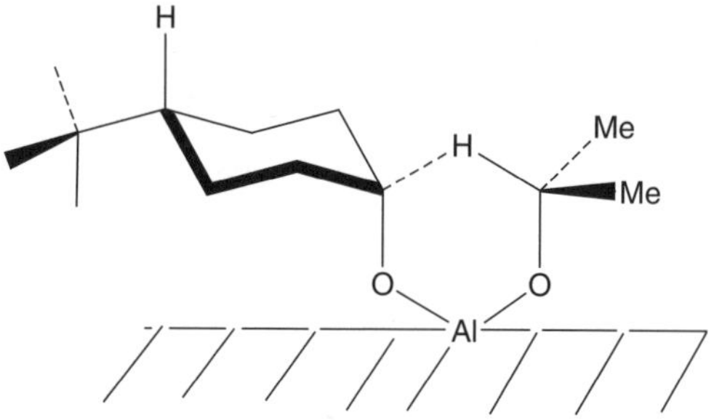

Scheme 5.41

The thermodynamically more stable *trans*-isomer, which is the sole product of homogeneous MPV reduction[61] in presence of propane-2-ol, was not detected in this case. This remarkable selectivity is attributed to the pore size of zeolite BEA.

5.4   Neat Reactions

A reaction under neat conditions, with or without catalyst, avoiding the use of organic solvents in the entire operation, including isolation and purification, is of great significance in the design of a green process for industrial application.

5.4.1   Reaction with Catalyst

5.4.1.1   *Enantioselective Epoxide Ring Cleavage*

Jacobsen and his co-workers[62] have showed a chromium salen complex (2 mol.%) and Me_3SiN_3 induced ring-opening reaction of cyclohexene oxide in enantioselective manner under solvent-free conditions (Scheme 5.42). To avoid the use of organic solvent the product was isolated by short-path distillation under reduced pressure directly from the reaction mixture.

5.4.1.2   *[3 + 2] Cycloaddition*

Imidazolines exhibit a wide range of pharmacological activities such as anti-hyperglycemic, antioxidant, anti-cancer, antitumor, *etc.*[63a,b] and thus synthesis of these compounds is of considerable interest. [3 + 2] Cycloaddition of aziridines with nitrile has been reported in neat in the presence of $Sc(OTf)_3$ to produce substituted imidazolines (Scheme 5.43) in high yields.[63c]

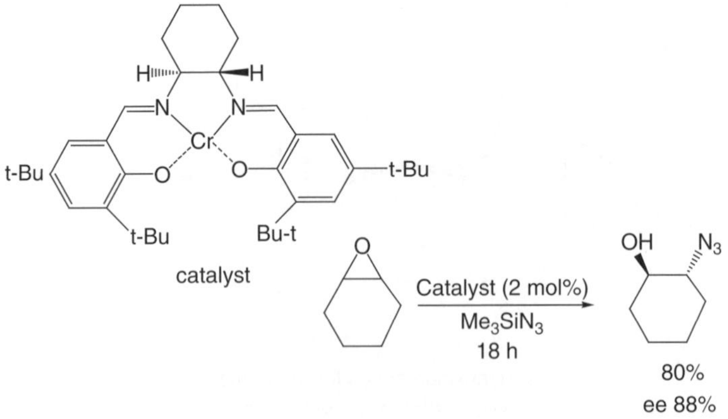

Scheme 5.42

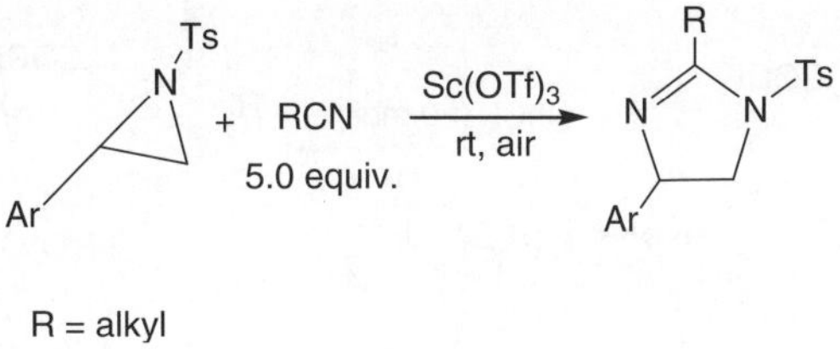

R = alkyl

Scheme 5.43

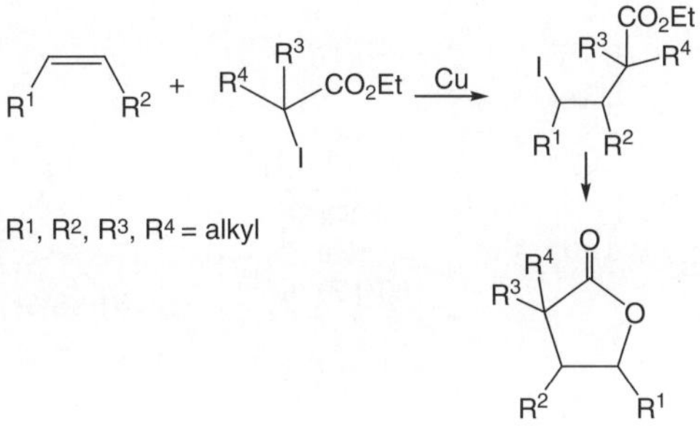

R¹, R², R³, R⁴ = alkyl

Scheme 5.44

5.4.1.3 Radical Reaction

A convenient procedure for the synthesis of substituted γ-butyrolactones has been developed by the reaction of alkenes and an α-iodoester through an electron-transfer process promoted by metallic Cu under solvent-free conditions at 130 °C under argon[64] (Scheme 5.44). To eliminate the use of solvent, the product was distilled directly from the reaction mixture.

5.4.1.4 Mukaiyama Aldol Reaction

Loh *et al* have demonstrated[65] the stereoselective synthesis of 1,3-amino alcohols based on the InCl₃-catalyzed Mukaiyama aldol reaction of a keto ester under solvent-free conditions. The high selectivity has been explained by the influence of a remote group present in the keto ester (Scheme 5.45).

5.4.1.5 Tsuji–Trost Reaction

Yamamoto and Patil have reported[66] the reaction of allyl alcohols with carbon pronucleophiles in the presence of a Pd(PPh₃)₄/carboxylic acid combined catalytic system, in neat conditions, leading to the corresponding allylated products in high yields in one step (Scheme 5.46).

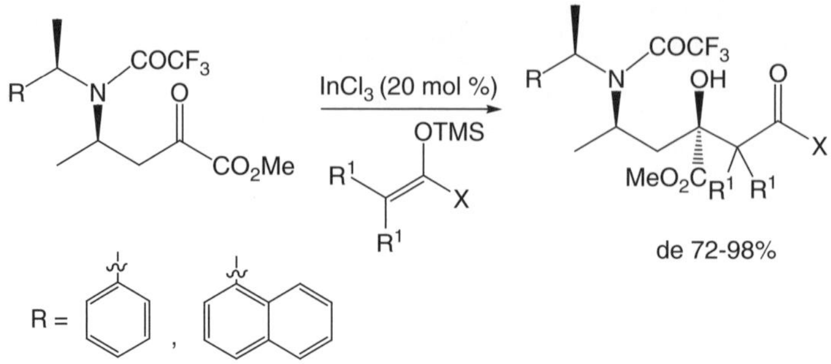

Scheme 5.45

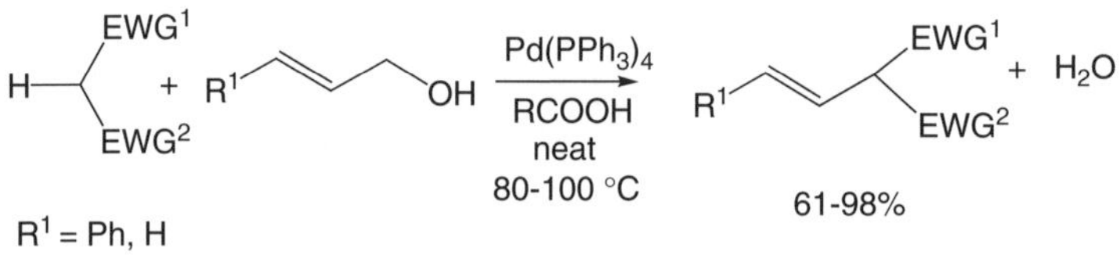

Scheme 5.46

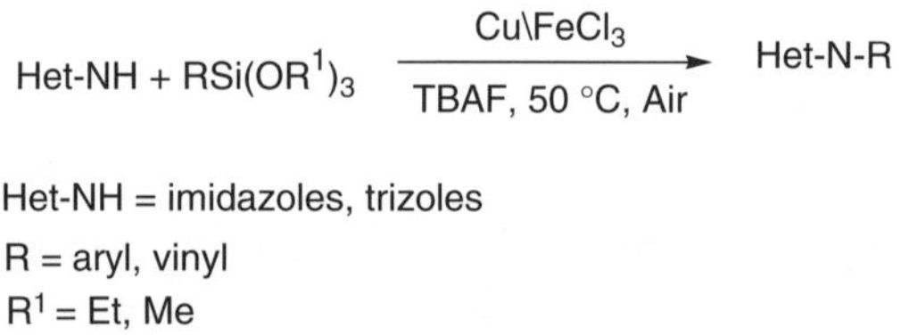

Scheme 5.47

5.4.1.6 N-Arylation

An efficient protocol for solvent-free N-arylation of N-heterocycles, including imidazoles and triazoles, by aryl trimethoxysilanes using $FeCl_3$/Cu as co-catalyst and TBAF as the base under air has been demonstrated by Li *et al.*[67] Various imidazoles and triazoles are compatible with this procedure (Scheme 5.47).

5.4.2 Reaction without Catalyst

5.4.2.1 Synthesis of Dihydropyrimidinones

Dihydropyrimidinone derivatives exhibit a wide range of pharmacological activities, *e.g.*, monastrol acts as a β-blocker while SQ 32926 exhibits anti-hypertensive properties.[68]

Various Lewis acids, such as $FeCl_3$, $InCl_3$, $BiCl_3$, $LiClO_4$, *etc.*, have been used to catalyze the century-old Biginelli reaction in the presence of different organic solvents. Recently, a few solvent-free reactions have been reported using $Yb(OTf)_3$, montmorillonite and ionic liquid as catalysts for the preparation of dihydropyrimidinones.

Interestingly, Ranu *et al.*[69] have reported an efficient reaction that involves simply stirring a mixture of aldehyde, diketone and urea at 100–105 °C for an hour without any catalyst or solvent (Scheme 5.48).

5.4.2.2 Synthesis of Dithiocarbamates

Dithiocarbamates have received considerable current attention due to their biological activities, pivotal role in agriculture and rubber industries, enzyme inhibition and radical polymerization among other uses.[70]

A one-pot solvent and catalyst free reaction of amine, CS_2 and alkyl halides at room temperature presents a very simple and green route to the synthesis of dithiocarbamates (Scheme 5.49). Significantly, higher yields were obtained in neat compared to those in organic solvents.[71] Certainly, this procedure is of much importance in pharmaceutical and agrochemical industries.

5.4.2.3 Synthesis of α-Aminophosphonate

α-Aminophosphonates are an important class of biologically active compounds and structural analogues to α-amino acids.[72a] They act as peptide

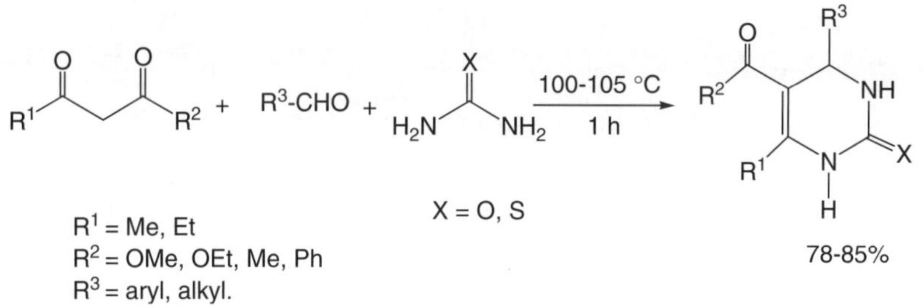

R^1 = Me, Et
R^2 = OMe, OEt, Me, Ph
R^3 = aryl, alkyl.

X = O, S

78-85%

Scheme 5.48

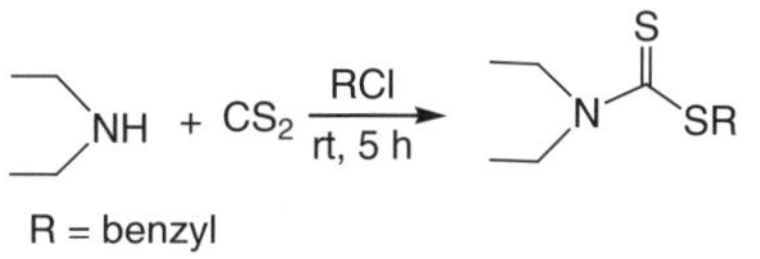

R = benzyl

Solvent	$ClCH_2CH_2Cl$	THF	CH_3CN	EtOH	ether	neat
Yield (%)	65	78	88	80	86	97

Scheme 5.49

mimic and enzyme inhibitor in addition to their role as versatile synthetic intermediates.

A general method for the synthesis of these compounds by a solvent-free, catalyst-free, one-pot, three-component condensation of carbonyl compound, amine and diethyl phosphate has been demonstrated by Ranu *et al.*[72b] (Scheme 5.50).

5.4.2.4   *[2 + 2] Cycloaddition*

A remarkable photochemical [2 + 2] addition of acenaphthylene and tetracyanoethylene in neat to produce the corresponding cycloadduct (Scheme 5.51) has been reported by Haga *et al.*[73] Interestingly, this reaction was inert in the presence of solvent.

5.4.2.5   *Reduction*

The $NaBH_4$ reduction of a pentacyclic cage diketone in EtOH afforded a 38 : 62 mixture of *endo-endo* and *exo-endo* diol.[74] In contrast, solid-state $NaBH_4$ reduction of diketone provided exclusively the corresponding *endo-endo* diol.[74] This indicates that the hydride transfer occurs exclusively at the *exo*-face of the carbonyl group in solid state reaction (Scheme 5.52).

5.4.2.6   *Photocyclization*

An interesting crystal-to-crystal asymmetric photoreaction of α,β-unsaturated thioamide to β-thiolactam (Scheme 5.53) with 94% ee in 96% yield of 58% conversion was reported by Sakamoto *et al.*[75]

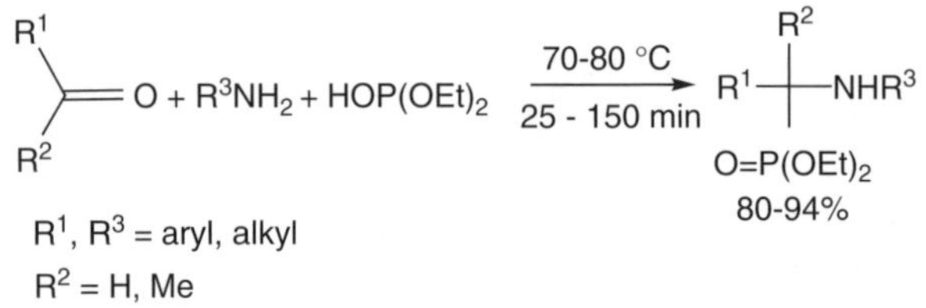

Scheme 5.50

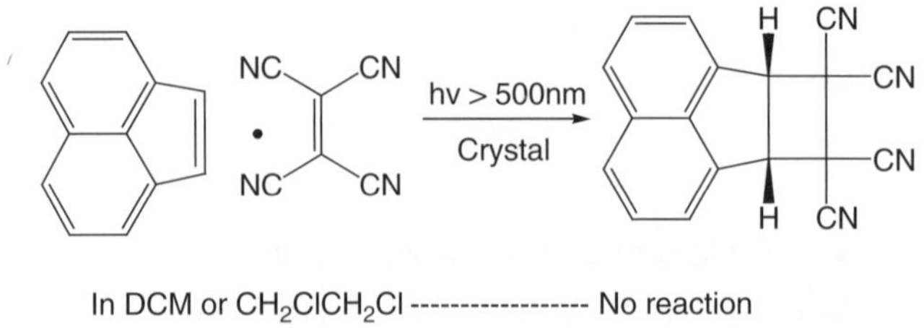

Scheme 5.51

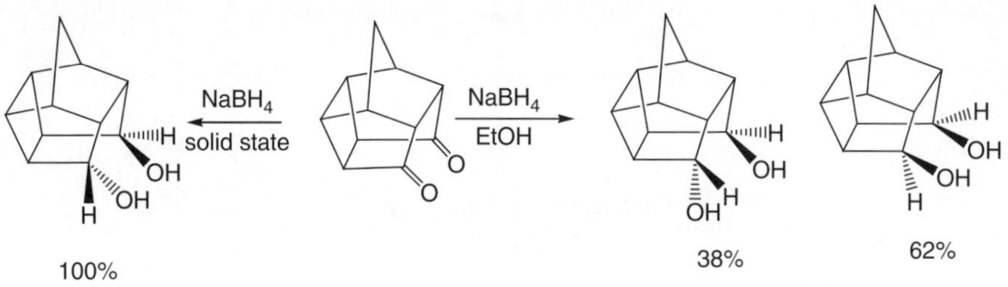

100%

Scheme 5.52

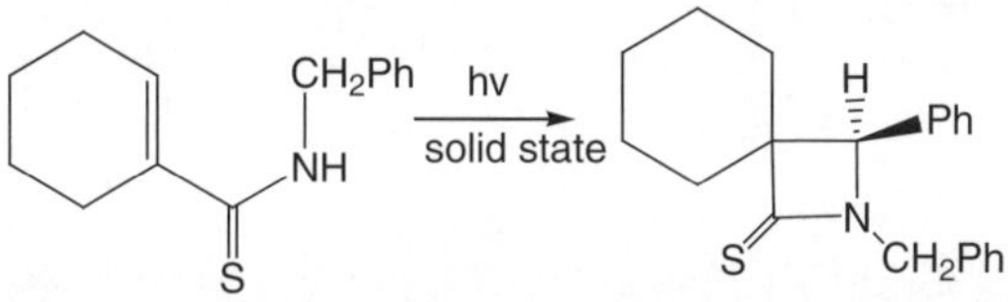

Scheme 5.53

5.5   Ionic Liquids

In recent times room temperature ionic liquids have been the subject of considerable interest since their introduction as green solvents for reactions.[76] Besides their usefulness as powerful reaction media, ionic liquids have been well recognized as efficient catalysts and successfully applied in many organic reactions.[77] Interestingly, ionic liquids are also found to steer several reactions in a particular direction, as outlined below.

In a remarkable observation, Earle's group[78] found that the reaction of toluene and nitric acid in three different ionic liquids leads to three completely different products in high yields (Scheme 5.54).

A similar dramatic influence of ionic liquid was also reported by Ranu's group[79] in the Michael reaction of active methylene compounds with methyl vinyl ketone and methyl acrylate in the presence of an ionic liquid, 1-butyl-3-methylimidazolium hydroxide ([bmIm]OH). Interestingly, open-chain 1,3-dicarbonyl compounds reacted with methyl vinyl ketone and chalcone to give the usual monoaddition products, whereas the same reactions with methyl acrylate or acrylonitrile provided exclusively bis-addition products (Scheme 5.55).

5.5.1   Synthesis of Organocarbonates

Chi and his group reported[80] the synthesis of symmetrical organic carbonates *via* alkylation of metal carbonates with various alkyl halides and sulfonate in 1-*n*-butyl-3-methylimidazolium hexafluorophosphate, [bmIm]PF_6. Besides providing

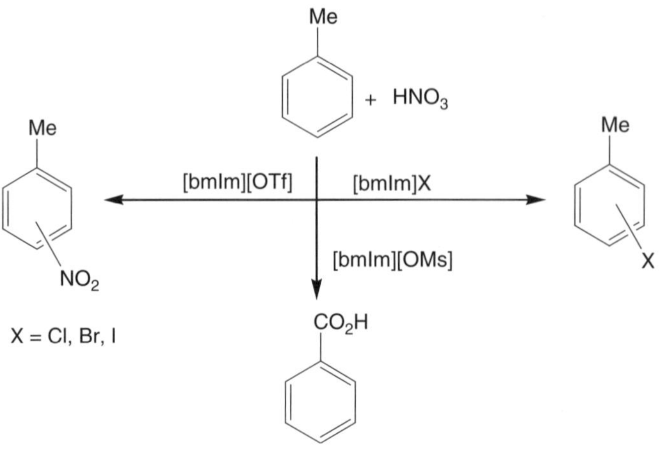

Scheme 5.54

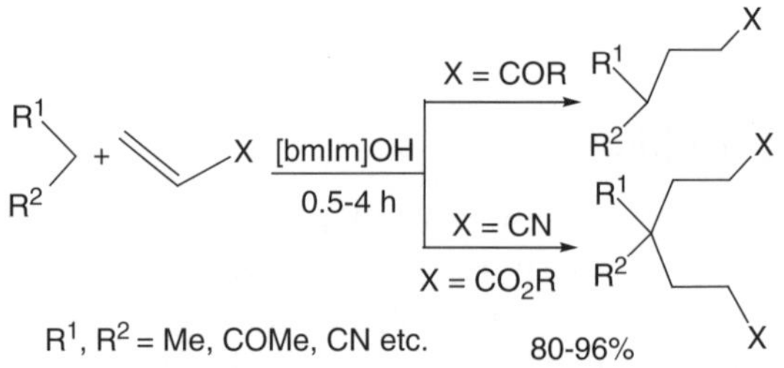

Scheme 5.55

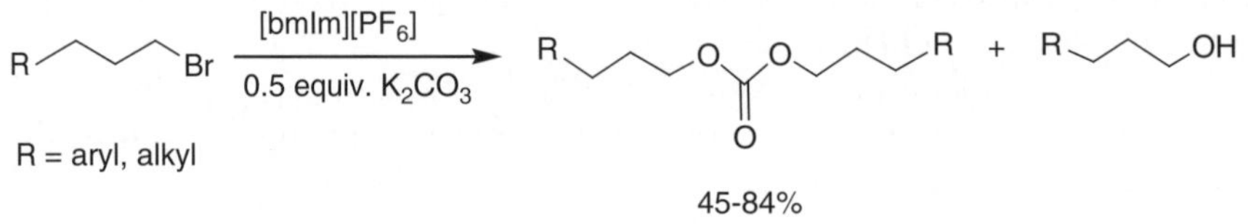

Scheme 5.56

good to excellent yields this methodology avoided highly toxic and harmful chemicals such as phosgene, carbon monoxide and pyridine[81] (Scheme 5.56).

5.5.2 Synthesis of 1,3-Dithianes

A simple, inexpensive ionic liquid, 1-methyl-3-pentylimidazolium bromide ([pmIm]Br), has been used as catalyst as well as reaction medium for the

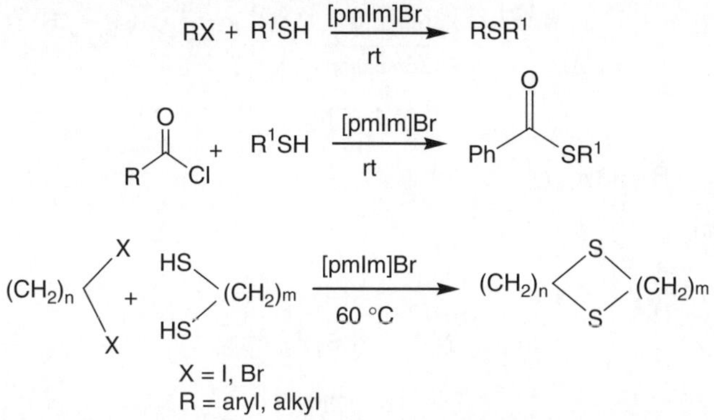

Scheme 5.57

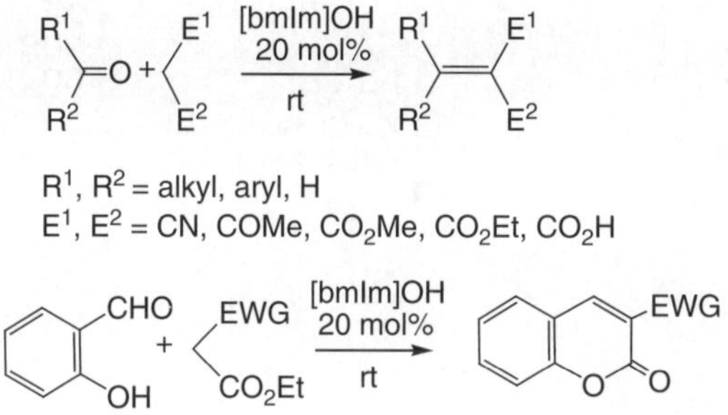

Scheme 5.58

conversion of thiols into thioethers and thioesters by reaction with alkyl or acyl halides, as demonstrated by Ranu *et al.*[82] This reaction has been successfully employed for the preparation of 1,3-dithianes (Scheme 5.57), a very useful chemical for organic transformation.

5.5.3 Knoevenagel Condensation

A basic ionic liquid, 1-butyl-3-methyl imidazolium hydroxide, [bmIm]OH, was found to catalyze the Knoevenagel condensation of aliphatic aldehydes and ketones with active methylene compounds efficiently in the absence of any organic solvent (Scheme 5.58). Coumarins have been obtained in one step from the reaction of *o*-hydroxy aldehydes following this procedure.[83]

5.5.4 Synthesis of Conjugated Dienes

Stereodefined conjugated dienes are of great importance in organic synthesis. A novel protocol involving an ionic liquid catalyzed rearrangement of

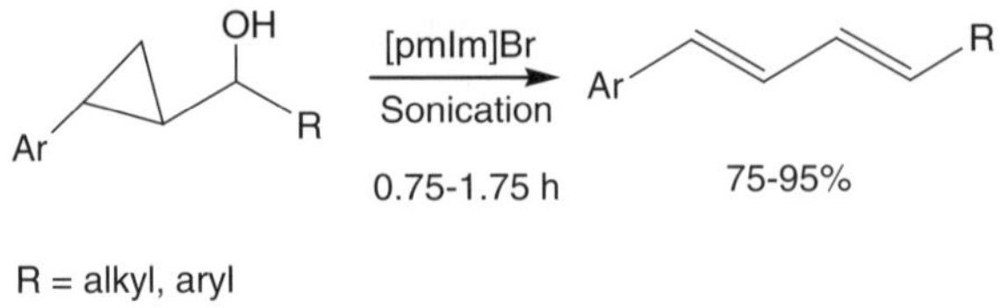
OH
Ar
R
[pmIm]Br
Sonication
0.75-1.75 h
Ar
R
75-95%
R = alkyl, aryl

Scheme 5.59

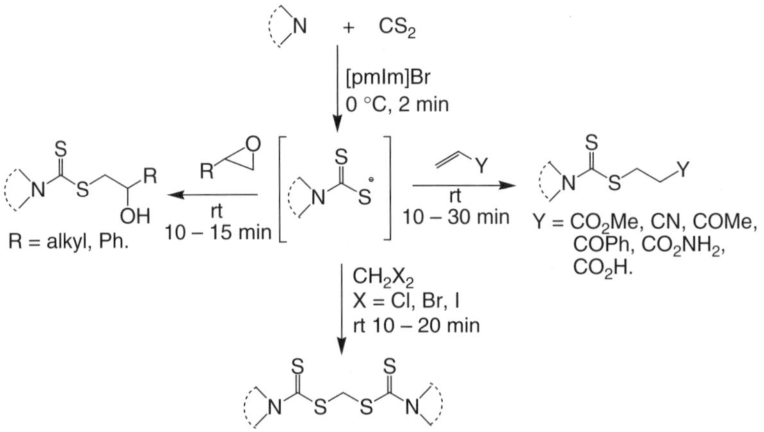
N + CS2
[pmIm]Br
0 °C, 2 min
R = alkyl, Ph.
rt
10 – 15 min
rt
10 – 30 min
Y = CO2Me, CN, COMe, COPh, CO2NH2, CO2H.
CH2X2
X = Cl, Br, I
rt 10 – 20 min

Scheme 5.60

cyclopropyl carbinol derivatives for the stereospecific synthesis of these molecules has been demonstrated by Ranu and his group[84] (Scheme 5.59).

5.5.5 Synthesis of Dithiocarbamates

The easily accessible neutral ionic liquid [pmIm]Br promoted a one-pot three-component condensation of an amine, carbon disulfide, and an activated alkene/dichloromethane/epoxide to produce the corresponding dithiocarbamates (Scheme 5.60) in high yields at room temperature.[85] The reactions were very fast in ionic liquids relative to those in other reaction media. These reactions did not require any additional catalyst or solvent. The ionic liquid was recovered and recycled for subsequent reactions.

5.5.6 Synthesis of Highly Substituted Pyridines

A basic ionic liquid, [bmIm]OH, efficiently promotes a one-pot, three-component condensation of aldehydes, malononitrile and thiophenols to produce highly substituted pyridines (Scheme 5.61) in high yields at room temperature.[86] These pyridines have much potential for therapeutic activity. This

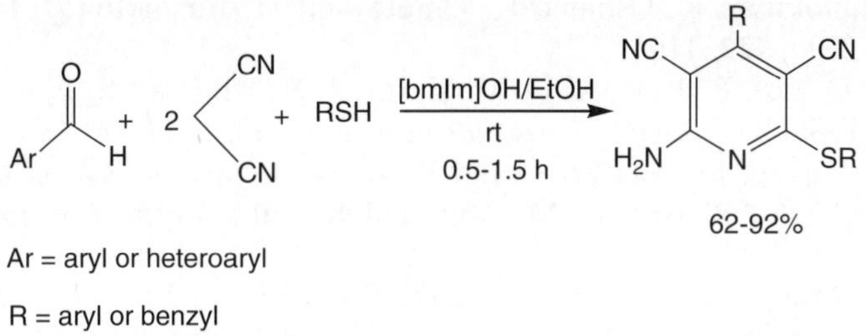

Ar = aryl or heteroaryl

R = aryl or benzyl

[bmIm]OH/EtOH, rt, 0.5-1.5 h

62-92%

Scheme 5.61

reaction does not lead to any side products and does not involve any hazardous organic solvents and toxic catalyst. The ionic liquid is recovered and recycled for subsequent reactions.

References

1. (a) S. Narayan, J. Muldoon, M. G. Finn, V. V. Fokin, H. C. Kolb and K. B. Sharpless, *Angew. Chem. Int. Ed*, 2005, **44**, 3275; (b) C.-J. Li, *Chem. Rev.*, 2005, **105**, 3095.
2. (a) A. Ben-Naim, *Hydrophobic Interaction,* Plenum Press, New York, 1980; (b) C. Tanford, *The Hydrophobic Effect*, 2nd ed., John Wiley & Sons, Inc, New York, 1980.
3. V. K. Aggarwal, D. K. Dean, A. Mereu and R. Williams, *J. Org. Chem.*, 2002, **67**, 510.
4. U. M. Lindstrom, *Chem. Rev.*, 2002, **102**, 2751.
5. (a) B. C. Ranu, *Eur. J. Org. Chem.*, 2000, 2347; (b) J. Auge, N. Lubin-Germain and J. Uziel, *Synthesis*, 2007, 1739.
6. (a) T.-H. Chan and M.-C. Lee, *J. Org. Chem.*, 1995, **60**, 4228; (b) T.-H. Chan and C.-J. Li, *Chem. Commun.*, 1992, 747.
7. X.-H. Yi, Y. Meng, X.-G. Hua and C.-J. Li, *J. Org. Chem.*, 1998, **63**, 7472.
8. (a) S. Kobayashi, *Synlett*, 1994, 689; (b) S. Kobayashi and I. Hachiya, *J. Org. Chem.*, 1994, **59**, 3590.
9. S. Guizzetti, M. Benaglia, L. Raimondi and G. Celentano, *Org. Lett.*, 2007, **9**, 1247 and references therein.
10. N. Mase, Y. Nakai, N. Ohara, H. Yoda, K. Takabe, F. Tanaka and C. F. Barbas III, *J. Am. Chem. Soc.*, 2006, **128**, 734.
11. N. Mase, K. Watanabe, H. Yoda, K. Takabe, F. Tanaka and C. F. Barbas III, *J. Am. Chem. Soc.*, 2006, **128**, 4966.
12. (a) D. C. Rideout and R. Breslow, *J. Am. Chem. Soc.*, 1980, **102**, 7816; (b) R. Breslow and U. Maitra, *Tetrahedron Lett.*, 1984, **25**, 1239.
13. L. A. van Royen, R. Mijngheer and P. J. De Clercq, *Tetrahedron*, 1985, **41**, 4667.
14. (a) H. Yorimitsu, K. Wakabayashi, H. Shinokubo and K. Oshima, *Tetrahedron Lett.*, 1999, **40**, 519; (b) H. Yorimitsu, T. Nakamura,

H. Shinokubo, K. Oshima, K. Omoto and H. Fujimoto, *J. Am. Chem. Soc.*, 2000, **122**, 11041.

15. (a) S.-O. Lee, R. Raja, K. D. M. Harris, J. M. Thomas, B. F. G. Johnson and G. Sankar, *Angew. Chem. Int. Ed.*, 2003, **42**, 1520; (b) C. D. Brooks, L.-C. Huang, M. McCarron and R. A. W. Johnstone, *Chem. Commun.*, 1999, 37; (c) R. Noyori, M. Aoki and K. Sato, *Chem. Commun.*, 2003, 1977.

16. B. C. Ranu, S. Bhadra and L. Adak, *Tetrahedron Lett.*, 2008, **49**, 2588.

17. Y. Usui, K. Sato and M. Tanaka, *Angew. Chem. Int. Ed.*, 2003, **42**, 5623.

18. A. Lubineau, J. Auge, N. Bellanger and S. Caillebourdin, *Tetrahedron Lett.*, 1990, **31**, 4147.

19. (a) G. Cardillo and C. Tomasini, *Chem. Soc. Rev.*, 1996, **25**, 117; (b) E. Hagiwara, A. Fujii and M. Sodeoka, *J. Am. Chem. Soc.*, 1998, **120**, 2474.

20. B. C. Ranu and S. Banerjee, *Tetrahedron Lett.*, 2007, **48**, 141.

21. G. L. Khatik, R. Kumar and A. K. Chakraborti, *Org. Lett.*, 2006, **8**, 2433.

22. B. C. Ranu and T. Mandal, *Synlett*, 2007, 925.

23. (a) Y. Sasson, Formation of carbon-halogen bond (Cl, Br, I) in *The Chemistry of Functional groups Supplement D2: The Chemistry of Halides, Pseudo Halides and Azides, Part 2*, S. Patai and Z. Rappoport, eds., John Wiley & Sons, Ltd., Chichester, 1995, pp. 535-620; (b) F. Diederich and P. J. Stang, *Metal-Catalysed Cross-Coupling Reactions,* Wiley-VCH Verlag GmbH, Weinheim, 1998; (c) X. Yang, A. Althammer and P. Knochel, *Org. Lett.*, 2004, **6**, 1665.

24. M. Sovak, *Radiocontrast Agents: Handbook of Experimental Pharmacology,* Springer, Berlin, 1993.

25. M. Jereb, M. Zupan and S. Stavber, *Chem. Commun.*, 2004, 2614.

26. (a) T.-P. Loh, S.B.K.W. Liung, K.-L. Tan and L.-L. Wei, *Tetrahedron*, 2000, **56**, 3227; (b) K. Manabe and S. Kobayashi, *Org. Lett.*, 1999, **1**, 1965.

27. T. Akiyama, J. Takaya and H. Kagoshima, *Tetrahedron Lett.*, 2001, **42**, 4025.

28. (a) E. J. Noga, G. T. Barthalmus and M. K. Mitchell, *Cell Biol. Int. Rep.*, 1986, **10**, 239; (b) P.N. Craig, In *Comprehensive Medicinal Chemistry, C.J. Drayton, (Ed),* Pergamon Press: New York, 1991, **vol. 8**.

29. (a) Y. Ju and R. S. Varma, *J. Org. Chem.*, 2006, **71**, 135; (b) Y. Ju and R. S. Varma, *Org. Lett.*, 2005, **7**, 2409.

30. Y. Wang, D. Dong, Y. Yang, J. Huang, Y. Ouyang and Q. Liu, *Tetrahedron*, 2007, **63**, 2724.

31. (a) B. M. Trost and D. L. Van Vranken, *Chem. Rev.*, 1996, **96**, 395; (b) J. Tsuji, *Transition Metal Reagents and Catalysts,* John Wiley & Sons, Inc, New York, 2000.

32. B. C. Ranu, K. Chattopadhyay and L. Adak, *Org. Lett.*, 2007, **9**, 4595.

33. S. U. Son, S. I. Lee, Y. K. Chung, S.-W. Kim and T. Hyeon, *Org. Lett.*, 2002, **4**, 277.

34. Y. Yokoyama, H. Hikawa, M. Mitsuhashi, A. Uyama and Y. Murakami, *Tetrahedron Lett.*, 1999, **40**, 7803.

35. H. Fujiwara, Y. Ogasawara, K. Yamaguchi and N. Mizuno, *Angew. Chem. Int. Ed.*, 2007, **46**, 5202.

36. (a) N. Zein, A. M. Sinha, W. J. McGahren and G. A. Ellestad, *Science*, 1988, **240**, 1198; (b) S. L. Iverson and J. P. Uetrecht, *Chem .Res. Toxicol.*, 2001, **14**, 175.

37. B. C. Ranu and K. Chattopadhyay, *Org. Lett.*, 2007, **9**, 2409.

38. B. C. Ranu, R. Dey and K. Chattopadhyay, *Tetrahedron Lett.*, 2008, **49**, 3430.

39. (a) E. Navarro, S. J. Alonso, J. Trujillo, E. Jorga and C. Perez, *J. Nat. Prod.*, 2001, **64**, 134; (b) G. A. Kraus and I. Kim, *Org. Lett.*, 2003, **5**, 1191; (c) H. Lu and G.-T. Liu, *Planta. Med*, 1992, **58**, 311.

40. M. Carril, R. SanMartin, I. Tellitu and E. Dominguez, *Org. Lett.*, 2006, **8**, 1467.

41. B.M. Trost and I. Fleming, In *Comprehensive Organic Synthesis,* Vol. 8, p. 417.

42. J. B. Peri, *J. Phys. Chem.*, 1965, **69**, 220.

43. M. Avalos, R. Babiano, J. L. Bravo, P. Cintas, J. L. Jimenez, J. C. Palacios and B. C. Ranu, *Tetrahedron Lett.*, 1998, **39**, 2013.

44. C. P. Wilgus, S. Downing, E. Molitor, S. Bains, R. M. Pagni and G. W. Kabalka, *Tetrahedron Lett.*, 1995, **36**, 3469.

45. (a) B. C. Ranu, S. Bhar and D. C. Sarkar, *Tetrahedron Lett.*, 1990, **32**, 2811; (b) B. C. Ranu, S. K. Guchhait, K. Ghosh and A. Patra, *Green Chem.*, 2000, **2**, 5.

46. (a) K. Mori, *Synlett*, 1995, 1097; (b) L. A. Paquette and H. C. Tsui, *J. Org. Chem.*, 1996, **61**, 142; (c) S. F. Martin, J. B. White and R. Wagner, *J. Org. Chem.*, 1982, **47**, 3190; (d) H. Waldmann, M. Weigerding, C. Dreisbach and C. Wandrey, *Helv. Chim. Acta.*, 1994, **77**, 2111.

47. (a) A. Togni, *Angew. Chem. Int. Ed. Engl.*, 1996, **35**, 1475; (b) W. G. Jary and J. Baungartner, *Tetrahedron Asymmetry*, 1998, **9**, 2081; (c) A. M. Giroud-Godquin and P. M. Maitlis, *Angew. Chem. Int. Ed. Engl.*, 1991, **30**, 375.

48. B. C. Ranu, U. Jana and A. Majee, *Green Chem.*, 1999, **1**, 33.

49. G. W. Kabalka, L. Wang and R. M. Pagni, *Synlett*, 2001, 676.

50. V. V. Veselovsky, A. S. Gybin, A. V. Lozanova, A. M. Moiseenkov and W. A. Smit, *Tetrahedron Lett.*, 1988, **29**, 175.

51. H. Kotsuki and T. Shimanouchi, *Tetrahedron Lett.*, 1996, **37**, 1845.

52. (a) T. H. Kang, K. Matsumoto, Y. Murakami, H. Takayama, M. Kitajima, N. Aimi and H. Watanabe, *Eur. J. Pharmacol.*, 2002, **444**, 39; (b) J. Ma and S. M. Hecht, *Chem. Commun.*, 2004, 1190; (c) M. M. Khafagy, A.H.F.A. El-Wahas, F. A. Eid and A. M. El-Agrody, *Farmaco*, 2002, **57**, 715.

53. G. Shanthi, G. Subbulakshmi and P. T. Perumal, *Tetrahedron*, 2007, **63**, 2057.

54. R.H.F. Manske and M. Kulka, In *Organic Reactions, R. Adams, (Ed),* John Wiley & Sons, Inc, New York, 1953, **Vol. 7**, p. 59.

55. B. C. Ranu, A. Hajra and U. Jana, *Tetrahedron Lett.*, 2000, **41**, 531.

56. (a) T.-L. Ho, *Tactics of Organic Synthesis,* John Wiley & Sons, Inc, New York, 1994; (b) M.E. Jung, In *Comprehensive Organic Synthesis, B.M. Trost and I. Fleming, (Eds),* **Vol. 4**, Pergamon, Oxford, 1991, p.1.

57. T. Kawabata, M. Kato, T. Mizugaki, K. Ebitani and K. Kaneda, *Chem. Eur. J.*, 2005, **11**, 288.

58. J.-F. Roudier and A. Foucaud, *Tetrahedron Lett.*, 1984, **25**, 4375.

59. S. M. Cricsery, *Chem. Brit.*, 1985, 473.

60. E. J. Creyghton, S. D. Ganeshie, R. S. Downing and H. van Bekkum, *Chem. Commun.*, 1995, 1859.

61. E. L. Eliel and R. S. Ro, *J. Am. Chem. Soc.*, 1957, **79**, 5992.

62. L. E. Martinez, J. L. Leighton, D. H. Carsten and E. N. Jacobsen, *J. Am. Chem. Soc.*, 1995, **117**, 5897.

63. (a) G. L. Bihan, F. Rondu, A. P. Tounian, X. Wang, S. Lidy, E. Touboul, A. Lamouri, G. Dive, J. Huet, B. Pfeiffer, P. Renard, B. Guardiola-Lamaitre, D. Manechez, L. Penicaud, A. Ktorza and J.-J. Godfroid, *J. Med. Chem.*, 1999, **42**, 1587; (b) J. V. Greenhill and P. Lue, *Prog. Med. Chem.*, 1993, **30**, 203; (c) J. Wu, X. Sun and H.-G. Xia, *Tetrahedron Lett.*, 2006, **47**, 1509.

64. J. O. Metzger and R. Mahler, *Angew. Chem. Int. Ed. Engl.*, 1995, **34**, 902.

65. T.-P. Loh, J.-M. Huang, S.-H. Goh and J. J. Vittal, *Org. Lett.*, 2000, **2**, 1291.

66. N. T. Patil and Y. Yamamoto, *Tetrahedron Lett.*, 2004, **45**, 3101.

67. R.-J. Song, C.-L. Deng, Y.-X. Xie and J.-H. Li, *Tetrahedron Lett.*, 2007, **48**, 7845.

68. (a) C. O. Kappe, *Tetrahedron*, 1993, **49**, 6937; (b) C. G. Rovnyak, K. S. Atwal, A. Hedberg, S. D. Kimball, S. Moreland, J. Z. Gougoutas, B. C. O'Reilly, J. Schwartz and M. F. Maleay, *J. Med. Chem.*, 1992, **35**, 3254.

69. B. C. Ranu, A. Hajra and S. S. Dey, *Org. Proc. Res. Dev.*, 2002, **6**, 817.

70. (a) E. D. Caldas, M. Hosana Conceicua, M. C. C. Miranda, L. Souza and J. F. Lima, *J. Agric. Food Chem.*, 2001, **49**, 4521; (b) A. W. Erian and S. M. Sherif, *Tetrahedron*, 1999, **55**, 7957; (c) T. F. Wood and J. H. Gardner, *J. Am. Chem. Soc.*, 1941, **63**, 2741; (d) W. Chen-Hsien, *Synthesis*, 1981, 622; (e) M. R. Wood, D. J. Duncalf, S. P. Rannard and S. Perrier, *Org. Lett.*, 2006, **8**, 553.

71. N. Azizi, F. Aryanasab and M. R. Saidi, *Org. Lett.*, 2006, **8**, 5275.

72. (a) H.V. Patel, K.V. Vyas and P.S. Fernandes, *Indian J. Chem.* 1990, **29B**, 836 and references therein; (b) B. C. Ranu and A. Hajra, *Green Chem.*, 2002, **4**, 551.

73. N. Haga, H. Nakajima, H. Takayanagi and K. Tokumaru, *J. Org. Chem.*, 1988, **63**, 5372.

74. A. P. Marchand and G. M. Reddy, *Tetrahedron*, 1991, **47**, 6571.

75. M. Sakamoto, M. Takahashi, K. Kamiya, K. Yamaguchi, T. Fujita and S. Watanabe, *J. Am. Chem. Soc.*, 1996, **118**, 10664.

76. (a) T. Welton, *Chem. Rev.*, 1999, **99**, 2071; (b) P. Wasserscheid and W. Keim, *Angew. Chem. Int. Ed.*, 2000, **39**, 3772; (c) J. S. Wilkes, *Green*

Chem., 2002, **4**, 73; (d) V. I. Parvulescu and C. Hardacre, *Chem. Rev.*, 2007, **107**, 2615.

77. (a) J. A. Boon, J. A. Levinsky, J. I. Pflug and J. S. Wilkes, *J. Org. Chem.*, 1986, **51**, 480 ; (b) J. R. Harjani, S. J. Nara and M. M. Salunkhe, *Tetrahedron Lett.*, 2002, **43**, 1127; (c) V. V. Namboodiri and R. S. Varma, *Chem. Comm.*, 2002, 342; (d) W. Sun, C.–G. Xia and H.–W. Wang, *Tetrahedron Lett.*, 2003, **44**, 2409; (e) K. Qiao and C. Yakoyama, *Chem. Lett.*, 2004, **33**, 472.

78. M. J. Earle, S. P. Katdare and K. R. Seddon, *Org. Lett.*, 2004, **6**, 707.

79. B. C. Ranu and S. Banerjee, *Org. Lett.*, 2005, **7**, 3049.

80. Y. R. Jorapur and D. Y. Chi, *J. Org. Chem.*, 2005, **70**, 10774.

81. (a) Gen. Patent 109,933 (Chemische Fabrik von Heydon) 1900 Friedl, 1901, 5.; Gen Patent 116,386 (Chemische Fabrik von Heydon) 1900 Friedl, 1904, 6, 1160; (b) H. Schnell, *Chemistry and Physics of Polycarbonates,* Interscience Publishers, New York, 1964, **vol. 9**, p. 91; (c) J.-P.C. R. Senet, *Acad. Sci. Paris Ser. IIc*, 2000, **3**, 505; (d) K. Kondo, N. Sonoda and S. Tsutsumi, *Tetrahedron Lett.*, 1971, **12**, 4885.

82. B. C. Ranu and R. Jana, *Adv. Synth. Catal.*, 2005, **347**, 1811.

83. B. C. Ranu and R. Jana, *Eur. J. Org. Chem.*, 2006, 3767.

84. B. C. Ranu, S. Banerjee and A. Das, *Tetrahedron Lett.*, 2006, **47**, 881.

85. B. C. Ranu, A. Saha and S. Banerjee, *Eur. J. Org. Chem.*, 2008, 519.

86. B. C. Ranu, R. Jana and S. Sowmiah, *J. Org. Chem.*, 2007, **72**, 3152.

Simple Reactions for the Synthesis of Complex Molecules

JANET L. SCOTT[a] AND CHRISTOPHER R. STRAUSS[b]

[a] School of Chemistry, Monash University, Wellington Road, Clayton, Victoria 3800, Australia*; [b] QUILL Centre, The Queen's University of Belfast, Northern Ireland BT9 5AG, UK, and Strauss Consulting Ltd, Box 1065, Kunyung LPO, Mt Eliza, Victoria 3930, Australia

6.1 Introduction

This chapter is concerned with green reactions, products and methodological advances facilitating them. To those ends, our work has attempted to simplify processes by combining as many tasks as possible into unit operations. Aspects include selection of media, reactants, catalysts, conditions and product isolation. Examples are presented along with details of some specific classes of relatively complex molecules that may be prepared. These include novel macrocycles, ligands and phenol/formaldehyde oligomers.

6.2 A New Platform for Phenol/Formaldehyde Based Chemicals and Materials

6.2.1 Phenol/Formaldehyde Products

Products of coupling reactions between phenolic derivatives and formaldehyde have been used to produce resins **1** that have been used widely as bulk chemicals

---

*Current address: Unilever Research & Development Port Sunlight, Bebington, CH63 3JW, UK.

RSC Green Chemistry Book Series
Eco-Friendly Synthesis of Fine Chemicals
Edited by Roberto Ballini
© Royal Society of Chemistry 2009
Published by the Royal Society of Chemistry, www.rsc.org

in the cookware, plywood, particleboard, abrasives and adhesives industries.[1] Calixarenes **2**, cyclic oligomers produced as by-products in phenol/formaldehyde reactions, were recognized about 60 years ago.[2] They now constitute a field on their own, with applications including ion scavenging, chromatography, controlled release and enzyme mimicry.[3]

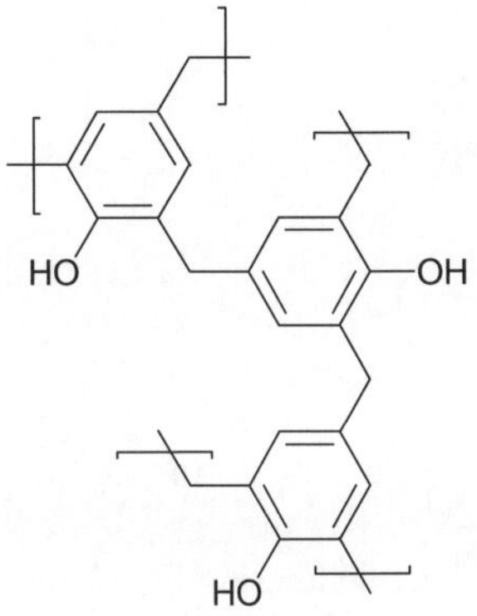

Traditional phenol/formaldehyde chemistry has depended upon empirical methods that limit the scope for ready introduction of structural diversity. Although random coupling of the aldehyde through the aryl 2-, 4- and 6-positions confers strength and rigidity to phenol/formaldehyde resins, it precludes the practical preparation of parent calixarenes from phenol directly. Instead, 4-substituted phenols, usually *p-tert*-butylphenol, are employed to generate the macrocyclic ring scaffolds. Calixarenes consisting of repeating methylene and phenolic units without para-substituents are normally obtained by dealkylation – a reaction often employing the undesirable reagent $AlCl_3$. In the parent macrocycles, chemo and regioselective derivatization of specific phenolic rings, among other identical moieties located within the same molecule, is not trivial. It is usually achieved by stepwise processes that can be difficult to control and that may afford complex mixtures. Isolation of the individual products in high yields can be problematic.

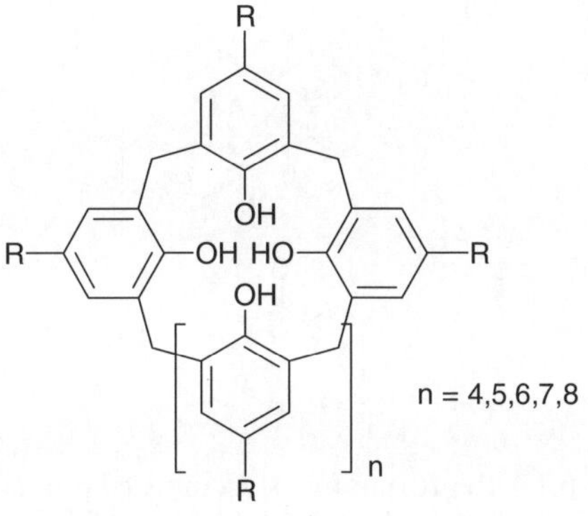

We have pursued an alternative array of somewhat related, novel molecules that could be more readily accessed and manipulated synthetically. The platform consists of structurally diverse oligomers, macrocycles and rod-like molecules that supplement Synthen and, in some cases, may eventually replace those available through phenol/formaldehyde chemistry. Consideration was given to the starting materials and reactions employed, a priority being the control of spatial aspects and conformation of the target molecules. Figure 6.1 shows representative compounds prepared by the strategies and methods outlined.

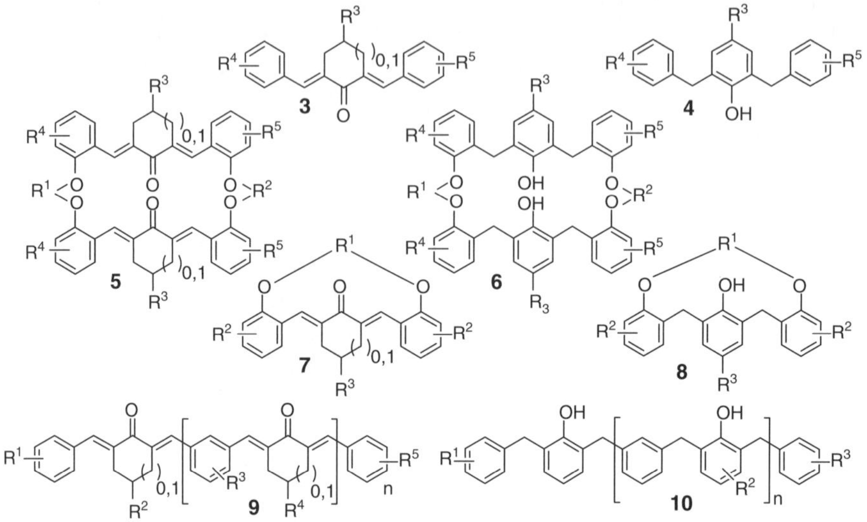

Figure 6.1 Representative structures of products accessible using the strategies outlined below include: bis-arylidinealkenones (**3**), 2,6-dibenzylphenols (**4**), "dienone" (**5** and **7**) and Horning-crown (**6** and **8**) macrocycles, rods (**9**) and flexible phenol/formaldehyde oligomers (**10**).

6.2.2 Building Block Approach

The synthetic methodology was based upon robust protocols and a hierarchical order of reactivity for various building blocks. Those aspects were exploited to produce rod-like molecules and macrocyclic rings, by sequential, or cascade, processes.[4] In one-pot reactions, successive steps were designed to occur in pre-determined sequences with predictable outcomes. First, a family of compounds was prepared through the *in situ* generation of dienone motifs **3**, **5**, **7** and **9**. Such molecules tended to have relatively rigid scaffolds. Formation of a dienone moiety usually was the key process that facilitated either chain extension or macrocyclic ring closure.[5] Dienone entities so formed were readily amenable to simple chemical transformations and functional group interconversions. Importantly, phenolic rings could be unmasked to afford entirely new families of molecules with increased conformational flexibility, **4**, **6**, **8** and **10**. This family of methylene-linked phenolic macrocyclic compounds was named Horning-crowns.[6]

The approach utilized molecular building blocks, each of which had a specific, directional connecting point. Plumbing terminology seems appropriate to illustrate the concept. Plumbers require rigid or flexible tubing, blank-off terminals, as well as connectors that may comprise rigid unions, flexible connectors, elbows, T-pieces and the like. A plumbing kit contains male (externally threaded) or female (internally threaded) forms of such items, along with flexible and rigid tubing. Accordingly, the plumber can exercise a high degree of control over the nature of connections and the shape, configuration and dimensions of the ultimate construction. Our synthetic approach attempted to emulate such control and outcomes at a molecular level.

6.2.3 Benzaldehyde-Cyclohexanone Chemistry as a Basis

Aryl aldehyde and α-methylene groups of cyclic ketones (although acyclic ketones also could serve in some cases) were the functionalities selected for making connections. They were arbitrarily assigned as female and male, respectively. Claisen–Schmidt condensation was the preferred reaction vehicle for connecting the building blocks (or the wrench, if one wished to over-do the plumbing analogy!). Relevant aspects of this regime, as exemplified by ben-zaldehyde (*i.e.*, female terminal)–cyclohexanone (*i.e.*, male–male rigid union) chemistry include the following:

- Under most conditions, 2,6-dibenzylidenecyclohexanone can be obtained exclusively, even when the concentration of starting ketone is in great excess.
- The exocyclic double bonds formed in 2,6-dibenzylidenecyclohexanone are exclusively trans toward the aryl ring and the carbonyl group of cyclohexanone, making the molecule essentially linear or rod-like and eliminating the risk of cis–trans isomerization in the condensation step.
- Rates for competing processes such as self-condensation of cyclohexanone and Cannizzaro or Tishchenko reactions of benzaldehyde are negligible compared with the Claisen–Schmidt condensation of aldehyde with ketone.
- If required, conditions can be varied to favor aldol addition over condensation.
- In contrast with Claisen–Schmidt condensation, aldol addition typically affords diastereoisomers of the mono-adduct, which may be dehydrated in a separate step to give (*E*)-2-benzylidenecyclohexanone.
- Condensations and additions are green reactions that can be carried out at moderate temperature (usually ambient) under acid or base catalysis.
- Aldol addition has a low activation energy, a potentially important consi-deration in dynamic covalent chemistry (reactions carried out reversibly, under conditions of equilibrium control).[7]
- (*E,E*)-2,6-Dibenzylidenecyclohexanones can undergo isoaromatization under a range of conditions to afford 2,6-dibenzylphenols.

A potential drawback (that we also found to be advantageous in appropriate circumstances) of Claisen–Schmidt condensation applied to ketones with two a-methylene groups is the proclivity for bis-condensation. "Male terminals" (retaining one active α-methylene group) are difficult to obtain. The need for building blocks including male terminals (*e.g.*, 2-benzylidenecyclohexanone) and flexible male–male connectors demanded methodology for their preparation in good yields and in one step. Literature methods regarded as unsatisfactory involved two-step addition–elimination processes or the Mukaiyama approach beginning from enol ether derivatives of the ketone.

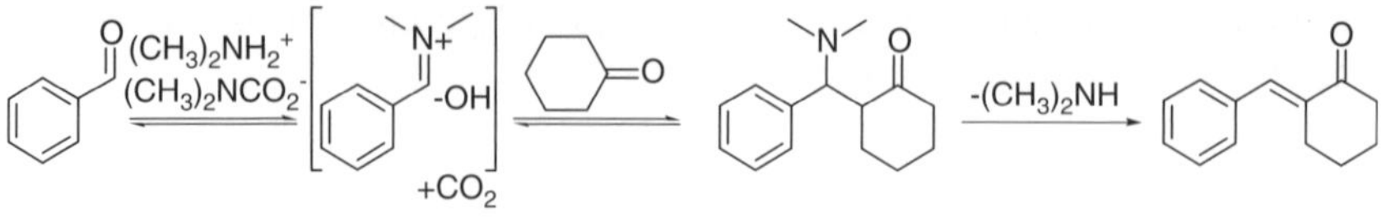

Scheme 6.1 Condensation of benzaldehyde and cyclohexanone in DIMCARB.

Mono-condensed α,β-unsaturated ketone products could be prepared directly in a liquid adduct of dimethylamine and carbon dioxide. These two gases react exothermically to yield a free-flowing liquid: a mixture of *N,N*-dimethylammonium, *N′,N′*-dimethylcarbamate and *N,N*-dimethylcarbamic acid (DIMCARB).[8] In contrast to its gaseous components, the liquid adduct is relatively stable up to 50 °C and has substantial ionic character.[9] Its polarity is comparable with that of ionic liquids,[10] which are interesting synthetic media owing to their typically negligible vapor pressure, solvent properties and usual low flammability. In contrast with ionic liquids, however, DIMCARB dissociates at 60 °C and the liberated CO_2 and Me_2NH gases can be re-associated and condensed, thereby facilitating recycling and reuse. Thus, in many respects DIMCARB could be regarded as a self-associated, "distillable" ionic liquid and several of its properties are conducive to green chemistry.

The pathway proposed for the reaction of aryl aldehydes and enolizable ketones in DIMCARB involved formation of a putative, transient iminium species that underwent Mannich reaction with the ketone. Subsequent elimination of dimethylamine from the Mannich adduct occurred slowly *in situ* to afford the 2-arylidene derivative (Scheme 6.1).[11]

6.2.4   Strategy and Building Blocks

The propensity for Claisen–Schmidt condensation to afford bis-adducts [*e.g.*, (*E,E*)-2,6-dibenzylidenecyclohexanone from benzaldehyde and cyclohexanone] virtually exclusively, even in the presence of excess ketone (as mentioned in Section 6.2.3), was the cornerstone of our strategy for the preparation of "symmetrical" adducts (*i.e.*, those with a two-fold axis through the C=O group as viewed in the plane of the page).[12] Such condensations occur sequentially, as mono-condensed products tend to be more reactive than the parent cycloalkanones under the reaction conditions, making them transient intermediates prepared for further reaction.

"Unsymmetrical" condensed products could be accessed by application of DIMCARB as both solvent and catalyst, yielding male terminals or flexible connectors (see below) and providing access to a multitude of different building blocks.

Each building block was designed to serve uniquely, as a terminal (male or female), as a rigid union (either male–male or female–female) or as a flexible connector (either male–male or female–female). Specifications and examples of each type of building block are presented for ease of explanation and illustrated in Figure 6.2: (a) female terminal – a non-enolizable aryl monoaldehyde, *e.g.*, benzaldehyde; (b) rigid female–female union – a conformationally rigid non-enolizable aryl dialdehyde, *e.g.*, benzene-1,3-dicarboxaldehyde; (c) flexible female–female connector – a molecule containing two non-enolizable aryl monoaldehyde moieties flexibly linked by a chain; (d) male terminal – a cyclic ketone with only one enolizable active methylene group, *e.g.*, 2-benzylidene-cyclohexanone; (e) rigid male–male union – a cyclic ketone having two eno-lizable methylene groups within the one ring system, *e.g.*, cyclohexanone, or alternatively, two rigidly linked cyclic ketones, each of which has only one active methylene group; and (f) flexible male–male connector – two flexibly linked cyclic ketones, each of which has only one active methylene group.

With specific functionality built-in, each building block qualified as only one type of fitting. Vast combinatorial diversity, however, was made possible by an extensive range of building blocks that could satisfy each type. For example, any aryl, fused aryl or heteroaryl monoaldehyde could be utilized as a female terminal and any cyclic ketone with active α,α'-methylene groups could con-stitute a rigid male–male union.

Building blocks for beginning or terminating sequences for open chain or rod-like molecules had only one aldehydic carbonyl group (female terminal) or only one active methylene group (male terminal). The construction of open chains or rods (but not macrocycles) could be started from either, provided that the selected building block was reacted with another possessing complementary functionality. Short chains or rods could be produced by connecting male and female terminals together in one Claisen–Schmidt condensation.

To enable two connections, *i.e.*, towards elaboration of oligomeric chains or rods or to participate in ring-forming processes, the building blocks possessed two non-enolizable aldehydic carbonyl functions (rigid female–female unions or flexible female–female connectors), or two potentially active methylene groups (rigid male–male unions or flexible male–male connectors); Figure 6.3 provides some examples.

Obviously, self-condensation, which could afford undesired dimerization, oligomerization or polymerization products, would have had the potential to short-circuit processes. Such circumstances were avoided by ensuring that no individual building block had an aldehydic as well as an active methylene function (*i.e.*, no individual union or connector possessed one male and one female fitting). Molecules designed to serve as elbows, T, X and H-pieces have also been produced, but will not be discussed herein. The geometry of the connections was determined by substitution patterns on aryl rings. Meta-sub-stituted aryl dialdehydes (rigid female–female unions) afforded linear connec-tions, after condensation. As the examples in Figure 6.2 suggest, not only the starting dialdehydes and cyclic ketones, but also some products of Claisen–Schmidt condensation could serve as unions. This versatility enhanced the range of building blocks available for preparing structurally diverse end pro-ducts and a small number are illustrated in Figure 6.3. Shape-selectivity, giving rise to elbows instead of unions, was obtained through the use of *o* or *p*-alde-hydic groups attached to monoaryl or fused aryl rings. Heteroatoms and

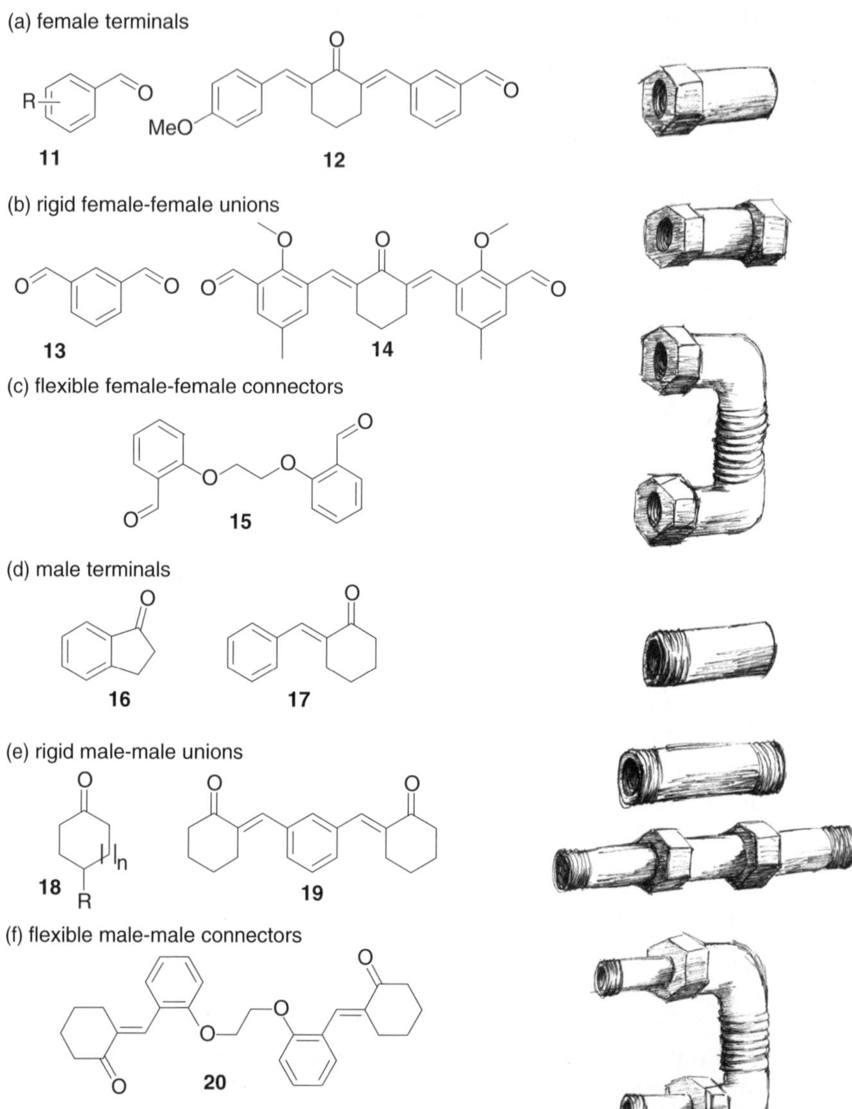

Figure 6.2 Examples of rigid and flexible terminals and unions: (a) female terminals *e.g.*, benzaldehyde (**11**) and 3-{(*E*)-[(*E*)-3-(4-methoxybenzylidene)-2-oxo-cyclohexylidene]methyl}benzaldehyde (**12**); (b) rigid female–female unions *e.g.*, benzene-1,3-dicarboxaldehyde (**13**) and 3,3′-(1*E*,1′*E*)-(2-oxo-cyclohexane-1,3-diylidene)bis(methan-1-yl-1-ylidene)bis(2-methoxy-5-methyl-benzaldehyde) (**14**); (c) flexible female–female connector *e.g.*, 2,2′-[ethane-1,2-diylbis(oxy)]dibenzaldehyde (**15**); (d) male terminals *e.g.*, 1-indanone (**16**) or (*E*)-2-benzylidenecyclohexanone (**17**); (e) rigid male–male unions *e.g.*, variously substituted cyclohexanones or cyclopentanones **18** and (2*E*,2′*E*)-2,2′-[1,3-phenylenebis(methan-1-yl-1-ylidene)]dicyclohexanone (**19**) (prepared by reaction of **13** with cyclohexanone in DIMCARB); and f) flexible male–male connector *e.g.*, (2*E*,2′*E*)-2,2′-[1,3-phenylenebis(methan-1-yl-1-ylidene)]dicyclohexanone (**20**) (prepared from **15** by reaction with cyclohexanone in DIMCARB).

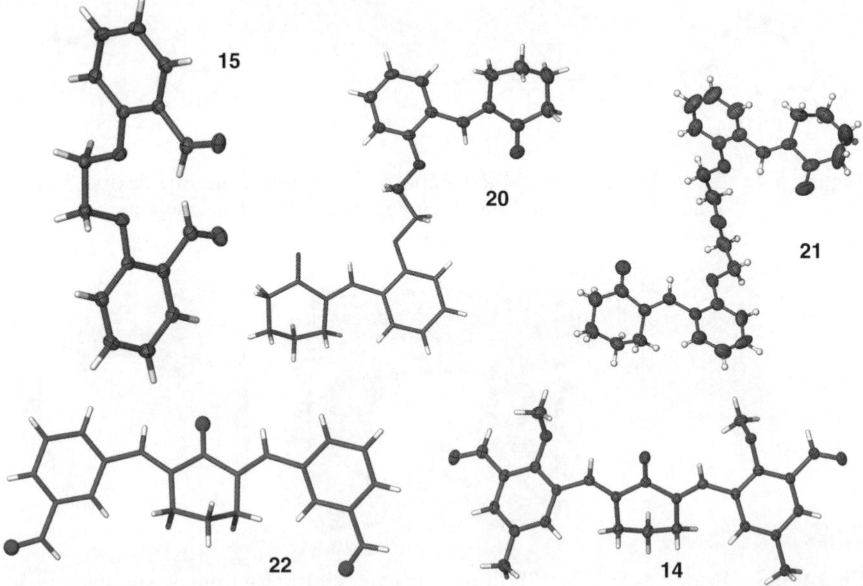

Figure 6.3   Molecular diagrams, derived from crystal structures, of examples of: a female–female flexible connector (**15**), two male–male flexible connectors (**20**) and (2*E*,2′*E*)-2,2′-{2,2′-[2,2′-oxybis(ethane-2,1-diyl)bis(oxy)]bis(2,1-phenylene)}bis(methan-1-yl-1-ylidene)dicyclohexanone (**21**), and rigid female–female unions 3,3′-(1*E*,1′*E*)-(2-oxocyclohexane-1,3-diylidene)-bis(methan-1-yl-1-ylidene)dibenzaldehyde **22** and **14**.

polyaryl rings also could be incorporated into building blocks. Such aspects of combinatorial design were considered important for the preparation of molecules with potential applications including non-linear optical materials and as connectors or spacers for the construction of larger molecules.[13]

6.2.5   Macrocyclic Dienone [1 + 1] Products

Flexible female–female connectors, having salicylaldehyde moieties as the terminal units and linked by etherification through the phenolic oxygen atoms, were cyclized *via* successive Claisen–Schmidt condensations with rigid male–male unions, *e.g.*, cyclohexanone or its derivatives.[5] If chains linking the aryl rings had ≥ 7 atoms, joined by single bonds, intramolecular cyclization afforded rings derived from one flexible female–female connector and one cyclohexanone (rigid *s*male–male union) – the so-called "[1 + 1]" products (Figure 6.4). Spatially, these macrocycles could be regarded as semi-circular or semi-elliptical, with the rigid, linear dibenzylidene cyclohexanone moiety making up the diameter and the flexible linker constituting the arc. Flexible linkers may consist of hydrocarbon chains, **23**, or polyether chains, **24**, amongst many other possibilities.

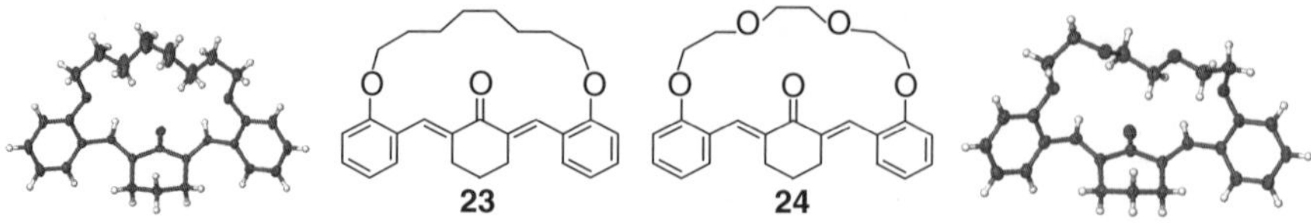

Figure 6.4   Examples of macrocyclic compounds prepared from one flexible female–female connector and one cyclohexanone rigid male–male union.

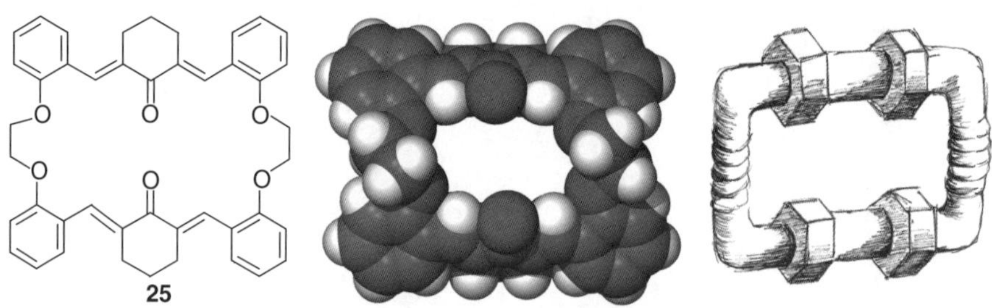

Figure 6.5   Example of a "[2 + 2]" dienone macrocycle formed from two flexible female–female connectors (**15**) and two rigid male–male unions (cyclohexanone).

6.2.6   Macrocyclic Dienone [2 + 2] Products

Conversely, if the flexible female–female connectors had chains containing <7 atoms joined by single bonds, ring closure afforded macrocycles derived from two cycloalkanone (male–male rigid union) moieties and two such female–female connectors, such as the example presented in Figure 6.5. When two pairs of identical male–male rigid unions and female–female flexible connectors were used, the macrocycle could be depicted two-dimensionally as rectangular or parallelo-gram-shaped, with the rod-like dibenzylidenecyclohexanone moieties making up one pair of parallel sides and the flexible linkers, the other pair. These were designated "[2 + 2]" dienone macrocycles in our shorthand. Each of these mac-rocycles had two axes of symmetry and was prepared from simple building blocks. By design, the single step ring-forming process involved four successive Claisen–Schmidt condensations, the first three of which were intermolecular and the fourth intramolecular. The original salicyl moiety became a 90° corner piece in the resultant macrocycle, as illustrated in **25**. Significantly, for organic molecules the available bond angles tend to be restricted to approximately 109°, 120° and 180°. This limits the scope for square and rectangular shapes and, even though a 90° angle can be obtained relatively readily through the use of metals in organome-tallic systems, efficient formation of rectangular molecules still poses a challenge.[14]

6.2.7   Trapezoidal Dienone Macrocycles

Macrocyclic products derived from two identical rigid male–male unions and two non-identical flexible female–female unions having linkers of differing

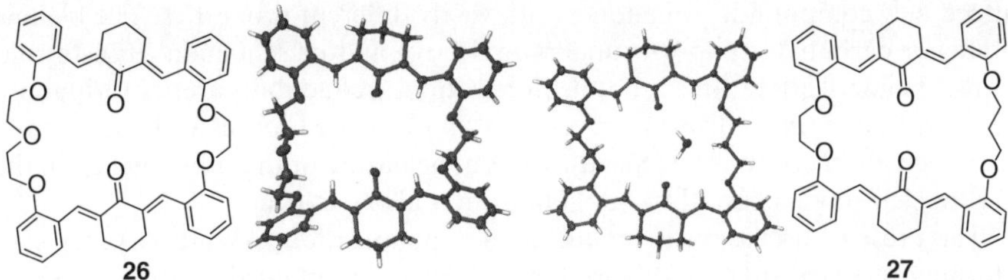

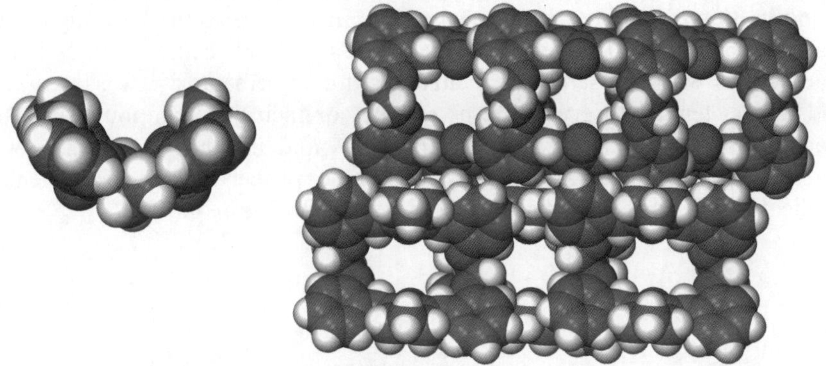

Figure 6.6 Trapezoidal macrocycles prepared by Claisen–Schmidt condensation of one flexible female–female connector with a flexible male–male connector. (Macrocycle **27** crystallizes as a hydrate with a hydrogen-bonded bridging water molecule between the phenolic moieties.)

Figure 6.7 Dienone [2 + 2] macrocycles fold at their flexible "hinges" and, in some cases, pack together to form solid-state structures containing channels filled with labile solvent molecules.

lengths but <7 atoms appeared, two dimensionally, as trapezoidal (Figure 6.6). They were prepared in one step by reaction of a flexible female–female connector (prepared using DIMCARB) with a flexible male–male connector having a tether of a different length.

When viewed three dimensionally, however, the flexible sides of all dienone macrocycles allowed some degree of folding. Depending upon the nature of the linking unit employed, the rings could assume shapes somewhat reminiscent of tortilla shells. Some symmetrical macrocycles had clefts, which allowed molecular stacking, as well as cavities that could be aligned (Figure 6.7). Their potential applications include membranes.

6.2.8 Horning-crown Macrocycles

A major advantage of the macrocyclic dienones obtained from building blocks described herein is the extensive potential for facile transformations that can afford new compounds, sometimes with vastly different properties. The olefinic bonds or carbonyl group can undergo selective and/or sequential transformations. Isomerization and photoisomerization, reduction, imine formation, reductive amination, disproportionation and dimerization can all be effected with high atom economy. Depending upon the nature of the substituents on the aryl nuclei, ring closure through ketalization also can occur.

The present discussion is limited to isoaromatization. About 60 years ago, Horning reported that 2,6-dibenzylphenol could be obtained by thermolysis of (2*E*,6*E*)-dibenzylidenecyclohexanone in the presence of a Pd catalyst. He coined the word "isoaromatization" to describe the process. Since then, others have attempted the same conversion, with variable degrees of success, under homogeneous catalysis[15] and with acid in the absence of metal.[16]

Isoaromatization of dienone macrocycles afforded Horning-crown macrocycles[6] – flexible macrocycles bearing structural elements reminiscent of those found in both calixarenes and crown ethers. In some cases the Horning-crown macrocycles exhibited solvent-dependent and switchable conformations.[17] For macrocycles with the same short linker, self-complementarity was observed, and dimers tended to crystallize as solvates or inclusion compounds.[18] This tendency was suppressed with longer linkers and in some Horning-crowns derived from trapezoidal macrocycles. These properties suggest potential applications in analysis, separation and detection (Figure 6.8).[17,18]

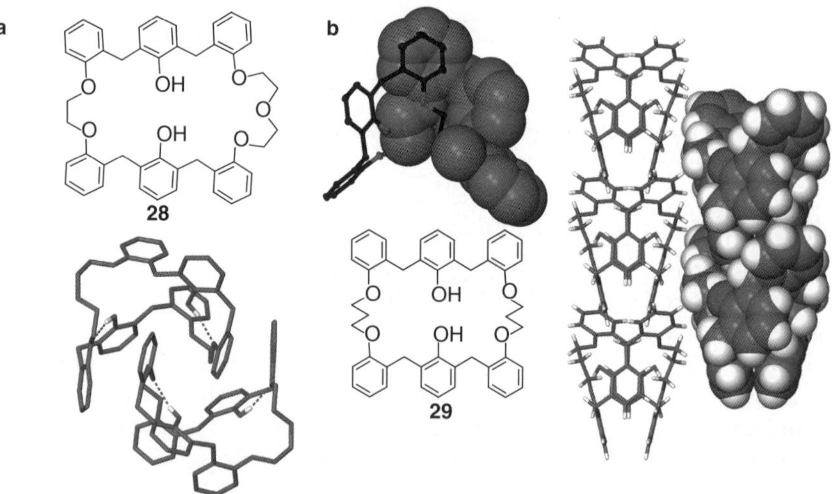

Figure 6.8 (a) Trapezoidal Horning-crown macrocycles with flexible tethers such as **28** may form dimeric globular structures, stabilized by intramolecular hydrogen bonds, which tend to pack with solvent molecules occluded in the crystals. (b) Longer tethers (symmetrical in **29**) allow the molecule to fold in on itself, with phenol to ether side-chain hydrogen bonds, and intramolecular $\pi\cdots\pi$ stacking, as well as either CH_2 to π or C–H$\cdots\pi$ interactions. These twisted conformers stack to form columns of macrocycles.

6.2.9 Rod-like Molecules and Phenol/Formaldehyde Oligomers

Rigid, rod-like molecules have attracted considerable recent interest, particularly in nanoscience and nanotechnology.[19] Here, the use of building blocks consisting of male and female terminals and rigid male–male and female–female unions facilitated the construction of zigzag rods. The products were characterized by a series of alternating cyclic ketonyl and cycloaryl moieties, any of which could possess heteroatoms (Figure 6.9). Rings were connected by methine groups that made up exocyclic trans double bonds at the α and α' carbons of the cyclic ketone units. Examples **30**, **31** and **33** may, if completely untwisted, possess a plane of symmetry, but no axis of symmetry, while **32** possesses a plane and an axis of symmetry, being prepared by classical Claisen–Schmidt condensation of rigid female union **13**, with a pair of rigid male terminals [in this case (*E*)-2-(4-methoxybenzylidene)cyclohexanone]. Through the flexible synthetic strategy described, all ring systems consisting of any one particular rod could be designed to differ from one another, if so desired, as illustrated in examples **30** and **31**. Rods with up to nine-ring systems (all or some of which could be unique if so desired) were prepared in a controlled and ordered fashion, linked by eight exocyclic methines, without the use of protective groups.

Isoaromatization of rods containing cyclohexyl residues led to the preparation of phenol/formaldehyde oligomers, such as examples **34** and **35**, thus illustrating the possibilities of preparing "designer" phenol/formaldehyde oligomers for modulating the properties of such resins.

6.3 A New Cascade Reaction for Aniline Derivatives

The foregoing discussion has focused upon condensations of non-enolizable aldehydes and ketones to afford green processes and products. That approach was extended to include a one step multicomponent preparation of N-mono-substituted or N,N-disubstituted aniline derivatives (Scheme 6.2). The

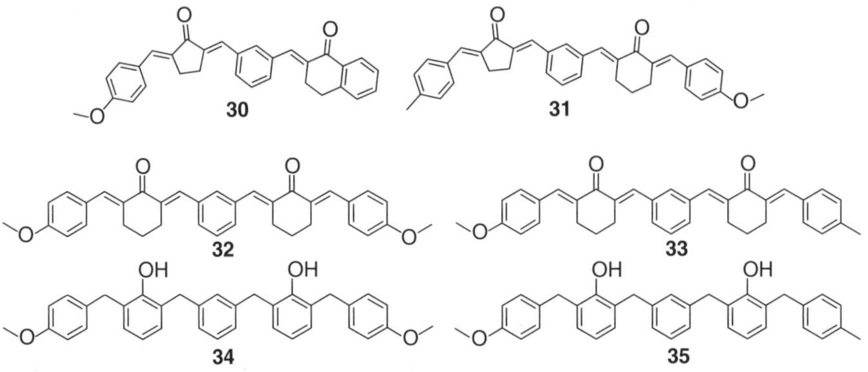

Figure 6.9 Rod-like molecules **30–32** and phenol/formaldehyde oligomers **34** and **35**.

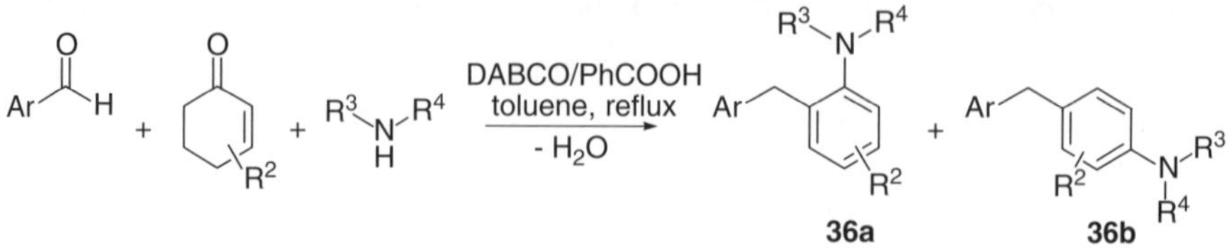

Scheme 6.2 A multicomponent reaction for the preparation of aniline derivatives. Primary amines tend to yield 2-substituted anilines **36a**, while sterically demanding secondary amines form predominantly 4-substituted anilines, **36b**.

reactants are an aldehyde without an enolizable carbonyl function, cyclohex-2-enone (or a derivative thereof) and an amine, which may be primary or secondary.[20] The simplest cases involve monoaldehydes and monoamines. With a dialdehyde or diamine, a second process analogous to the first can occur, so great complexity can be introduced in one step. The mechanistic pathway was investigated and established. Despite a large number of potentially competing equilibria, aromatization appeared to be the only non-reversible reactive step. Thus, the thermodynamically most stable product could slowly accumulate. In other work, a final irreversible aromatization was similarly exploited for the synthesis of phenylenediamines.[21]

The outcome of the present multicomponent process, regarding the cyclohexene ring, entailed removal of the keto function and replacement of it with two single bonds, one of which was attached to the carbon atom of a methylene group and the other to a nitrogen atom, followed by aromatization of the ring. The carbonyl oxygen atom of the aryl aldehyde was lost, with that function becoming transformed into a methylene group, while the amine starting material became N-arylated by what was the cyclohexene ring and at the position previously occupied by the keto group. Accordingly, in one cascade, several functional group interconversions and transformations were generated. This process offers opportunities for the introduction of combinatorial diversity into the subject molecules, while satisfying many of the principles and objectives of green chemistry.

In our hands, the regioselectivity was usually high, but it could become variable without careful attention to the conditions. Catalysts and rates of addition of enones influenced product distributions. Regioselectivity, though, was dictated mainly by the relative bulk of the participating amine, with primary and secondary amines mainly affording 2- and 4-substituted anilines, respectively.

With derivatives of benzaldehyde, the reaction proceeded regardless of electron-donating or electron-withdrawing properties of the substituents. A diverse range of aldehydes was employed, extending to heterocyclic compounds such as furfural and aryl dialdehydes. Benzene-1,3-dicarboxaldehyde, with cyclohexenone and benzylamine, reacted to produce a bis-anilino derivative, 2,2′-[1,3-phenylenebis(methylene)bis(*N*-benzylaniline)]. The yield (65%) was

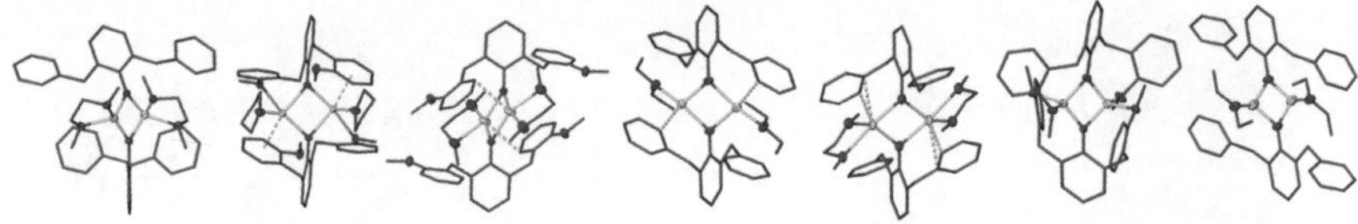

Figure 6.10 Group I and II metal phenolates with 2,6-dibenzylphenolate ligands. Oxygenated solvents are also co-ordinated to the metal centers.

respectable in light of the potential for competing processes and the number of transformations involved in the assembly of a somewhat complex molecule in one step.

The new reaction could be performed simply, under relatively uniform conditions. It was quite convergent and predictable, regardless of the diverse array of starting materials that could be employed. Participating amines, for example, included primary examples, *e.g.*, benzylamine, aniline, 2-amino-methylpyridine, methyl aminoacetate and tryptamine, or secondary, *e.g.*, di-(2-methoxyethyl)amine, morpholine, amphetamine and di-*n*-butylamine. As mentioned above, cascade and/or multicomponent reactions can be useful tools for green chemistry, if they proceed in good yields with high atom economy and generate little waste.

This multicomponent reaction was used to construct much larger ligand structures, including "scorpion" ligands,[22] and recent advances extend to the preparation of ligand structures from seven components in a single step!

6.4 Properties of Some of the Above Products

Various of the classes of molecules described above are amenable to metal ion complexation. 2,6-Dibenzylphenols have been employed in the preparation of group I and II metal phenolates (Figure 6.10).[23–25] as well as lanthanoid aryloxides.[26]

The cascade reaction for the preparation of anilines is particularly well suited to the synthesis of potential ligands, as is illustrated by the "scorpion ligand" *N,N*-4-tris(pyridin-2-ylmethyl)aniline (**37**), which forms a range of Cu complexes with widely varying solid-state structures.[22] Clearly, from the metal co-ordination geometries illustrated in Figure 6.9, Horning-crown macrocycles may fulfil similar roles.

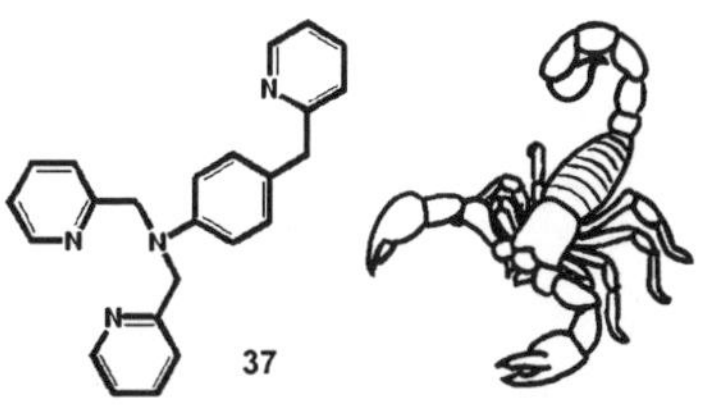

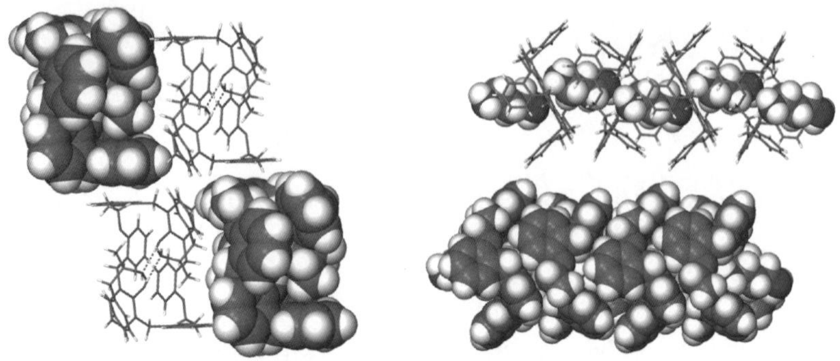

Figure 6.11   Horning-crown macrocycle (derived from isoaromatization of **25**) inclusion complexes with α-methylphenethylamine and propylenediamine contain ball-like structures (left) and sheath/core structures (right), respectively.

The ability to produce structurally varied products, by selection of building blocks, without modification to the synthetic protocols described is of particular utility in the preparation of libraries of ligands. This has facilitated investigation of the effect of subtle variation of ligand structure on the properties of the complexes produced.

Crystallization of the conformationally restrained (E,E,E',E')-[2 + 2] dienone macrocycles to form porous materials has been alluded to above, but light-mediated E/Z isomerization can result in globular structures, where the molecular cavity is collapsed. Such switchability in response to the environment is exhibited in a different manner by the Horning-crowns, which show greater or lesser degrees of self-association in different solvents.[17]

The conformational flexibility of the Horning-crowns leads to a range of supramolecular architectures, including varied, guest dependent packing of the dimers referred to above.[18] Guest molecules bearing hydrogen bond donor/acceptor groups, such as amines, disrupt the Horning-crown intramolecular hydrogen bonding, providing diverse solid-state structures, such as host–guest "balls" and insulated core/sheath structures (Figure 6.11).

These flexible macrocycles also formed host–guest inclusion complexes with buckminsterfullerenes, C_{60} and C_{70}, utilizing various π system interactions to encase the large, globular guest molecules in columnar or ball and socket configurations.[27]

The extended rod structures and flexible phenol/formaldehyde oligomers offer interesting opportunities in the development of nonlinear optical (NLO) materials and Bakelite modifiers, respectively.

6.5   Conclusion

The application of apparently "simple" reactions to yield molecular structures of relative complexity has been demonstrated. The protocols described give access to significant domains of "molecular space," some of which hitherto have been poorly explored. Direct approaches, avoiding protection/deprotection strategies and having great flexibility with respect to combination of building blocks, fulfill various of the strictures of green chemistry and allow rapid generation of molecular diversity.

References

1. For recent reviews, see: (*a*) A. Gardziella, L. A. Pilato, A. Knop, *Phenolic Resins: Chemistry, Applications, Standardization, Safety, and Ecology*, Springer, New York, 2000; (*b*) D. H. Solomon, G. G. Qiao, M. J. Caulfield in *Chemistry of Phenols* (Ed.: Z. Rappoport), Chichester, England, 2003, pp. 1455–1506.
2. For reviews, see: (*a*) V. Böhmer, *Angew. Chem.*, 1995, **107**, 785; V. Böhmer, *Angew. Chem. Int. Ed. Engl.*, 1995, **34**, 713; (*b*) A. Ikeda, S. Shinkai, *Chem. Rev.*, 1997, **97**, 1713; (*c*) C. D. Gutsche, *Calixarenes Revisited, Monographs in Supramolecular Chemistry* (Ed.: J. F. Stoddart), The Royal Society of Chemistry, Cambridge, England 1998; (*d*) *Calixarenes in Action*, (Eds.: L. Mandolini, R. Ungaro), Imperial College Press, London, 2000; (*e*) *Calixarenes 2001*, (Eds.: Z. Asfari, V. Böhmer, J. M. Harrowfield, J. Vicens, M. Saadioui), Kluwer, Dordrecht, 2001.
3. Many excellent examples have been published. Selected recent examples include: (*a*) L. Baldini, A. Casnati, F. Sansone, R. Ungaro, *Chem. Soc. Rev.*, 2007, **36**, 254; (*b*) A. J. Petrella, C. L. Raston, *J. Organomet. Chem.*, 2004, **689**, 4125; (*c*) A. Wei, *Chem. Commun.*, 2006, 1581; S. J. Dalgarno, J. L. Atwood, C. L. Raston, *Chem. Commun.*, 2006, 4567; (*d*) R. Meyer, T. Jira, *Curr. Anal. Chem.*, 2007, **3**, 161; (*e*) E. da Silva, A. N. Lazar, A. W. Coleman, *J. Drug Delivery Sci. Tech.*; 2004, **14**, 3; (*f*) C. D. Gutsche, *Acc. Chem. Res.*, 1983, 16, 161; (g) F. Perret, A. N. Lazar, A. W. Coleman, *Chem. Commun.*, 2006, 2425.
4. M. A. Giarrusso, L. T. Higham, U. P. Kreher, R. S. Mohan, A. E. Rosamilia, J. L. Scott, C. R. Strauss, *Green Chem.*, 2008, **10**, 842.
5. L. T. Higham, U. P. Kreher, C. L. Raston, J. L. Scott and C. R. Strauss, *Org. Lett.*, 2004, **6**, 3257.
6. L. T. Higham, U. P. Kreher, C. L. Raston, J. L. Scott and C. R. Strauss, *Org. Lett.*, 2004, **6**, 3261.
7. S. J. Rowan, S. J. Cantrill, G. R. L. Cousins, J. K. M. Sanders and J. F. Stoddart, *Angew. Chem. Int. Ed.*, 2002, **41**, 898.
8. U. P. Kreher, A. E. Rosamilia, C. L. Raston, J. L. Scott and C. R. Strauss, *Molecules*, 2004, **9**, 387.
9. J. Zhang, A. I. Bhatt, A. M. Bond, A. G. Wedd, J. L. Scott and C. R. Strauss, *Electrochem. Commun.*, 2005, **7**, 1285.
10. J. Zhang, A. I. Bhatt, A. M. Bond, D. R. MacFarlane, J. L. Scott, C. R. Strauss, P. I. Iotov, S. V. Kalcheva and R. Shaw, *Green Chem.*, 2006, **8**, 261.
11. U. P. Kreher, A. E. Rosamilia, C. L. Raston, J. L. Scott and C. R. Strauss, *Org. Lett.*, 2003, **5**, 3107.

12. C. E. Garland and E. E. Reid, *J. Am. Chem. Soc.*, 1925, **47**, 2333.
13. (a) P. F. H. Schwab, M. D. Levin and J. Michl, *Chem. Rev.*, 1999, **99**, 1863; (b) P. F. H. Schwab, J. R. Smith and J. Michl, *Chem. Rev.*, 2005, **105**, 1197.
14. C. J. Kuehl, T. Yamamoto, S. R. Seidel and P. J. Stang, *Org. Lett.*, 2002, **4**, 913.
15. (a) Z. Aizenshtat, M. Hausmann, Y. Pickholtz, D. Tal and J. Blum, *J. Org. Chem.*, 1977, **42**, 2386; (b) Y. Pickholtz, Y. Sasson and J. Blum, *Tetrahedron Lett.*, 1974, 1263.
16. (a) A. Jha, M. P. Padmanilayam and J. R. Dimmock, *Synthesis*, 2002, 463; (b) J.-M. Conia and P. Amice, *Bull. Soc. Chim. Fr.*, 1968, **8**, 3327.
17. L. T. Higham, U. P. Kreher, R. J. Mulder, C. R. Strauss and J. L. Scott, *Chem. Commun.*, 2004, 2264.
18. L. T. Higham, U. P. Kreher, J. L. Scott and C. R. Strauss, *CrystEngComm*, 2004, 484.
19. (a) P. F. H. Schwab, M. D. Levin and J. Michl, *Chem. Rev.*, 1999, **99**, 1863; (b) P. F. H. Schwab, J. R. Smith and J. Michl, *Chem. Rev.*, 2005, **105**, 1197.
20. A. E. Rosamilia, J. L. Scott and C. R. Strauss, *Org. Lett.*, 2005, **7**, 1525.
21. L. T. Higham, K. Konno, J. L. Scott, C. R. Strauss and T. Yamaguchi, *Green Chem.*, 2007, **9**, 80.
22. A. Almesåker, S. A. Bourne, G. Ramon, J. L. Scott and C. R. Strauss, *CrystEngComm*, 2007, **9**, 997.
23. M. L. Cole, L. T. Higham, P. C. Junk, K. M. Proctor, J. L. Scott and C. R. Strauss, *Inorg. Chim. Acta*, 2005, **358**, 3159.
24. M. L. Cole, G. B. Deacon, C. M. Forsyth, P. C. Junk, K. M. Proctor, J. L. Scott and C. R. Strauss, *Dalton Trans.*, 2006, 3338.
25. M. L. Cole, G. B. Deacon, C. M. Forsyth, P. C. Junk, K. M. Proctor, J. L. Scott and C. R. Strauss, *Polyhedron*, 2007, **26**, 244.
26. M. L. Cole, G. B. Deacon, P. C. Junk, K. M. Proctor, J. L. Scott and C. R. Strauss, *Eur. J. Inorg. Chem.*, 2005, 4138.
27. M. Makha, J. L. Scott, A. N. Sobolev, C. R Strauss, C. L. Raston, *Cryst. Growth & Des.*, 2008, submitted.

High Pressure: a Clean Activation Method for Sustainable Organic Synthesis

LUCIO MINUTI

Dipartimento di Chimica, Università degli Studi di Perugia, via Elce di Sotto 8, 06123 Perugia, Italy

7.1 Introduction

High-pressure technology has been used in recent decades in an increasing number of applications in many areas of chemistry, particularly in synthetic organic chemistry. With this technology organic reactions in which reagents and/or products are heat sensitive and/or are unreactive at ambient pressure can be carried out under mild conditions. High pressure is a clean non-destructive physical activation mode for organic reactions and is a valuable tool in green organic chemistry because it is energy-saving and can markedly improve both reaction yields and reaction selectivities, *i.e.* stereoselectivity, regioselectivity and chemoselectivity, *avoiding the production of wastes and the employment of metallic catalyst and/or high temperatures.* Nowadays, this technique is commonly used in the laboratory, and in industry, since even a modest increase in pressure may have marked effects on reactions in solution. Using high-pressure chemistry, all reactions that have a negative volume of activation are activated. Most organic reactions have a negative activation volume and are, therefore, influenced by high pressure in various ways depending on the value of the activation volume.

RSC Green Chemistry Book Series
Eco-Friendly Synthesis of Fine Chemicals
Edited by Roberto Ballini
© Royal Society of Chemistry 2009
Published by the Royal Society of Chemistry, www.rsc.org

The aim of this chapter is not to discuss the general principles or provide an extensive survey of the applications of high pressure since there are excellent books and reviews on this topic, [1-6] rather, examples will be presented that show how this technique influences organic reactions and how it can be a useful tool for developing environmentally benign processes.

The term "high pressure" is used to indicate a range between 0.1–2 GPa. The relationship between pressure and the rate constant of the reaction is expressed as follows:

$$\frac{\delta \ln k}{\delta P} = -\frac{\Delta V^{\neq}}{RT}$$

where $\Delta V^{\neq}$ is the activation volume, defined as the difference between the volume occupied by the transition state and that occupied by the reactants ($\Delta V^{\neq} = V_{TS} - V_R$). If $\Delta V^{\neq}$ is negative, the rate constant will increase with increasing pressure. Similarly the effect of pressure on the reaction equilibria is expressed by:

$$\frac{\delta \ln K}{\delta P} = -\frac{\Delta V}{RT}$$

where ΔV is the molar volume.

The first part of this chapter deals with the effects of high pressure on cycloaddition reactions, particularly the Diels–Alder reaction, which is the most important cycloaddition reaction. The second part will illustrate applications of pressure to nucleophilic substitutions, condensations and other reactions (miscellaneous reactions), such as Mannich, Heck, ene, S_EAr, Wittig, Horner–Wadsworth–Emmons and multicomponent Strecker reactions.

Descriptions of high-pressure equipment needed to perform the experiments are discussed in detail by Matsumoto and Acheson.[1]

7.2 Cycloadditions

7.2.1 Diels–Alder Reaction

Diels–Alder reactions[7] have large negative volumes of activation and intermolecular reactions have larger activation volume values than those of the intramolecular ones. Diels–Alder reactions may be pericyclic, in which bond-forming and bond-breaking processes are concerted in the six-membered transition state, or stepwise-cycloadditions in which the two σ-bonds are formed in two separate steps. Pericyclic processes have the largest activation volumes. Some monographs and reviews dealing with high-pressure cycloadditions, especially Diels–Alder reactions, have been published.[8-16]

The synthetic approach to aklavinone, the aglycone of aclacinomycin A, developed by Li and Wu,[17] is an interesting example of a pressure-promoted cycloaddition reaction. A thermal reaction between *p*-benzoquinone (**1**) and the electron-poor dienic ester **2** gives a mixture of the heat-sensitive cycloadduct **3** and its isomeric hydroquinone **4** in low yield. When the Diels–Alder reaction is carried out at 1.5 GPa at room temperature, the enedione cycloadduct **3** is obtained in good yield (60%) (Scheme 7.1) without significant contamination by the hydroquinone side-product.

Cycloadditions between 3-substituted coumarins **5** and dienes **6** allow various novel coumarin derivatives to be prepared regioselectively in high yields[18] (Scheme 7.2).

The reactions are carried out either in water or under high pressure – the best yields are achieved with the latter.

The reactions with (*E*)-piperylene occur *endo*-diastereoselectively; an inverse diastereoselectivity is observed with $X = CO_2H$. Extensive decarboxylation of the cycloadducts occurs in the reactions carried out in water.

Hydrophenanthrenes are known to be useful intermediates in the total synthesis of diterpenes and steroids. The Diels–Alder reaction of heteroannular bicyclic dienone **8** with (*E*)-piperylene (**9a**) and 2,3-dimethyl-1,3-butadiene (**9b**)

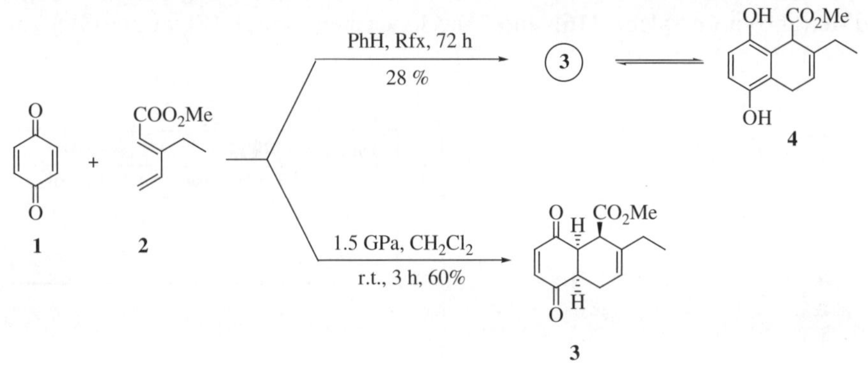

Scheme 7.1

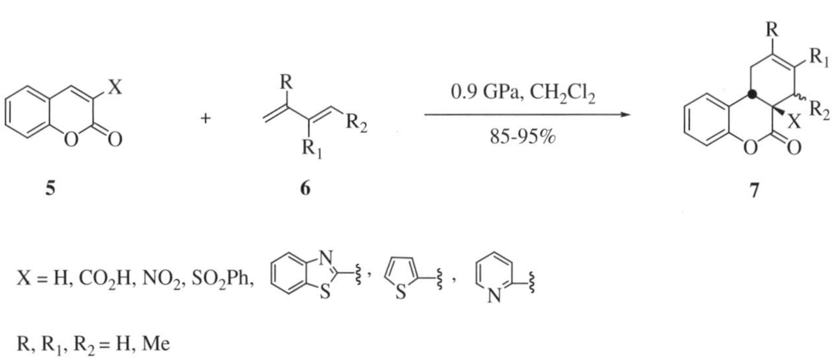

$X = H, CO_2H, NO_2, SO_2Ph,$

$R, R_1, R_2 = H, Me$

Scheme 7.2

is an interesting example of high-pressure controlled regioselective and diastereoselective preparation of hydrophenanthrenones.[19] The different ratios of compounds **10/11** produced by cycloadditions performed at ambient- or under high-pressure, and the formation of the unexpected trans-compound **11** can be explained by (i) Diels–Alder reactions at ambient pressure are thermodynamically controlled and (ii) the *anti-endo* adducts **10** are converted into the short-lived *syn-endo* adducts **12** which tautomerize to **11**. The formation of trans-compounds **11** by induced post-cycloaddition isomerization makes this method very useful in organic synthesis (Scheme 7.3).

Another approach to hydrophenanthrenones and mono- and di-substituted phenanthrenes, based on Diels–Alder reaction between Danishefsky diene (**13**) and 1-nitronaphthalene (**14**), has been developed by Mancini *et al.*[20] When 1-nitronaphthalene (**14**) interacts with the electron-rich diene **13** under high pressure, the cycloaddition occurs at a lower temperature and, after aromatization, allows 2-hydroxyphenanthrene (**15**) to be prepared (Scheme 7.4).

Another application of high-pressure cycloaddition is the one-pot preparation[21] of 3,4-dihydro-1-(2*H*)-anthracenone (**18**), a useful intermediate in the synthesis of phenacenes.[22] Compound **18** can be obtained by reacting $\alpha,\alpha,\alpha',\alpha'$-tetrabromo-*o*-xylene (**16**) and 2-cyclohexenen-1-one (**17**) at 0.9 GPa; no

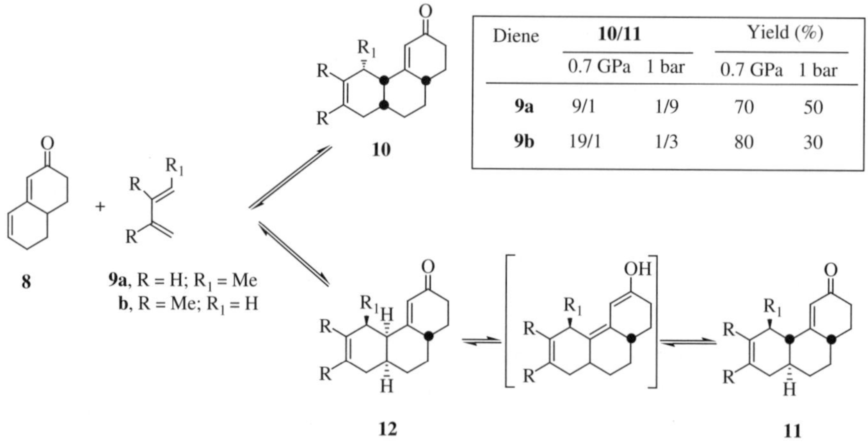

Diene	10/11		Yield (%)	
	0.7 GPa	1 bar	0.7 GPa	1 bar
9a	9/1	1/9	70	50
9b	19/1	1/3	80	30

Scheme 7.3

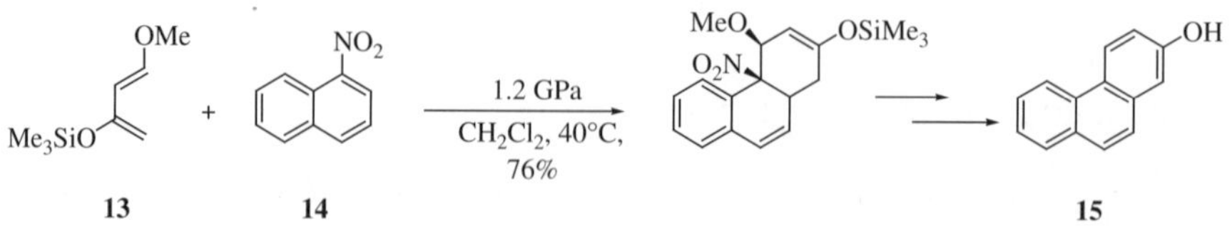

Scheme 7.4

reaction occurs under thermal and/or catalysed conditions at normal pressure because of the low reactivity[23] of cycloalkenone **17** (Scheme 7.5).

This one-pot procedure gives a greater yield and is more convenient than the previously reported four-step synthesis.

Small helical molecules are of interest because of their photophysical and chiroptical properties. Helicenophanes, *i.e.* helical molecules containing the [2.2]paracyclophane moiety, have been studied due to the peculiar structural properties of the cyclophane framework. Enantiopure compounds have extraordinarily high specific rotation values. Some helical cyclophanes have been prepared[24,25] by cycloadditions between [2.2]paracyclophane-based dienes **19** and **20** with dienophiles **1** and **21** (Scheme 7.6).

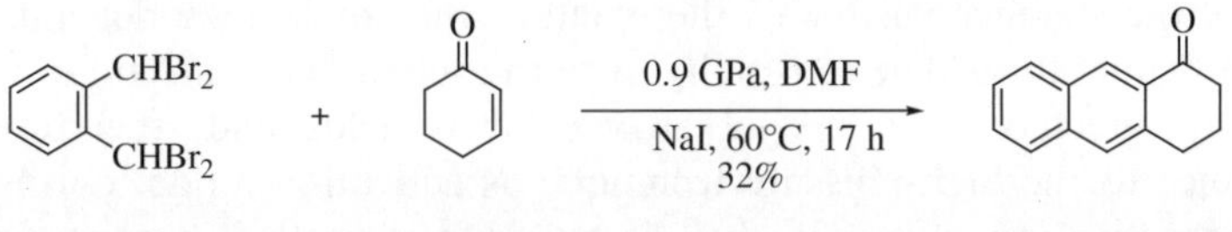

16   +   **17** $\xrightarrow[\text{NaI, 60°C, 17 h} \atop \text{32\%}]{\text{0.9 GPa, DMF}}$ **18**

Scheme 7.5

22 $\xleftarrow[\text{1 GPa, 40°C, 18h} \atop 20\%]{\boxed{19}}$ **1** $\xrightarrow[\text{0.6 GPa, 35°C,} \atop \text{CH}_2\text{Cl}_2, 61\%]{\boxed{20}}$ **23**

24 $\xleftarrow[\text{1 GPa, 40°C,} \atop \text{CH}_2\text{Cl}_2, 70\%]{\boxed{19}}$ **21** $\xrightarrow[\text{0.65 GPa, 35°C,} \atop \text{CH}_2\text{Cl}_2, 76\%]{\boxed{20}}$ **25**

24 $\xrightarrow[84\%]{\text{DDQ}}$ **26**

(S)-(+)-**19**   **20**

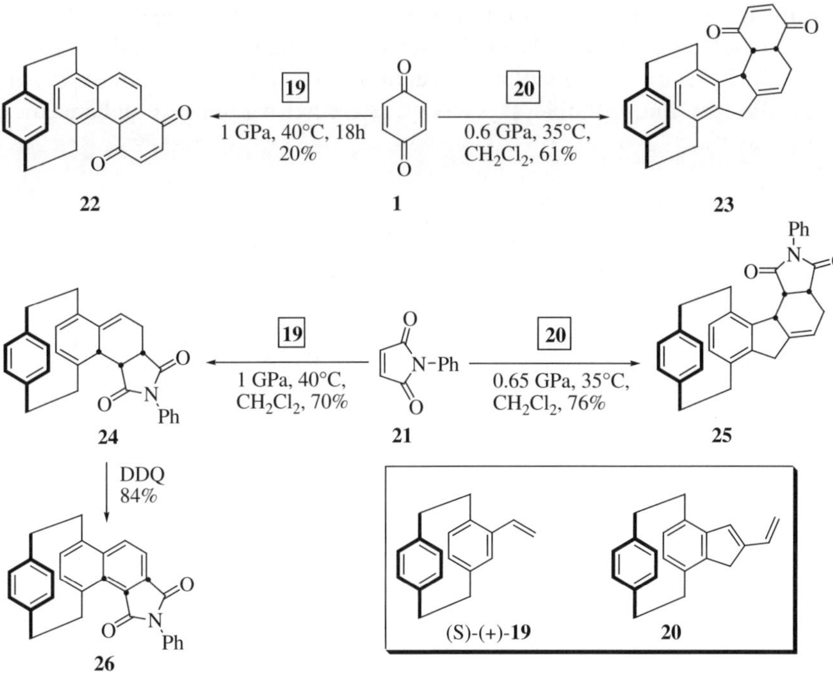

Scheme 7.6

Dienes **19** and **20** only react with *N*-phenylmaleimide (**21**) when the reactions are activated by high pressure; cycloadditions with the more reactive dienophile **1** also occur under thermal conditions, but in lower yield. The reactions are totally *anti*-(with respect to the unsubstituted benzene ring of the para-cyclophane unit) *endo*-diastereoselective. DDQ oxidation of cycloadduct **24** led to the aromatized compound **26** while the aromatization of **23** and **25** failed. Interestingly, the cycloaddition between **1** and **19** led to the aromatized compound **22**; this clearly indicates that 1,4-benzoquinone also acts as an oxidant.

Heterocyclic five-membered compounds such as furan and thiophene are slightly reactive dienes because they are aromatic. Since furan has a lower aromatic character, it is more reactive and undergoes thermal Diels–Alder reaction[26] with maleic anhydride and *N*-phenylmaleimide. Whereas the cycloaddition with the maleic anhydride affords the *exo*-adduct at room temperature, the stereochemistry of the reaction with maleimide depends on the reaction temperature. The sensitivity of both cycloadducts and furan to acidic conditions precludes the use of strong Lewis acids and, therefore, gives importance to the high-pressure technique as activation mode. Owing to the strong deactivating effect of the electron-donor methyl group, citraconic anhydride (**27**) only reacts with furan (**28**) at high pressure to afford, *exo*-selectively and in high yield (97%), cycloadduct **29**, an intermediate in the synthesis of palasonin[27] (Scheme 7.7).

Thiophene (**30**) does not usually undergo Diels–Alder reaction under thermal and Lewis acid catalysed conditions but it has been reported[28] that under high pressure (2.0 GPa) it reacts with maleic anhydride (**31**) in reasonable yield (47%) (Scheme 7.8). It was recently observed that the reaction yield is strongly improved (77%) if the cycloaddition reaction is carried out in perfluorohexane at 0.8 GPa; the fluorophobic effect plays a crucial role in rate enhancement

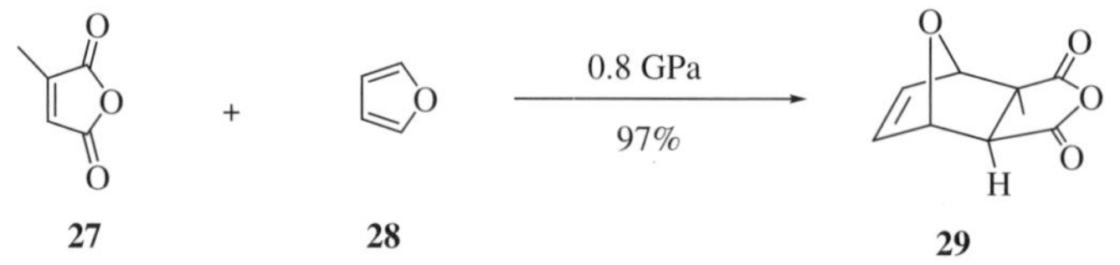

Scheme 7.7

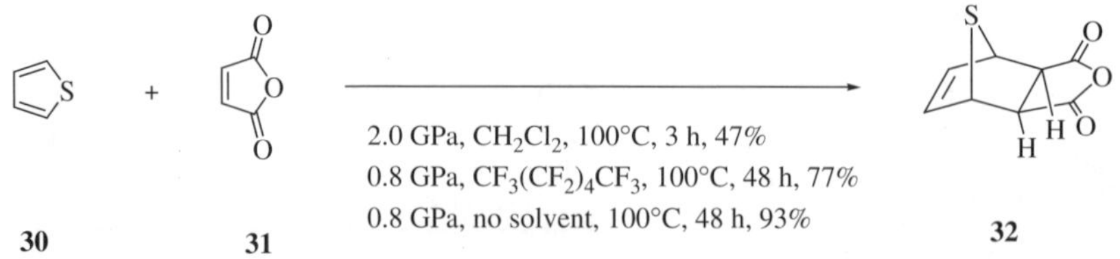

Scheme 7.8

(Scheme 7.8). A higher reaction yield (93%) was obtained when the reaction was carried out under pressure and solvent-free conditions (Scheme 7.8). The reactions were totally *exo*-diastereoselective.

Thiophene (**30**) also undergoes Diels–Alder reactions activated by high pressure with maleimides **33** in high yield.[29] These reactions (Scheme 7.9) are almost non-diastereoselective since mixtures of *exo*- and *endo*- isomers **34** and **35** are obtained.

3-Substituted furans **36** undergo regioselective and *exo*-diastereoselective Diels–Alder cycloaddition with various cycloalkenones **37** under high pressure and at low temperatures to give the corresponding 7-oxabicyclo[2.2.1]-heptene derivatives **38** in good yield.[30] The reaction does not occur at ambient pressure or at high temperatures due to the low reactivities of both furans and cycloalkenones.[31] Lewis acid activation was precluded because furans and cycloadducts are sensitive to acidic conditions. The cycloadducts are thermally labile and generally unstable; they are oxidized (*m*-CPBA) to the more stable corresponding sulfones (**39**, Scheme 7.10).

Multi-ring heteroaromatic compounds containing furan and thiophene rings have been prepared by pressured cycloaddition of 2-vinylbenzo[*b*]furan (**40a**) and 2-vinylbenzo[*b*]thiophene (**40b**) with 3-nitro-2-cyclohexen-1-one (**41**) and 2-inden-1-one (**42**), respectively, generated *in situ*.[32]

Cycloadducts **43**, obtained by regioselective and *exo*-diastereoselective reactions, are then converted into tetracyclic aromatic ketones **45** (Scheme 7.11).

The cycloadditions of 2-inden-1-one (**42**) generated *in situ* are regioselective and cycloadducts **46** are converted into pentacyclic heterohelicenes **47**.

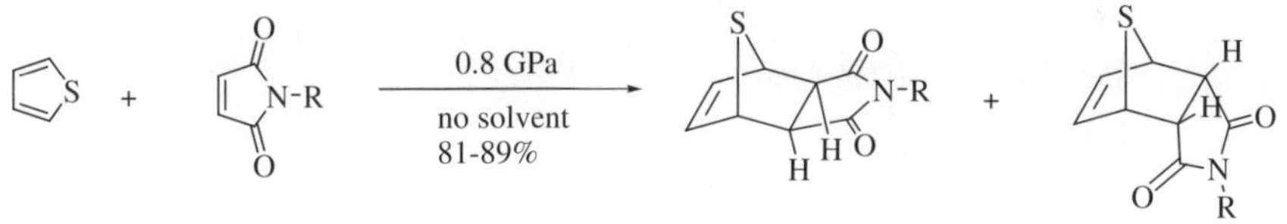

Scheme 7.9

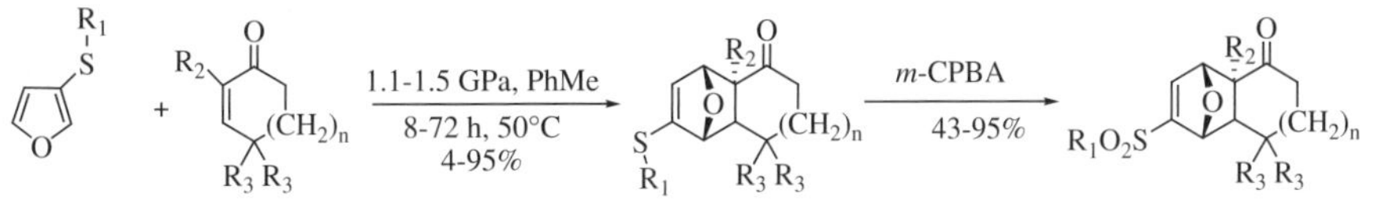

Scheme 7.10

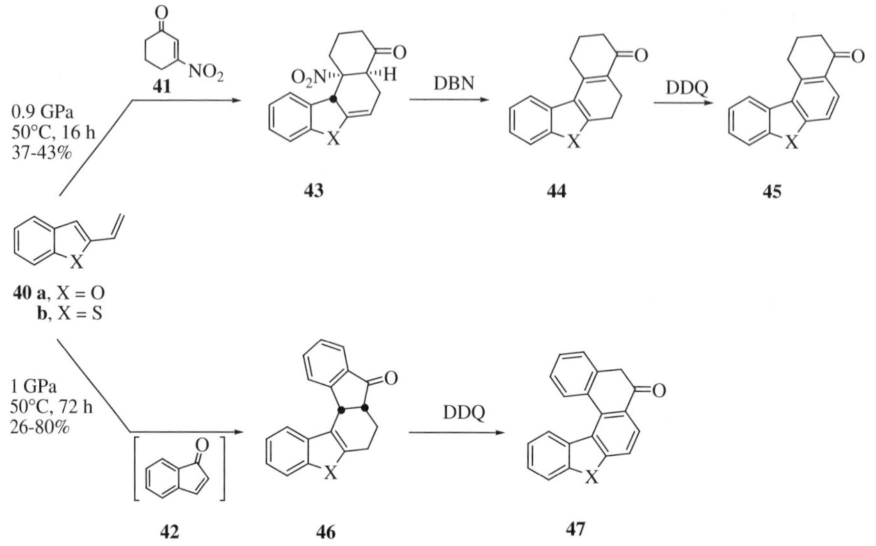

41

0.9 GPa
50°C, 16 h
37-43%

43

DBN

44

DDQ

45

40 a, X = O
  b, X = S

1 GPa
50°C, 72 h
26-80%

42

46

DDQ

47

Scheme 7.11

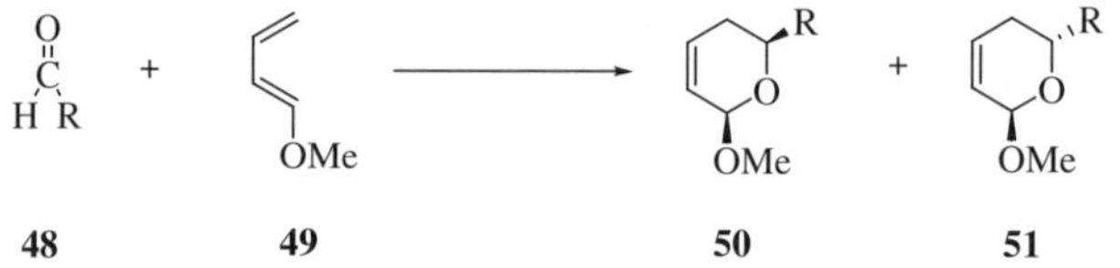

48

+

OMe

49

50

OMe

+

OMe

51

R	P (GPa)	Yield (%)	$50/51^{a}$
Me	2	65	2.3 (0.45)
$n\text{-}C_5H_{11}$	1.4	28	3.5
Ph	1.9	80	3 (1)
2-Fu	1.95	73	2.7

Scheme 7.12

Dihydropyrans are useful intermediates in the total synthesis of mono-saccharides and have been prepared by hetero-Diels–Alder reaction[33–35] of simple alkyl- and arylaldehydes **48** with electron-rich (*E*)-1-methoxy-1,3-butadiene (**49**) in reasonable to good yield. As expected *endo*-diastereoselectivity is predominant because, as usual, pressure strongly favours *endo*-addition (Scheme 7.12).

Jurczak[36] has reported an interesting approach to optically active 6-substituted-2-methoxy-5,6-dihydropyrans. The approach is based on a high-pressure cata-lysed hetero-Diels–Alder reaction of activated diene **49** with non-activated heterodienophile **52** (Scheme 7.13). These authors studied the effects of catalyst

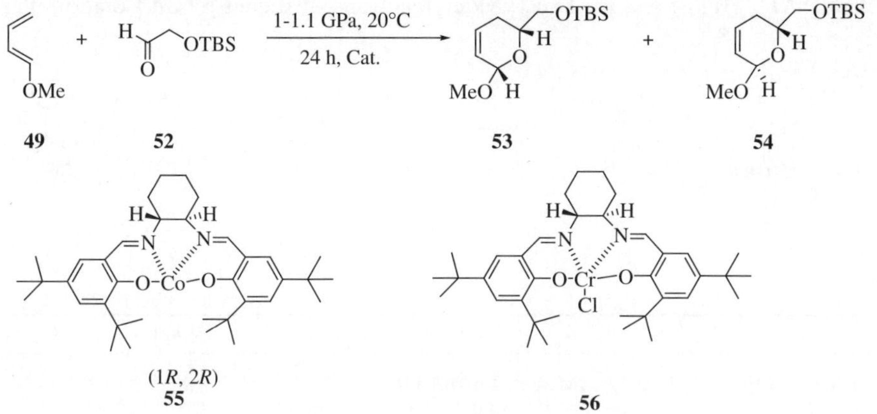

Scheme 7.13

and solvent on reaction yield and enantioselectivity for the cycloadditions carried out at 1–1.1 GPa and at room temperature. Commercially available (salen)Co complex **55** and chromium complex **56** were effective catalysts in the reactions carried out under high pressure. The highest enantioselectivities were observed in reactions carried out in the presence of **55** (up to 94% ee); in contrast, higher yields (up to 90%) but somewhat lower enantioselectivities (84–87%) were obtained with catalyst **56**.

Synthetic approaches to dihydropyrans by inverse electron demand (LUMO diene-controlled) Diels–Alder reaction have also been reported. Boger and co-workers[37,38] investigated the reactions of heterodienes **57** and electron-rich alkenes **58**. Table 7.1 summarizes some of the results. All reactions are totally regioselective and although thermal and catalysed cycloadditions preferentially proceed through an *endo*-transition state the pressure-promoted cycloadditions are more *endo*-diastereoselective and give higher reaction yields. This study is an interesting example of controlling reaction stereochemistry by high pressure.

Enol ethers have often been used as the olefinic component in inverse electron demand Diels–Alder reactions. De Meijere and co-workers[39] reported a novel approach to the synthesis of racemic 2-deoxy-*arabino*-hexopyranoside skeleton by a pressure-promoted hetero-Diels–Alder reaction between the cyclopropanated enol ether **61** and β,γ-unsaturated α-ketoesters **62** (Scheme 7.14).

Mixtures of cis- and trans-diastereoisomers **63** and **64** were obtained and the cis–trans ratios, as well as reaction yields, were dependent on reaction time and reaction temperature. The highest cis–trans ratios were obtained when the cycloadditions were carried out at 1.0 GPa, 25 °C in CH_2Cl_2.

Cis- and trans-diastereoisomers were separated and transformed into spiro-*arabino*-hexopyranoside-cyclopropane.

An example of an all-carbon inverse electron demand Diels–Alder reaction is given in the short synthesis of Corey's lactone **69** that was reported recently by Markò *et al.*[40]

Table 7.1 High-pressure Diels–Alder reactions of dienes **57** and dienophiles **58**.

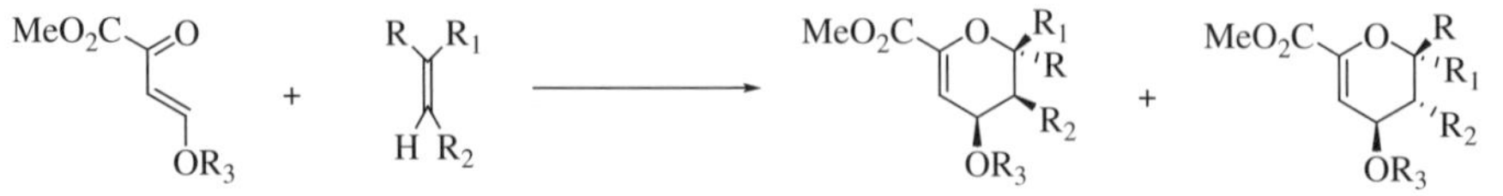

R	R_1	R_2	R_3	*Reaction conditions*	**59/60**	*Yield (%)*
H	OEt	H	Me	PhMe, 110 °C	1.8	48
				TiCl$_4$, CH$_2$Cl$_2$, –78 °C	0.3	61
				1.3 GPa, neat, 24 °C	5.7	82
H	OBn	Me	Me	PhH, 110 °C	3	11
				EtAlCl$_2$, CH$_2$Cl$_2$, –78 °C	6	46
				0.95 GPa, CH$_2$Cl$_2$, 24 °C	19	50
H	OEt	H	Ph	PhH, 80 °C	4	73
				EtAlCl$_2$, CH$_2$Cl$_2$, –78 °C	0.4	94
				0.62 GPa, neat, 24 °C	9	86
OMe	OMe	Me	Ph	CH$_2$Cl$_2$, 40 °C	–	–
				0.62 GPa, CH$_2$Cl$_2$, 24 °C	–	65
OMe	OMe	OMe	Ph	1 GPa, CH$_2$Cl$_2$, 24 °C	1	72

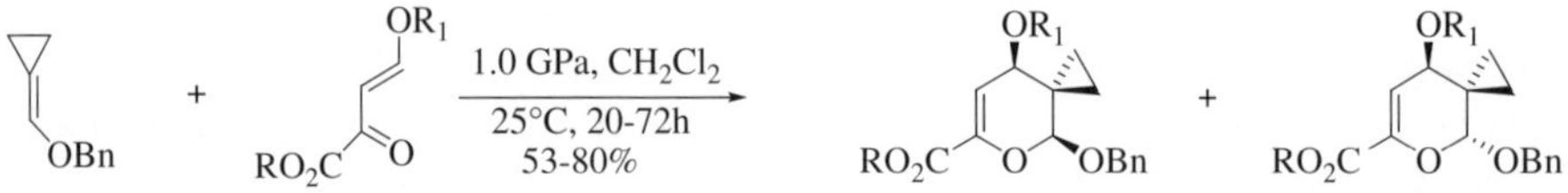

a, R = R$_1$ = Et

b, R = Me, R$_1$ = Bn

Scheme 7.14

This approach is based on the rearrangement of the bicyclic seleno ester **67** prepared by pressure-promoted inverse electron demand cycloaddition of carbomethoxypyrone **65** and vinyl selenide **66** (Scheme 7.15). The cycloaddition reaction occurs regioselectively and *endo*-diastereoselectively.

Pressure activation is crucial, due to the weaker electron-donating ability of selenium compared to that of the sulfur- or oxygen-containing analogues. Lactone **69** is a versatile and key intermediate in the synthetic strategy developed by Corey for the synthesis of prostaglandins.

A high-pressure technique has also been applied successfully to intramolecular Diels–Alder reactions.

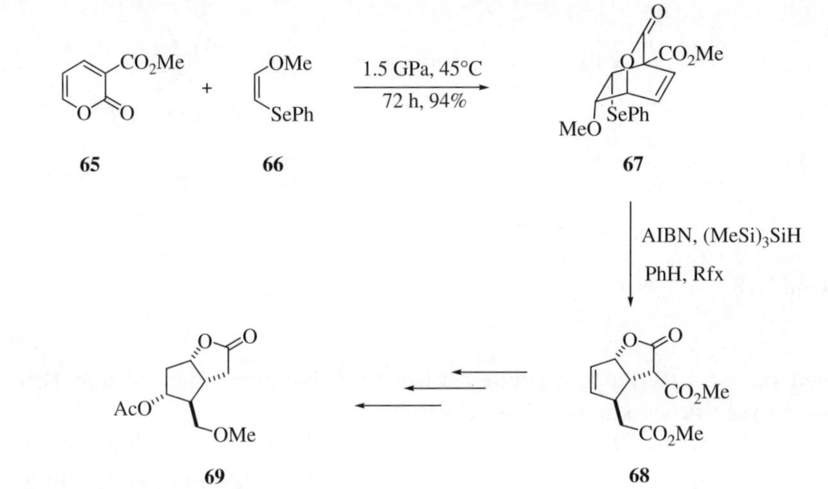

Scheme 7.15

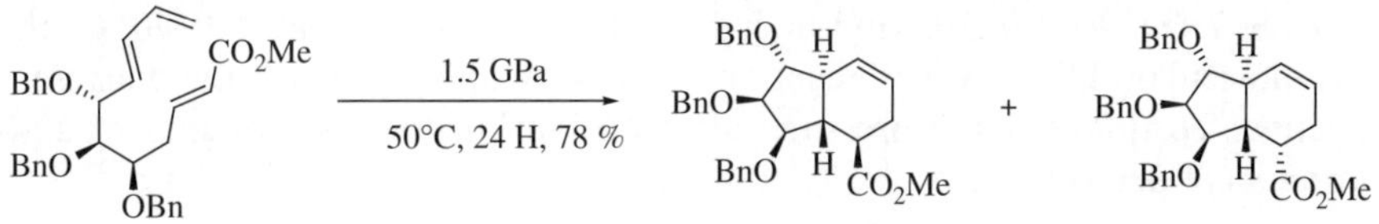

Scheme 7.16

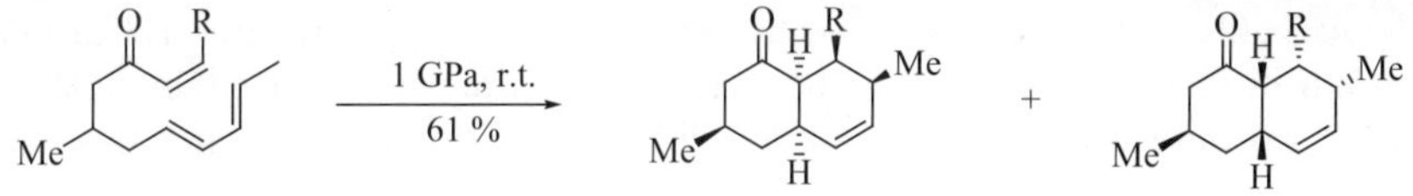

Scheme 7.17

The synthesis of highly oxygenated chiral cyclopentanes reported by Jarosz *et al.*[41] is an interesting example of a pressure-promoted intramolecular process. Whereas $AlCl_3$-catalyzed intramolecular cycloaddition of trienoate **70** led to a 2 : 1 mixture of stereoisomers **71** and **72**, the reaction performed under pressure (1.5 GPa) led stereoselectively to cycloadduct **71** (Scheme 7.16). The cycloadduct was then converted into the desired chiral cyclopentanones.

A similar strategy was followed by Konoike *et al.*[42] for the enantioselective total synthesis of (+)-6-*epi*-mevinolin and its analogues. Intramolecular cycloaddition of chiral (*Z*)-trienone **73** led to a mixture of *cis*-decalins **74** and **75** (Scheme 7.17); the most satisfactory result was achieved when the reaction was

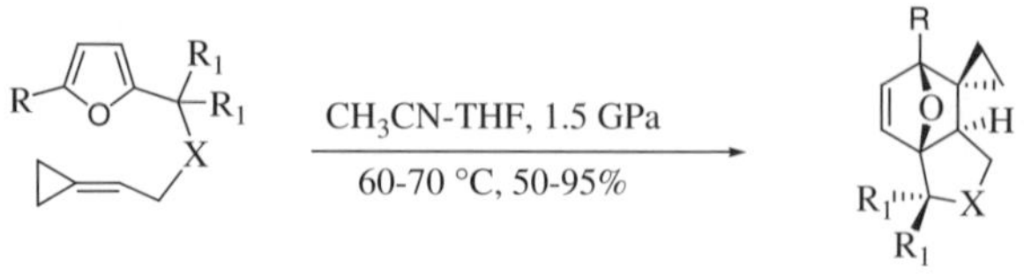

$$\text{76} \quad \xrightarrow[\text{60-70 °C, 50-95\%}]{\text{CH}_3\text{CN-THF, 1.5 GPa}} \quad \text{77}$$

X = O, NMe$_2$; R = H, Me, OMe; R$_1$ = H, Me, = O

Scheme 7.18

carried out at 1.0 GPa and room temperature. Notably, thermal and Lewis acid-catalysed processes gave less selectivity.

Furyl derivatives **76**, with an allylether or allylamine-type linkage to a methylenecyclopropane framework, readily undergo high pressure-promoted intramolecular cycloaddition[43] to give spirocyclopropane tricyclic products **77**. No cycloaddition reaction occurred at ambient pressure and the products were mostly tar and polymers. Lewis acid catalysis was only marginally successful (Scheme 7.18). At 1.0 GPa and a slightly elevated temperature (60–70 °C) the intramolecular Diels–Alder reaction occurs readily and is *exo*-diastereoselective. To quantify the pressure effect on the kinetics the volumes of activation were determined.

7.2.2   [3 + 2] Cycloaddition

[4 + 2] Cycloadditions and [3 + 2] reactions are associated with a negative volume of activation and, therefore, the application of high pressure usually enhances the reaction rate.[2,4] The influence of pressure on 1,3-dipolar cycloaddition reactions of organic azides was recently reported by Klärner *et al.*[44] To date no cycloaddition of azides to α-alkylnyl-enamines has been reported.

The cycloaddition reactions of α-ethynylamine **78** with azides **79** have been examined under ambient- and high-pressure conditions. The results reported in Scheme 7.19 clearly show that the reaction yield depends on the pressure at which the reaction is carried out.

In recent years there has been growing interest in multiprocesses such as tandem and multicomponent reactions since they allow bonds, rings and stereogenic centers, to be formed. These reactions are powerful means to rapidly construct complex molecules. The combination of multiple reactions in simple operations is one of the fundamental objectives for enhancing synthetic efficiency. Tandem processes are very useful for achieving high synthetic efficiency since complex molecules can be rapidly constructed from simple ones.[45–47] A good example of this is the one-pot preparation of nitroso-acetals by high pressure-promoted tandem [4 + 2]/[3 + 2] cycloadditions of nitroalkanes **81** and ethyl vinyl ether (**82**)[48] (Scheme 7.20).

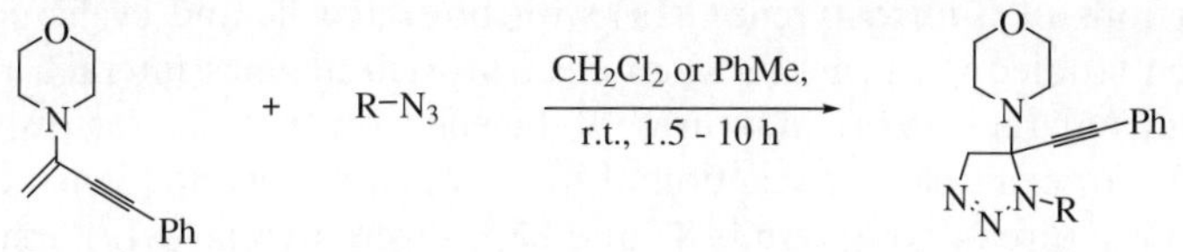

a = 4-NO$_2$-C$_6$H$_4$; **b** = EtOCO; **c** = Ph

Azide	Pressure	Product	Yield (%)
79a	1 bar	**80a**	37
	O.72 GPa		54
79b	1 bar	**80b**	–
	O.78 GPa		85
79c	1 bar	**80c**	–
	O.67 GPa		31

Scheme 7.19

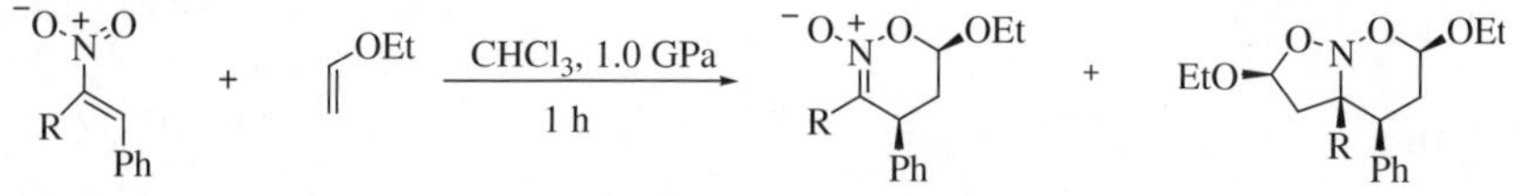

81   **82**   **83**   **84**

a, R = H
b, R = Me

Scheme 7.20

The reactions performed at 1.0 GPa gave almost quantitatively 1 : 6 (R=H) and 1 : 20 (R=Me) mixtures of mono adduct **83** and tandem-adduct **84**. Activation by pressure allowed the long reaction times to be reduced and circumvented the use of large excesses of reagents.[49]

An interesting example of pressure-promoted tandem reactions is the one-pot synthesis of bicyclic- and tricyclic-*N*-oxy-β-lactams **85**,[50] which are compounds that strongly resemble the biologically active oxamazins.

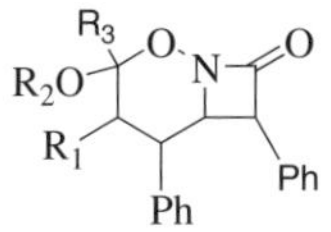

a, R$_1$ = H, R$_2$ = Et, R$_3$ = H
b, R$_1$ = H, R$_2$ = PMB, R$_3$ = H
c, R$_1$-R$_3$ = (CH$_2$)$_4$, R$_2$ = Me
d, R$_1$-R$_2$ = (CH$_2$)$_2$, R$_3$ = H

85

The reactions of β-nitrostyrene (**81a**) with both acyclic and cyclic enol-ethers have been studied. In general, when electron-rich alkenes interact at 1.5 GPa with β-nitrostyrene (**81a**), mixtures of bicyclic or tricyclic regioisomers are obtained. For example, the reaction of **81a** with enol ether **86** (Scheme 7.21) led to a 7 : 3 mixture of compounds **87** and **88**. β-Nitrostyrene (**81a**) first reacts as an electron-poor diene in an inverse electron demand Diels–Alder reaction with the enol ether, and then as an electron-poor dipolarophile with the formed monoadduct in a 1,3-dipolar cycloaddition.

The main product **87** was then converted into the bicyclic lactam **85b** by base treatment.

Pressure-promoted three-component tandem cycloadditions [4 + 2]/[3 + 2] have also been investigated.[48] The components were a nitroalkene, an electron-rich enol ether and an electron-poor alkene: β-nitrostyrene (**81a**) (or 1-phenyl-2-nitropropene), ethyl vinyl ether (**82**) and methyl acrylate (**89**).

The components were chosen based on the fact that (i) nitronates react faster with electron-poor alkene than with electron-rich alkenes[49–51] and (ii) methyl acrylate reacts as an electron-poor dipolarophile because it does not react with the enol ether under high pressure conditions. The reaction of **81a**, **82** and **89** afforded a 2.5 : 1 mixture of two main products, **90** and **91**, in 62% yield (Scheme 7.22).

Nitroalkene **81a** reacts as an electron-poor diene with an electron-rich alkene in an inverse electron-demand hetero-Diels–Alder reaction to give a nitronate,

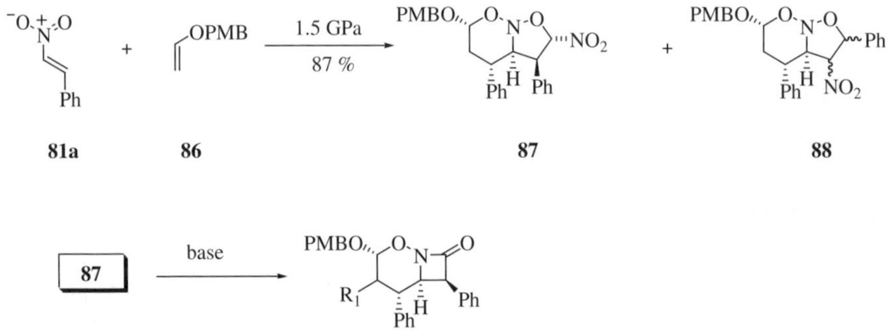

81a   **86**   **87**   **88**

87   base   **85b**

Scheme 7.21

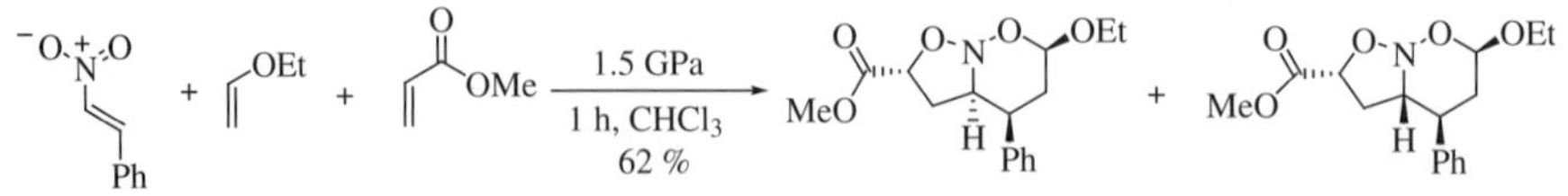

81a   **82**   **89**   **90**   **91**

Scheme 7.22

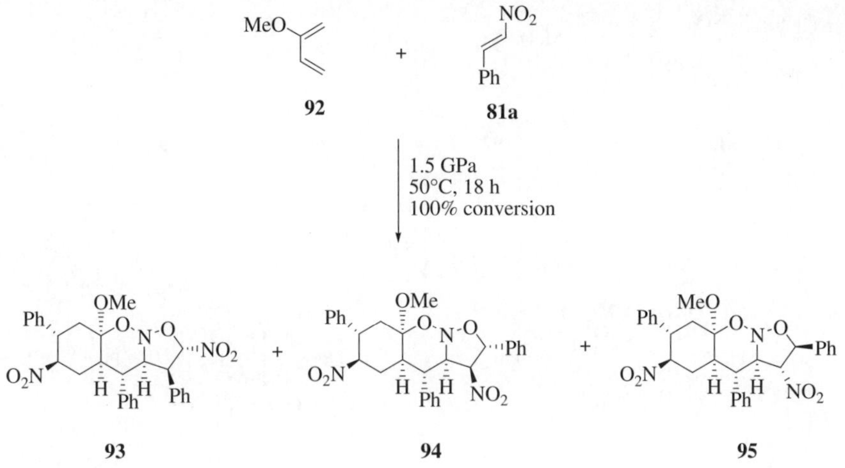

Scheme 7.23

which then undergoes 1,3-dipolar cycloaddition. The powerful effect of high pressure results in the formation of the nitroso-acetals without the need of stoichiometric amounts of a Lewis acid catalyst, as previously reported.

A four-component [4 + 2]/[4 + 2]/[3 + 2] cycloaddition reaction, reported by Scheeren *et al.*,[52] is another application of multicomponent reactions. The interaction of 2-methoxy-1,3-butadiene (**92**) (1 equiv.) with β-nitrostyrene (**81a**) (3 equiv.) at 1.5 GPa led to a mixture of nitroso-acetals **93–95**. In this reaction, β-nitrostyrene (**81a**) first reacts as a dienophile in the normal electron-demand Diels–Alder reaction, it then reacts as a heterodiene in the inverse electron-demand hetero-Diels–Alder reaction and, finally, it acts as a dipolarophile in the [3 + 2] cycloaddition (Scheme 7.23).

A [4 + 2]/[4 + 2]/[3 + 2] cycloaddition using four different components has also been investigated.[52] The interaction of β-nitrostyrene (**81a**), diene **92**, styrene (**96**) and *N*-phenylmaleimide (**97**) under 1.5 GPa pressure in dichloromethane solution for 16 h at room temperature (Scheme 7.24) afforded a mixture of diastereomeric nitroso-acetals **98** and **99**. In this case, **92** reacted as a diene, *N*-phenylmaleimide as a dienophile, β-nitrostyrene as the heterodiene and styrene as a dipolarophile.

7.3 Nucleophilic Substitutions

7.3.1 Aliphatic Nucleophilic Substitutions

High pressure usually accelerates these reactions and affects their selectivities.

S_N1- and S_N2-solvolyses of neutral molecules can be distinguished by measuring the reaction rates under high pressure. It has been demonstrated that the activation volumes for both the S_N1- and S_N2-solvolyses of alkyl halides are

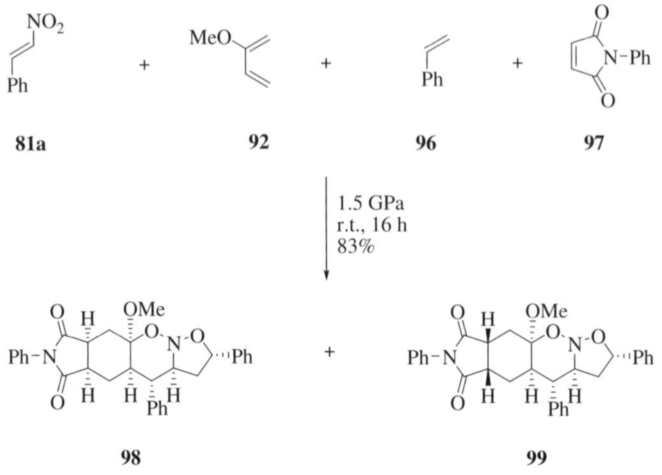

Scheme 7.24

negative at low pressure and increase with increasing pressure; they become positive for S_N1 reactions at high pressure. For example, the activation volume for the solvolysis of methyl iodide in glycerol solution (S_N2) is $-4.5\,cm^3\,mol^{-1}$ at 1.1 GPa and it remains negative up to 7.0 GPa.[53] In contrast, the activation volume for the solvolysis of *tert*-butyl chloride in glycerol solution (S_N1) is about $-4\,cm^3\,mol^{-1}$ at 0.4 GPa; it becomes positive at pressures above 1.6 GPa. Thus, S_N2 reactions are favoured at higher pressures.

The epoxide ring-opening reaction with certain nucleophiles in the absence of acid or basic catalysts is slow at normal pressure and generally requires large excess amounts of reagents, long reaction times and high temperature. Under these severe conditions other functionalities in the substrate can give side-reactions. Under high pressure, various epoxides have been hydrolysed nearly quantitatively into diols with water without using any catalyst.[54] Excellent chemoselectivity has been observed (Scheme 7.25); in runs 3 and 4, the ester group survived completely.[54]

Epoxide ring-opening reactions with various nitrogen heterocycles (indoles, pyrroles, pyrazoles) have been efficiently promoted under mild conditions with high pressure or silica gel-catalysed conditions.[55]

Whereas the silica gel-catalysed reactions of epoxides with indoles **101** are more efficient than the high-pressure activated reactions for pyrrole **102**, pyrazoles **103** and imidazoles **104**, the best results were obtained with reactions carried out under high pressure.[55b] The reaction between epoxides **100** and indoles **101** occurred regioselectively and only 3-substituted indoles were produced (Scheme 7.26).

Using optically active epoxides, chiral N-alkylated pyrazole and imidazole derivative ligands have been prepared by high pressure reactions with **103** or **104**; their catalytic efficiency as chiral ligands in the enantioselective addition of diethylzinc to benzaldehyde has been tested.[55c,e]

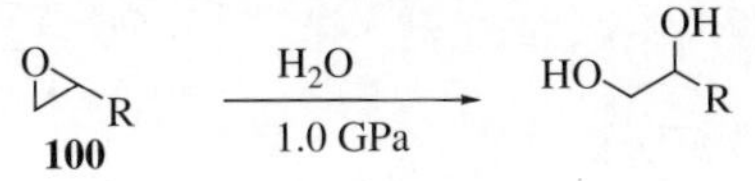

Run	R	T (°C)	t (h)	Yield (%)
1	Ph	60	24	92
2	C_5H_{11}	80	24	95
3	CH_2OCOPh	80	46	76
4	$CH_2CH{=}CHCO_2Et$	80	46	89

Scheme 7.25

Aminolysis of epoxides is also promoted by high pressure or silica gel catalysis. For example, N-(β-hydroxyalkyl)glycine esters have been prepared in high yields by nucleophilic substitution at 1.0 GPa or under silica gel catalysed conditions of various epoxides with a stoichiometric amount of *tert-* butyl glycinate.[56]

At atmospheric pressure, the conversion of sterically hindered primary and secondary alcohols into the corresponding phenyl sulfide derivatives with a Bu_3P–PhSSPh system occurs in low yields (0–56%); at 1.0 GPa pressure, the same reactions have given a 65–100% yield.[57a] Thus, this reaction has been used in the seven-step enantioselective synthesis of (–)-solenopsin B.[57b]

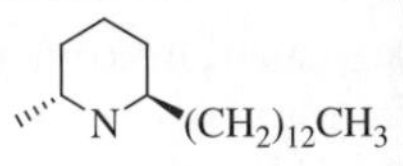

The hydrolysis of hydroxy, amino or unsaturated esters is usually accompanied by side reactions, loss of chirality or isomerization. Various esters have been selectively hydrolysed at room temperature and in high yields under high pressure in the presence of N-methylmorpholine or ${}^i Pr_2NEt$.[58] For example, the hydrolysis of the β,γ-unsaturated diester **105** into β-hydroxyester **106** occurred chemoselectively in 100% yield at high pressure in methanol solution; subsequent high-pressure induced hydrolysis of **106**, in CH_3CN/H_2O (60 : 1) as medium, afforded the desired hydroxy acid **107** as a single product (Scheme 7.27). The usual aqueous procedures produced a mixture of products.

Similarly, at 1.0 GPa the L-serine derivative **108** undergoes quantitative hydrolysis to the acid **109** enantioselectively and chemoselectively; hydrolysis does not occur at atmospheric pressure (Scheme 7.27).

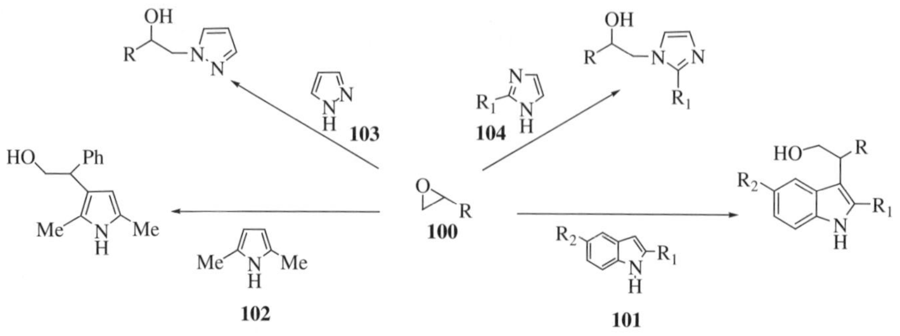

Epoxide **100** reacts with substrates **101**–**104** to give the products shown (compounds **101**, **102**, **103**, **104**).

Epoxide 100	Substrate	Conditions[a]	Yield (%)	Ref.
R = Ph	**101** $R_1 = R_2 = H$	85 °C, 24 h	13	55a
		SiO_2, 21 °C, 7 d	88	55b
		1.0 GPa, 42 °C, 24 h	56	55a,b
R = Ph	**101** $R_1 = Me$, $R_2 = H$	SiO_2, 22 °C, 7 d	70	55b
		1.0 GPa, 42 °C, 24 h	44	55a,b
R = Ph	**101** $R_2 = Me$, $R_1 = H$	SiO_2, 22 °C, 7 d	68	55b
		1.0 GPa, 42 °C, 24 h	57	55b
R = CH_2OPh	**101** $R_1 = R_2 = H$	SiO_2, 22 °C, 10 d	10	55b
		1.0 GPa, 65 °C, 3 d	16	55b
		1.0 GPa, SiO_2, 65 °C, 3 d	29	55b
		1.0 GPa, Yb(OTf), 60 °C, 42 h	66	55d
R = Ph	**102**	SiO_2, 22 °C, 2 d	30	55b
		1.0 GPa, 42 °C, 24 h, CH_2Cl_2	41	55b
R = Ph	**103**	SiO_2, 22 °C, 7 d	53	55b
		1.0 GPa, 65 °C, 3 d	61	55b,c,e
R = C_6F_5	**103**	1.0 GPa, 65 °C, 3 d	76	55c,e
R = Ph	**104** $R_1 = H$	SiO_2, 22 °C, 7 d	47	55b
		1.0 GPa, 65 °C, 3 d	59	55b,c,e
R = C_8H_{17}	**104** $R_1 = H$	SiO_2, 22 °C, 10 d	56	55b
		1.0 GPa, 65 °C, 3 d	72	55b
R = CH_2OPh	**104** $R_1 = H$	SiO_2, 22 °C, 10 d	50	55b
		1.0 GPa, 65 °C, 3 d	67	55b
R = Ph	**104** $R_1 = Me$	SiO_2, 22 °C, 8 d	36	55b
		1.0 GPa, 65°, 3 d	53	55b,c,e

[a]Unless otherwise noted, all reactions were performed in MeCN.

Scheme 7.26

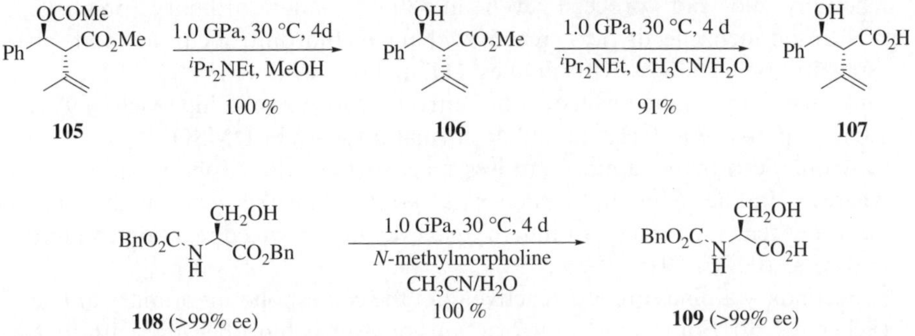

105   1.0 GPa, 30 °C, 4d / ^{i}Pr$_2$NEt, MeOH / 100 %   **106**   1.0 GPa, 30 °C, 4 d / ^{i}Pr$_2$NEt, CH$_3$CN/H$_2$O / 91%   **107**

108 (>99% ee)   1.0 GPa, 30 °C, 4 d / *N*-methylmorpholine / CH$_3$CN/H$_2$O / 100 %   **109** (>99% ee)

Scheme 7.27

$$O_2N-C_6H_4-X \ (\mathbf{110}) \ + \ (Ar)RNH_2 \ (\mathbf{111}) \ \xrightarrow[\text{50-100 °C}]{\text{0.72-1.0 GPa}} \ O_2N-C_6H_4-NHR(Ar) \ (\mathbf{112}) \ + \ HX$$

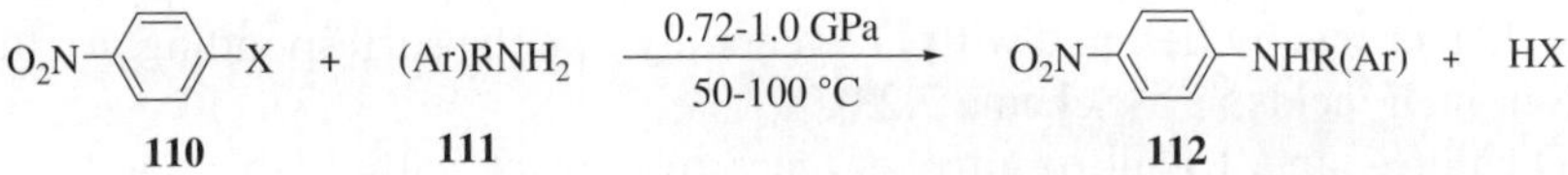

X	(Ar)R	Yield (%)	X	(Ar)R	Yield (%)
Cl	Pr	93	I	Pr	28
	Bu	76		Bu	35
	Hex	65		^{t}Bu	0
	^{i}Pr	26			
	^{t}Bu	2	F	^{t}Bu	89
	Ph	0		Ph	100
	p-CH$_3$O-C$_6$H$_4$	2			
			OTf	Ph	35

Scheme 7.28

7.3.2 Aromatic Nucleophilic Substitution (S_NAr)

Nucleophilic substitution of aromatic halides with amines is generally more difficult than the substitution reaction of aliphatic halides, and under ordinary pressure is limited to the reactions of some halides that have strong electron-attracting substituents, with amines of strong nucleophilicity.

These S_NAr reactions have a negative volume of activation ($\Delta V^{\neq} = -30$ to $-40\,\mathrm{cm^3\,mol^{-1}}$) and are therefore accelerated under high pressure. Thus, N-substituted aromatic amines have been synthesized by S_NAr of aryl triflate and aromatic halides that contain electron-withdrawing substituents with primary and secondary amines under hyperbaric conditions.[59]

Reactions of *p*-nitrochlorobenzene (**110**) with primary aliphatic amines under 0.72 GPa pressure have given the corresponding *p*-nitroanilines in 2–93% yield, depending upon the bulkiness of amines **111** (Scheme 7.28). The same reactions did not proceed even at 80 °C under ordinary pressure.[59a] *p*-Nitroiodobenzene is less reactive than *p*-chloronitrobenzene and gives substitution products with amines **111** in low yields (0–35%).[59a] However, bulky *tert*-butylamine reacted with *p*-nitrofluorobenzene in high yield (89%) at 1.0 GPa pressure in CH_3CN and at normal pressure in DMSO.[59c]

Aromatic primary amines are less reactive than the aliphatic amines, and whereas aniline does not react with *p*-nitrochlorobenzene, with *p*-nitrofluorobenzene it gave a quantitative yield even under forced reaction conditions (1.0 GPa, 100 °C, 50 h).[59a,c]

Secondary amines are less reactive than the corresponding primary amines[59] (Scheme 7.29). For example, the reactions of *p*-nitrochlorobenzene with Et_2NH and $^iPr_2NH$ gave the corresponding N-substituted anilines in low yields (39% and 0%, respectively) at 0.8 GPa. However, when $^iPr_2NH$ reacted with *p*-nitrofluorobenzene under 1.0 GPa pressure there was a good product yield (68%). Cyclic secondary amines such as morpholine, piperidine and pyrrolidine have higher reactivities, and with **113** generally give the corresponding products **115** in high yields[59a–c] (Scheme 7.29).

Dichloro- and trichloro-nitrobenzenes also react with Et_2NH and cyclic amines (morpholine, pyrrolidine) at 0.6 GPa pressure to give mono-, di-, and trisubstitution products. For example, 2,3-dichloronitrobenzene (**116a**) and 2,5-dichloronitrobenzene (**116b**) reacted with these amines to give the

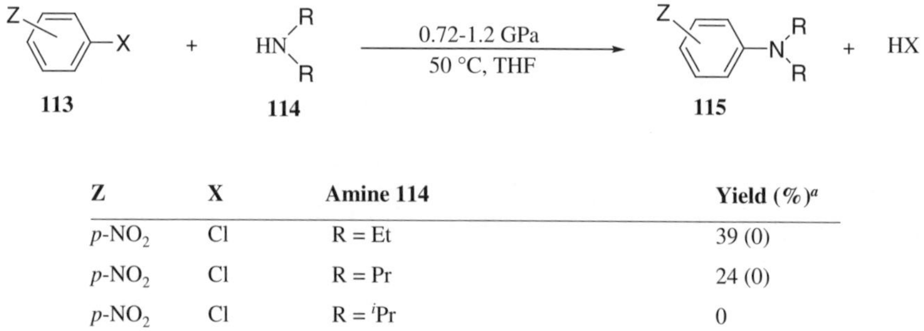

Z	X	Amine 114	Yield (%)[a]
p-NO$_2$	Cl	R = Et	39 (0)
p-NO$_2$	Cl	R = Pr	24 (0)
p-NO$_2$	Cl	R = ^{i}Pr	0
p-NO$_2$	OTf	R = ^{i}Pr	23
p-NO$_2$	F	R = ^{i}Pr	68
p-NO$_2$	Cl	Morpholine, piperidine, pyrrolidine	100
o-NO$_2$	Cl	Morpholine, pyrrolidine	100
m-NO$_2$	F	Morpholine	77 (0)
m-NO$_2$	F	Piperidine, pyrrolidine	100
p-CN	F	Morpholine	88 (69)
p-CH$_3$CO	F	Morpholine	68

[a]Yields in parentheses refer to reactions under 1 atm at 50–80 °C.

Scheme 7.29

corresponding *ortho*-substitution products **118a,b** in nearly quantitative yield; 3,4-dichloronitrobenzene (**116c**) reacted with the same amines selectively at the *para* position to the nitro group, again in high yield. In contrast, 2,4-dichloronitrobenzene (**116d**) gave a mixture of *ortho*-substitution, *para*-substitution and disubstitution products **118d**; the ratio of these three types of products depends on the amount and reactivity of the amine (Scheme 7.30).[59a]

High pressure (0.6 GPa) S_NAr of 2,3,5,6,-tetrachloronitrobenzene (**119**) with various primary and secondary amines **120** gives nitro-group-substitution products **121** together with mono-, and dichloro substitution products **122–127** (Scheme 7.31). The ratio of substitutions of the nitro group and chlorine atom is affected by the bulkiness of the amines.[60a,b]

The reaction of tetrachloronitrobenzene (**119**) with diamines under 0.6 GPa pressure predominantly gives mixtures of 1 : 1-substitution, bridgehead 2 : 1-substitution and cyclization products.[60c]

Finally, the nucleophilic substitution reactions of mono-, di- and trichloronitrobenzenes **128** with N-methyl substituted cyclic tertiary amines **129**

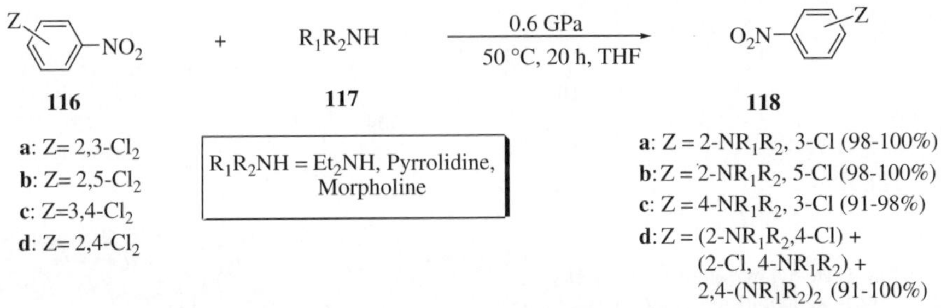

Scheme 7.30

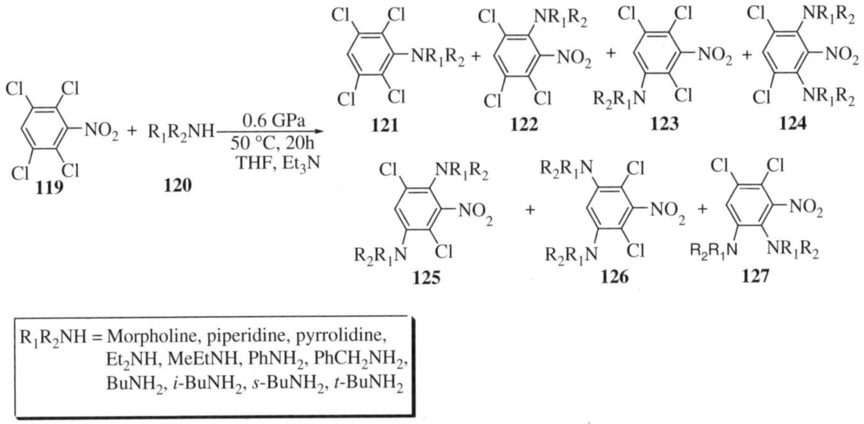

Scheme 7.31

under high pressure give demethylation products **130** and/or ring-opening products **131** and **132** (Scheme 7.32).[61] For example, **128** reacts with 1-methylpyrrolidine (**129a**) under high pressure to give the corresponding products of demethylation **130** and ring opening **131** and **132** initiated by the S_NAr reactions at the *para* and/or *ortho* positions to the nitro group; however, the reactions with 1-methylpiperidine (**129b**) and 4-methylmorpholine (**129c**) under the same conditions only give demethylation products **130**.[61a,b]

The high-pressure-promoted S_NAr reaction of aromatic diamines with fluoronitrobenzenes provides a new, practical access to aniline oligomers that are useful in the field of materials.[62] Linear and bent aniline trimers **135** have been prepared in good yields by carrying out reactions of 1,3- or 1,4-phenylendiammines **133** with 2-nitro- or 4-nitro-fluorobenzenes **134** at 1.5 GPa. Hydrogenation of products **135** over 10% Pd/C gives amine substituted aniline trimers **136**. Aniline tetramers, pentamers and hexamers **137** have been prepared by successive high-pressure S_NAr reaction followed by nitro group reductions (Scheme 7.33).[62]

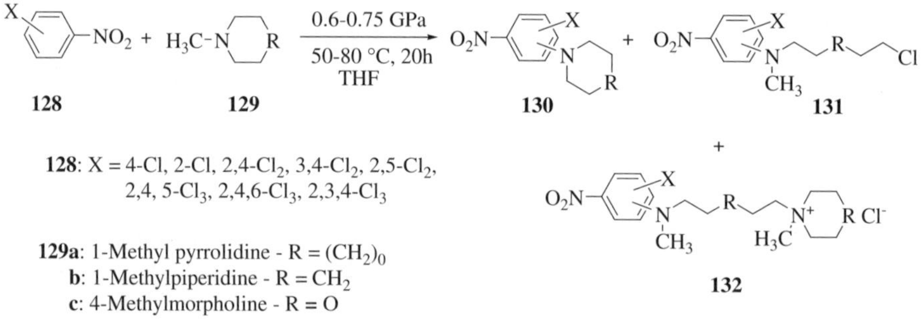

128: X = 4-Cl, 2-Cl, 2,4-Cl$_2$, 3,4-Cl$_2$, 2,5-Cl$_2$,
2,4, 5-Cl$_3$, 2,4,6-Cl$_3$, 2,3,4-Cl$_3$

129a: 1-Methyl pyrrolidine - R = (CH$_2$)$_0$
  b: 1-Methylpiperidine - R = CH$_2$
  c: 4-Methylmorpholine - R = O

Scheme 7.32

135 R = H, NO$_2$ (86-100%)

137 R = H, NH$_2$
n = 2-4, (65-68%)

136 R = H, NH$_2$ (97-100%)

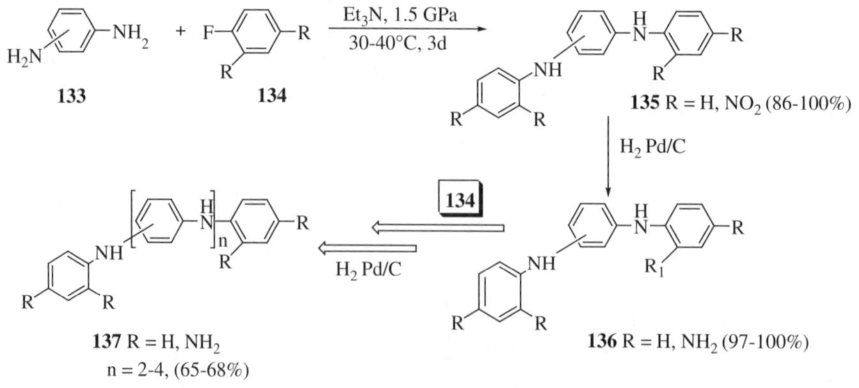

Scheme 7.33

Various heteroaromatic alkylamines, such as (di)alkylamino derivatives of pyridine, quinoline and pyrimidine, which are difficult to prepare at normal pressure, have been obtained in good to high yields by high-pressure S_NAr reactions of the corresponding heteroaromatic halides with various amines.[63] 4-(Di)alkylaminopyridine derivatives **140** have been synthesized *via* the high-pressure-promoted (0.8 GPa) S_NAr of 4-chloropyridine hydrochloride **(138)** with primary and secondary amines **139** (Scheme 7.34).[63d]

The use of high pressure is also essential for preparing racemic and enantiomerically pure C4 amino pyrazolo[3,4-*d*]pyrimidines **143** in satisfactory yield by S_NAr reaction of 4-amino-6-chloro-1-phenylpyrazolo[3,4-*d*]pyrimidine **(141)** with the appropriate 2-aminopropan-amides **142** in racemic or enantiomerically pure form. At 1.5 GPa pressure the (*S*)-enantiomer **143** has been prepared in good yield and shows a high affinity and selectivity for the adenosine A_1 receptor. The same reactions at atmospheric pressure have been unsuccessful[64] (Scheme 7.35).

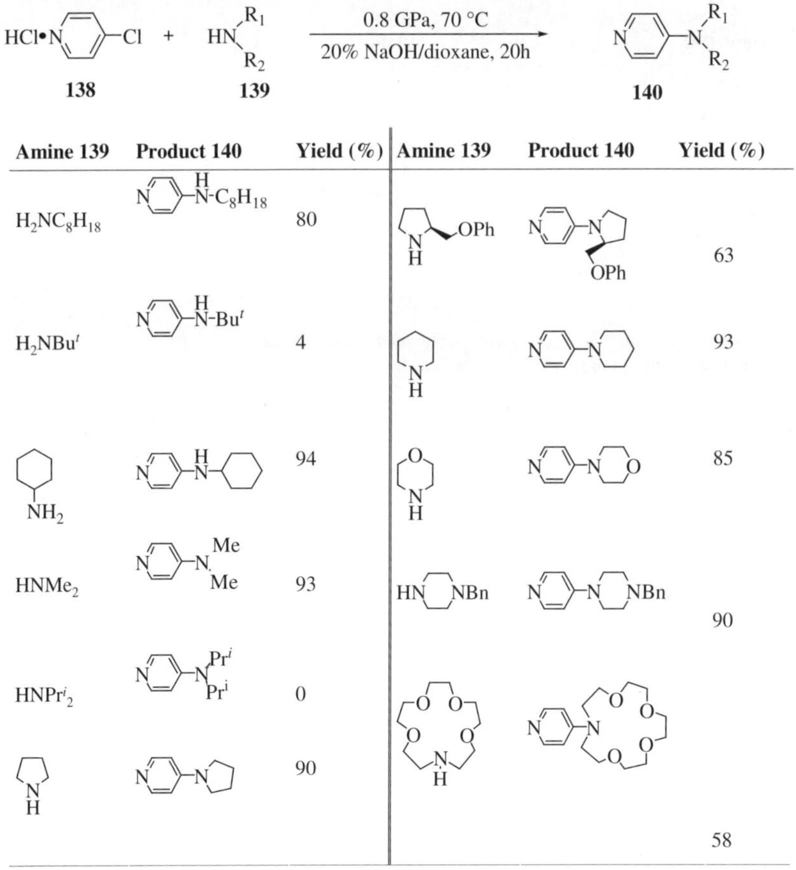

Amine 139	Product 140	Yield (%)	Amine 139	Product 140	Yield (%)
$H_2NC_8H_{18}$		80			63
H_2NBu^t		4			93
		94			85
$HNMe_2$		93			90
$HNPr^i_2$		0			
		90			58

Scheme 7.34

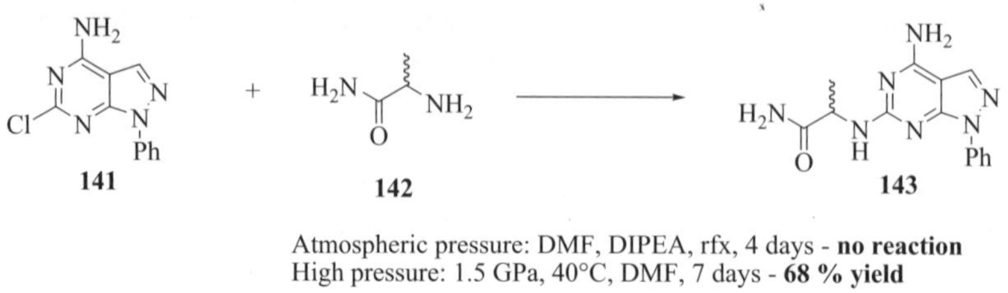

Atmospheric pressure: DMF, DIPEA, rfx, 4 days - **no reaction**
High pressure: 1.5 GPa, 40°C, DMF, 7 days - **68 % yield**

Scheme 7.35

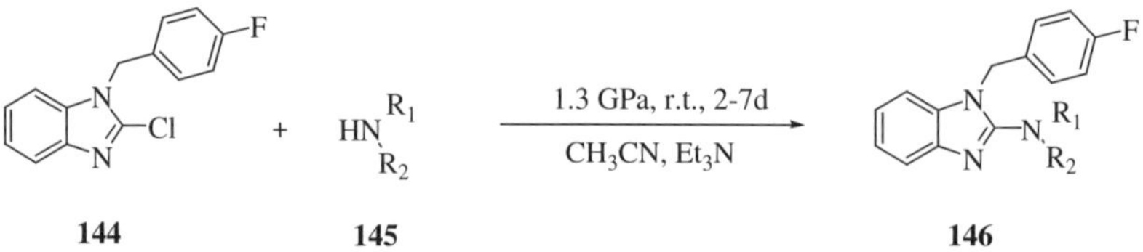

Amine 145	Yield (%)
R_1 = H, R_2 = ^{n}Bu	91 (51)[a]
R_1 = H, R_2 = Bn	83
R_1 = H, R_2 = ^{i}Pr	66 (No reaction)[a]
R_1 = H, R_2 = ^{t}Bu	No reaction
R_1 = R_2 = Et	80
R_1 = H, R_2 = p-CH$_3$OC$_6$H$_4$	No reaction
R_1-R_2 = morpholine	(90)[a]

[a]Values in parenthesis refer to yields obtained under thermal conditions (normal pressure, Rfx, n-butanol).

Scheme 7.36

Similarly, the high-pressure amination of *N-p*-fluorobenzyl-2-chlorobenzimidazole (**144**) provides access to various norastemizole analogues **146** that have a wide range of biological activities[65] (Scheme 7.36).

As shown in Scheme 7.36, primary and secondary amines under high pressure give high yields of adduct, except for anilines and sterically-hindered amines.

Finally, a new type of crown ether, such as some double-armed diaza-crown ethers directly connected to various heteroaromatic rings, has been obtained by high-pressure S_NAr reactions of heteroaromatic halides with the corresponding unsubstituted crown ethers.[66]

7.4   Condensation Reactions

Aldol reactions are one of the most important carbon–carbon bond-forming reactions in organic synthesis. These reactions are generally activated by base or Lewis acid catalysts and are a valuable tool for preparing optically active β-hydroxy carbonyl compounds by using various asymmetric catalysts. Non-metallic chiral catalysts, the so-called "chiral organocatalysts", such as L-proline have been used to develop an environmentally-friendly asymmetric methodology. However, these reactions usually require drastic conditions such as high temperatures, polar solvents and long reaction times. Aldol reactions and other related condensation reactions, involving bond-forming processes and ionic intermediates, are characterized by large negative activation volumes and are, therefore, greatly accelerated by high pressure.[67]

The L-proline-catalysed asymmetric aldol condensation between acetone (**147**) and various aldehydes has been efficiently promoted by applying high pressure (0.2 GPa). In these processes acetone is used as both reactant and solvent. As shown in Scheme 7.37, all of the reactions give the desired aldol products in high yield, except the reaction between acetone and isovaleraldehyde; the highest ee values are obtained with cyclohexyl- and naphthylaldehydes[68] (Scheme 7.37).

The high-pressure technique even promotes the diastereoselective catalytic asymmetric Henry (nitro-aldol) reaction of optically active α-aminoaldehydes with nitro-derivatives, which is one of the most valuable methods for ready access to non-natural 3-amino-2-hydroxy acids and 1,3-diamino-2-hydroxy compounds.[69]

Matsumoto *et al.* have reported the first example of a diastereoselective nitro-aldol reaction of optically active *N,N*-dibenzyl α-amino aldehydes **150** with nitroalkanes **151** under high pressure without a catalyst (Scheme 7.38).[69b] The reactions occurred at 0.8 GPa (room temperature, 12 h) and gave mixtures of diastereomers **152** and **153** in good yields and with high optical purity; under atmospheric pressure **150** did not react with **151**. Very high diastereoselectivity (**152**/**153** = 99) was observed in the reaction of **150a** with 2-nitropropane (**151c**).

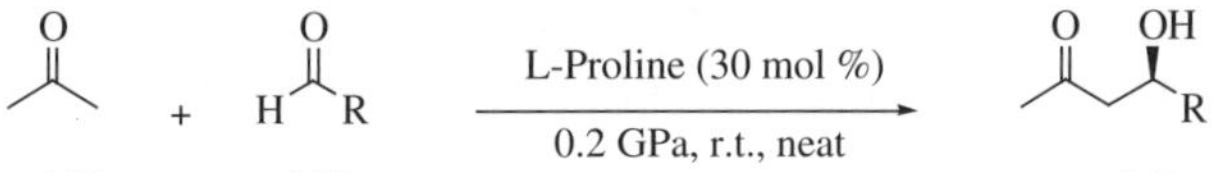

R	% Yield (% ee)
C_6H_5	90 (72)
p-$NO_2C_6H_4$	89 (64)
p-BrC_6H_4	89 (70)
o-ClC_6H_4	92 (70)
Naphthyl	95 (92)
Cyclohexyl	84 (92)
^iPr	32 (62)

Scheme 7.37

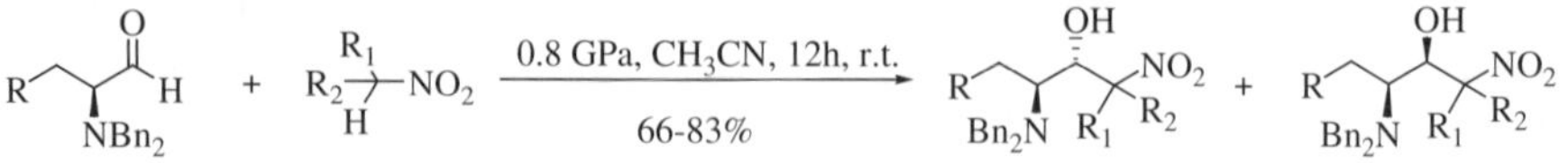

150a, R = Ph   **151 a**, R$_1$ = R$_2$ = H   **152**   **153**
   b, R = Me   **b**, R$_1$ = Me, R$_2$ = H
   c, R = ^{i}Pr   **c**, R$_1$ = R$_2$ = Me
   d, R = ^{i}Bu   **152/153** = 2.44-99
                                               ee (**152**) = 90-99%

Scheme 7.38

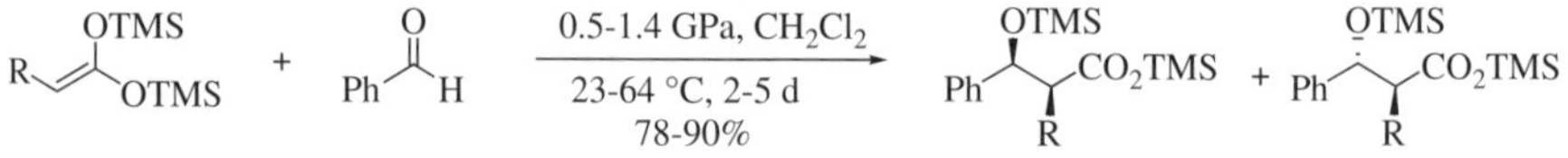

154   **155**   syn **156**   anti **157**

R = Me, Et, iPr   **156/157** = 1.63-5.25

Scheme 7.39

Other important aldol condensations are the Mukaiyama-type aldol reactions of silyl enol ethers with aldehydes that usually require catalyst activation. Yamamoto reported that such reactions under high pressure proceed: (i) without catalyst even at room temperature, (ii) without isomerization of the formed adducts and (iii) with a reversed *syn/anti* stereoselectivity compared with that of the TiCl$_4$-catalysed reactions.[70]

Dumas *et al.* noted the good yields and *syn* diastereoselectivities obtained in a high-pressure aldol reaction of bis-silyl ketene acetals **154** with benzaldehyde (**155**) (Scheme 7.39). The *syn* aldol **156** was obtained with a diastereoselectivity that was significantly correlated with the steric bulkiness of the R-substituent in the acetals **154**. The preference for *syn* bis-silyl aldols **156** has been attributed to the reaction pathway that involves compact transition states in which steric interactions between the R substituent of **154** and the phenyl group of benzaldehyde are minimized. The authors also studied the condensation of unsaturated bis-silyl ketene acetal as a model for the synthesis of retinoid compounds.[71]

Michael condensations, together with the Diels–Alder cycloaddition and the aldol reactions, are the most powerful and useful bond-forming reactions in synthetic organic chemistry. Like the nitro-aldol (Henry) additions, nitroalkanes are particularly appropriate reagents in Michael reactions; they act as α-hydrogen donors. Nitroalkanes react easily with typical Michael acceptors such as α,β-unsaturated aldehydes or ketones under base or Lewis acid catalysis.[72]

Jenner has investigated the effect of kinetic pressure on some nitro Michael and Henry reactions and has concluded that, although the kinetic effect is less than that reported for [4 + 2] cycloadditions, high pressure is a valuable tool

R	Experimental conditions[a]	Time (d)	Yield (%)
COCH$_3$	A	1	50
	B	2	95
COOCH$_3$	A	1	81
	B	12	0

[a]Experimental Conditions: A, Yb(OTf)$_3$ + H$_2$O in CH$_3$CN, 0.8 GPa, 60 °C; B, Yb(OTf)$_3$-SiO$_2$, 1 atm, r.t.

Scheme 7.40

for performing both nitro Michael and Henry reactions, because it improves the yields as well as the selectivity.[73] Theoretical investigations of the influence of pressure on the selectivity of the Michael addition of diphenylmethaneamine to stereogenic crotonates have also been reported.[74]

Michael addition reactions of β-ketoesters with various α,β-unsaturated carbonyl compounds have been performed using a Yb(OTf)$_3$ catalyst on either silica gel surface or in combination with 0.8 GPa pressure (Scheme 7.40).

Although the reactions on silica gel at normal pressure are relatively slow compared with the high pressure reactions, this method is preferred for conjugated enone systems. In contrast, acrylic esters react most successfully under high pressure conditions.[75] For example, the reaction of **159** with methyl vinyl ketone (**158** R=COCH$_3$) at normal or under high pressure gives comparable yields. In contrast, cyclohexanone **159** reacts with acrylic ester (**158** R=COOCH$_3$) to give a high yield at 0.8 GPa while there is no reaction at normal pressure.

Pressure creates interesting possibilities for controlling the course of a Michael reaction. The aza-Michael reaction of methyl 4-*tert*-butylcyclohexylidene ester **161** with cyclic and acyclic amines gives various products, depending on the experimental conditions and the nature of the amine.

Under thermal activation conditions, ester **161** reacts with benzylamine to give a mixture of α-amino-β,γ-unsaturated ester **164** and amide **165**, but under 1.1 GPa pressure ester **161** gives the spiro-aziridines **162** and **163** in high yield and good diastereoselectivity. These are part of an attractive class of compounds that are formed by hetero-Michael addition of amine onto the unsaturated ester **161**, followed by intramolecular nucleophilic substitution of the bromine atom (Scheme 7.41).[76]

The high-pressure aza-Michael addition of mono- and diamines on α,β-unsaturated diesters provide a practical route for preparing some β-aminoesters and novel bridged β-aminodiesters.[77]

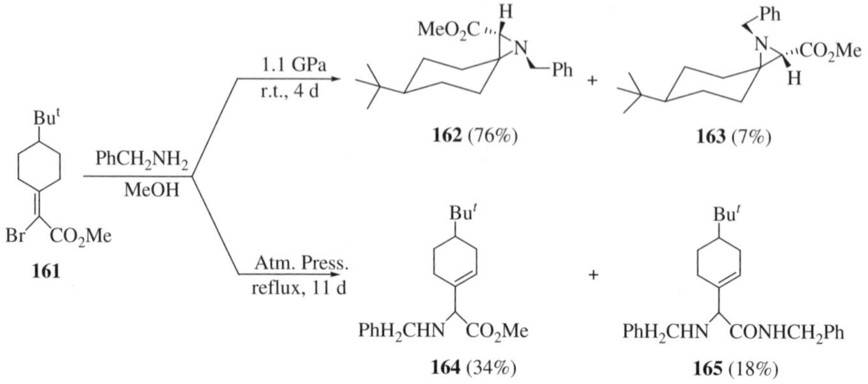

Scheme 7.41

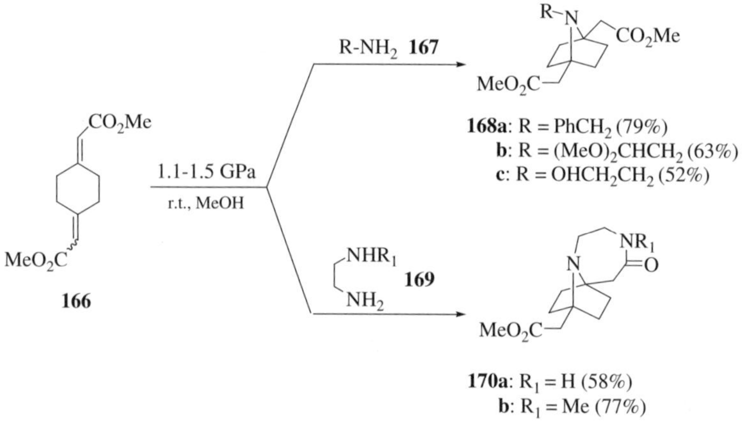

Scheme 7.42

The hyperbaric addition of primary amines **167** to diester **166** leads directly to bridgehead β-aminodiesters **168** in good yields; the combination of **166** and diamines **169**, under the same conditions, affords the tricyclic derivatives **170** in a single step (Scheme 7.42).[77]

The high-pressure technique is also very useful for promoting the conjugated addition of 1-methylindole (**171**) to α,β-unsaturated ketones **172**. The more reactive and less sterically hindered electrophiles usually give the expected 3-akylated indoles in good yields at normal pressure in the presence of $Yb(OTf)_3$ catalyst. When there is significant steric hindrance, as in the reaction between **171** and enones **172**, the reaction time is greatly reduced and yields are significantly improved when pressure (1.3 GPa) is applied (Scheme 7.43).[78]

A recent study shows fair yields, with a high degree of regio-, diastereo- and enantioselectivity, in the high pressure-mediated Michael addition of chiral imines to methyl and phenyl crotonates[79] (Scheme 7.44).

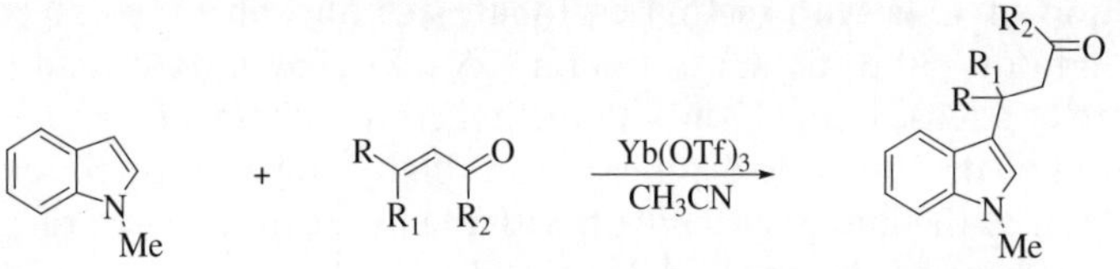

Enone 172	Time	Pressure	Yield (%)
$R = R_1 = R_2 = Me$	7 d	Atmospheric	38
	2 d	1.3 GPa	72
$R = {}^{i}Pr; R_1 = H; R_2 = Me$	4 d	Atmospheric	77
	18 h	1.3 GPa	84
$R = Ph; R_1 = H; R_2 = Me$	7 d	Atmospheric	10
	26 h	1.3 GPa	67
2-Cyclohexen-1-one	7 d	Atmospheric	37
	2 d	1.3 GPa	56
3-Methyl-2-cyclohexen-1-one	7 d	Atmospheric	3
	7 d	1.3 GPa	11

Scheme 7.43

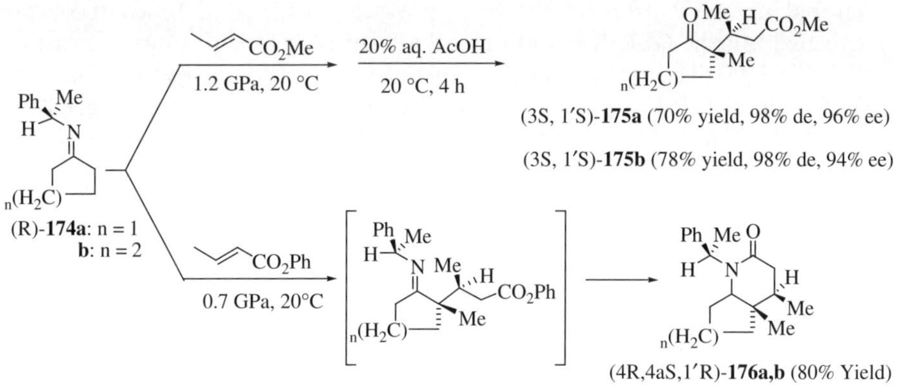

Scheme 7.44

High-pressure condensation of imines (R)-**174a,b** with methyl crotonate gives the corresponding adducts, which when submitted to hydrolysis (20% aqueous AcOH, THF) afford ketoesters $(3S,1'S)$-**175a,b** in 70 and 78% yields, respectively, and 94–98% diastereomeric and enantiomeric excesses.

The reaction of **174a** with methyl crotonate at atmospheric pressure requires forced condition (30 d, 68 °C) to obtain (3*S*,1′*S*)-**175a** in low yield (34%); all attempts to condense imine **174b** with methyl crotonate under standard thermal conditions failed. Under high pressure or high temperature conditions, the condensation of the imines (*R*)-**174a,b** with phenyl crotonate invariably affords enantiopure bicyclic lactams (4*R*,4a*S*,1′*R*)-**176a,b**. These are obtained by N-heterocyclization of the corresponding transient Michael adducts;[79] under atmospheric pressure (120 °C, 5 d) the yields of **176a** and **176b** were 50% and 60%, respectively.

High pressure has also been used in the ketalization of ketones and oxy-Michael/ketalization of conjugated enones without using any catalyst (Scheme 7.45).

At 0.8 GPa pressure and in the presence of trimethyl orthoformate as water scavenger, various saturated cyclic and acyclic ketones **177** were converted nearly quantitatively into the corresponding dimethyl ketals **178** under essentially neutral conditions. Under the same reaction conditions, several α,β-unsaturated cyclic and acyclic ketones **179** have given the oxy-Michael/ketal derivatives **180** in good to excellent yields[80] (Scheme 7.45).

7.5 Miscellaneous Reactions

Many other important organic reactions, such as the Mannich,[81] Heck,[82] ene,[83] S_EAr[84] and Wittig[85] reactions are accelerated by high pressure.

For example, pressure has an effect on Friedel–Crafts benzoylation.

The benzoylation yield of deactivated fluorobenzene (**181**, R=F) increases ten-fold on going from atmospheric pressure to 0.5 GPa. In contrast, for activated toluene (**181**, R=Me) the yield decreases as the pressure increases[86] (Scheme 7.46). These results can be explained in terms of competition between the favourable effect of pressure which allows the transition states between ArH and the 1 : 1 acyl chloride complex (TS1) or the acyllium (TS2) to be reached, and the unfavourable effect that pressure has on the subsequent HCl

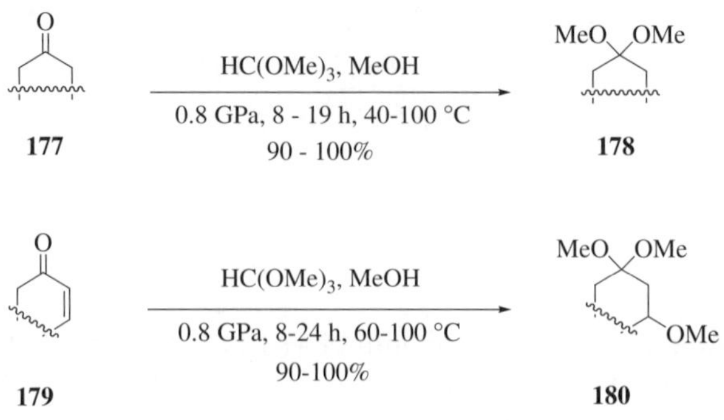

177   HC(OMe)₃, MeOH   0.8 GPa, 8 - 19 h, 40-100 °C   90 - 100%   178

179   HC(OMe)₃, MeOH   0.8 GPa, 8-24 h, 60-100 °C   90-100%   180

Scheme 7.45

elimination step. Since activated toluene does not need any activation to reach TS1 or TS2, the limiting step of the reaction becomes HCl elimination, which is increasingly disfavoured as the pressure increases.

However, the combined use of high pressure (1.0 GPa) and chiral (salen) Co(II) complex (R,R)-**186** as catalyst allows the synthesis of optically active furfuryl alcohols (R)-**187** and (S)-**188** from 2-methylfuran (**184**) in good yield (up to 70%) and with enantioselectivity (up to 76% ee)[87] (Scheme 7.47). Under atmospheric pressure, the reaction of **184** and **185** in the presence of the catalyst (R,R)-**186** gives a negligible yield (15%) and low enantioselectivity (44% ee).

The high-pressure technique is also very useful when organometalloid reagents are used for difficult organic synthesis.

When pressure (1.5 GPa) is applied at room temperature on the reaction of $[Pt_2(\mu\text{-}S)_2(dppp)]$ (**189**) with an excess of α-α'-dichloro-o-xylene (**190**) a product mixture is obtained from which 3,8-dibenzo-1,6-dithiacyclodecane **192** (35%) has been isolated. Apart from its pharmacological value, it could be used in molecular sensors and conductive polymers. The one-pot formation of **192** at 1.5 GPa is a much more convenient synthetic strategy than the alternative multistep synthesis of this macrocyclic dithioether[88] (Scheme 7.48).

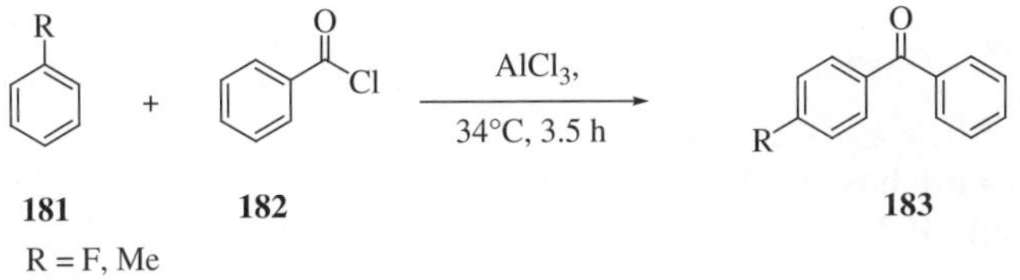

Scheme 7.46

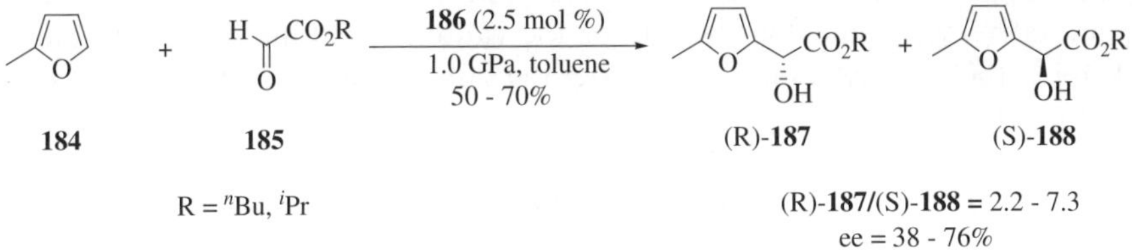

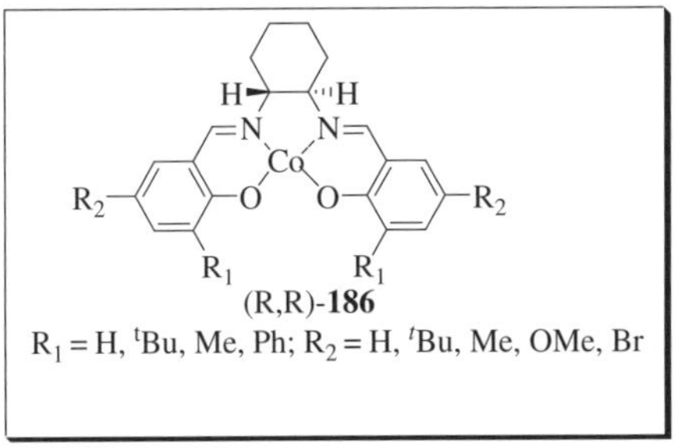

Scheme 7.47

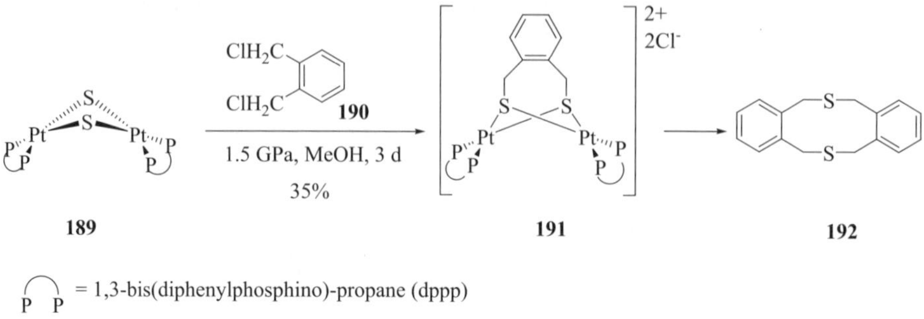

Scheme 7.48

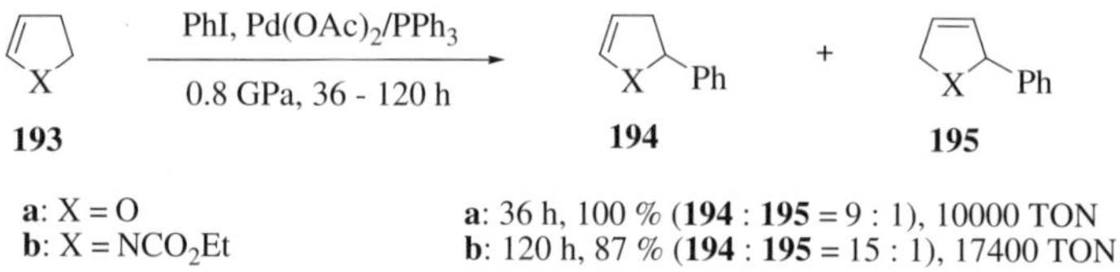

Scheme 7.49

a: X = O
b: X = NCO$_2$Et

a: 36 h, 100 % (**194** : **195** = 9 : 1), 10000 TON
b: 120 h, 87 % (**194** : **195** = 15 : 1), 17400 TON

High pressure has also been used in the palladium-catalysed arylation of 2,3-dihydrofuran (**193a**) as tool to increase the reaction yields and the lifetime of the catalyst, which is reflected in the turnover number (TON). At normal pressure, a 9 : 1 mixture of products **194a** and **195a** is obtained in low yield (up to 27%) with a maximum TON for catalyst [Pd(OAc)$_2$/PPh$_3$] of about 250–280. Under high pressure (0.8 GPa) the 9 : 1 mixture of **194a/195a** is obtained in quantitative yield and the lifetime of the catalyst is dramatically prolonged (TON > 10000)[89] (Scheme 7.49). Under the same conditions similar results are obtained for 2,3-dihydropyrrole **193b**.

Under high-pressure (0.8 GPa) the alkenylation of carbonyl compounds **196** with phosphonate **197** [Horner–Wadsworth–Emmons (HWE) reaction] is accomplished efficiently (50–90% yields) at room temperature in the presence of triethylamine without further activation by Lewis acid. In contrast, reactions carried out at atmospheric pressure give negligible yields (0–8%)[90] (Scheme 7.50).

Based on this protocol, a domino process has been developed that combines the HWE-reaction with a Michael addition. A mixture of an aldehyde **196**, a phosphonate **197** and a nucleophile (various piperidines, secondary amines or thiols) in the presence of triethylamine at 0.8 GPa allows a one-pot synthesis of β-amino- or β-thio-esters with the general structure **199** (Scheme 7.50).[91]

Another application of high pressure in combinatorial chemistry has been reported by Matsumoto and Jenner.[91] For the first time, uncatalysed, high-pressure (0.6 GPa) three-component Strecker synthesis of α-amino nitriles **202** was carried out in high yields by reacting aniline (**200**) with various ketones **201**

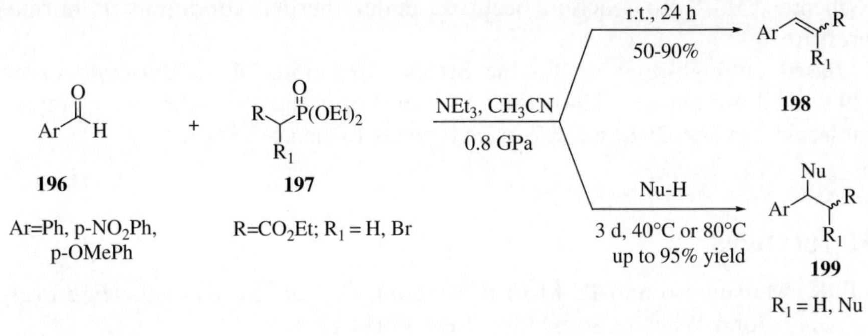

196 Ar=Ph, p-NO$_2$Ph, p-OMePh

197 R=CO$_2$Et; R$_1$ = H, Br

r.t., 24 h
50-90%
198

Nu-H
3 d, 40°C or 80°C
up to 95% yield
199
R$_1$ = H, Nu

NEt$_3$, CH$_3$CN
0.8 GPa

Scheme 7.50

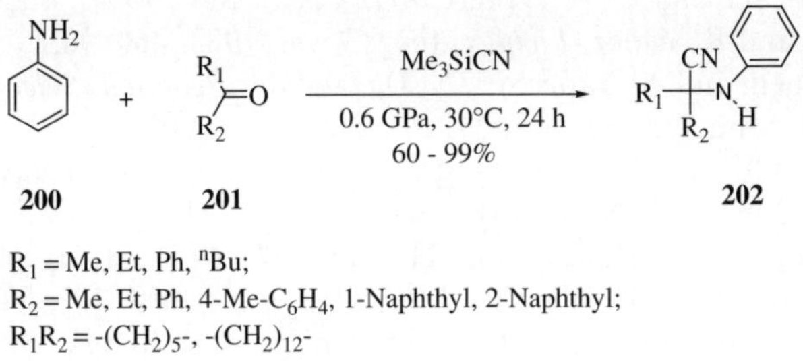

200

201

Me$_3$SiCN
0.6 GPa, 30°C, 24 h
60 - 99%

202

R$_1$ = Me, Et, Ph, ^{n}Bu;
R$_2$ = Me, Et, Ph, 4-Me-C$_6$H$_4$, 1-Naphthyl, 2-Naphthyl;
R$_1$R$_2$ = -(CH$_2$)$_5$-, -(CH$_2$)$_{12}$-

Scheme 7.51

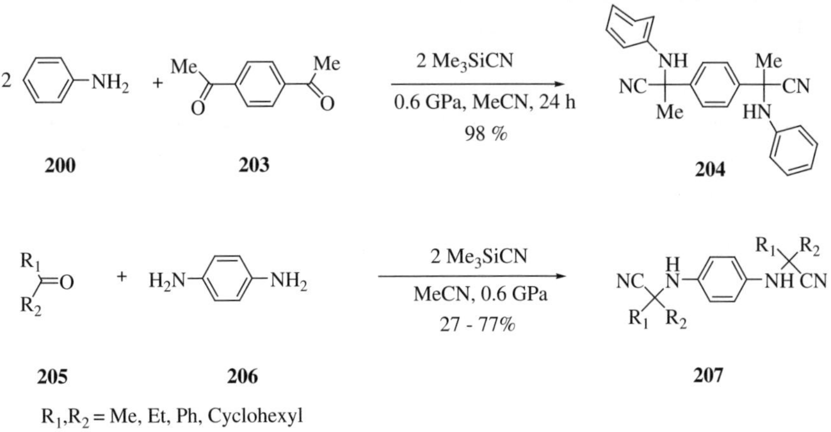

200

203

2 Me$_3$SiCN
0.6 GPa, MeCN, 24 h
98 %

204

205

206

2 Me$_3$SiCN
MeCN, 0.6 GPa
27 - 77%

207

R$_1$,R$_2$ = Me, Et, Ph, Cyclohexyl

Scheme 7.52

(Scheme 7.51);[91] no reaction occurred under thermal conditions at normal pressure.

Based on high-pressure double Strecker reactions of 1,4-diacetylbenzene (**203**) or 1,4-diamines (**206**), the same authors[91] have synthesized complex molecules such as **204** and **207** in high yields (Scheme 7.52).

References

1. K. Matsumoto and R. Morrin Acheson, *Organic Synthesis at High Pressure*, John Wiley & Sons, Inc., New York, 1991.
2. R. van Eldik and F.-G. Klärner, *High Pressure Chemistry-Synthetic, Mechanistic, and Supercritical Applications*, Wiley-VCH, Weinheim, 2002.
3. T. Asano and W. J. le Noble, *Chem. Rev.*, 1978, **78**, 407.
4. W. J. le Noble and H. Kelm, *Angew. Chem. Int. Ed. Engl.*, 1980, **19**, 841.
5. N. S. Isaacs, *Tetrahedron Report N. 299*, 1991, **47**, 8463.
6. J. R. McCabe and C. A. Ackert, *Acc. Chem. Res.*, 1974, **7**, 251.
7. O. Diels and K. Alder, *Liebigs Ann. Chem.*, 1928, **460**, 98.
8. F. Fringuelli and A. Taticchi, *The Diels–Alder Reaction. Selected Practical Methods*, John Wiley & Sons, Ltd, New York, 2002.
9. F. Fringuelli and A. Taticchi, *Dienes in the Diels–Alder Reaction*, John Wiley & Sons, Ltd, New York, 1990.
10. K. Matsumoto, H. Hamana and H. Iida, *Helv. Chim. Acta*, 2005, **88**, 2033.
11. F.-G. Klarner and F. Wurche, *J. Prakt. Chem.*, 2000, **342**, 609.
12. M. Ciobannand and K. Matsumoto, *Liebigs Ann./Recuil*, 1997, 623.
13. L. Minuti, A. Taticchi and L. Costantini, *Recent Res. Devel. Organic Chem.*, 1999, **3**, 105.
14. G. Jenner, *Tetrahedron*, 2002, **58**, 5185.
15. V. Pindur, G. Lutz and C. Otto, *Chem. Rev.*, 1993, **93**, 741.
16. K. Matsumoto and A. Sera, *Synthesis*, 1985, 999.
17. (a) T. Li and Y. L. Wu, *J. Am. Chem. Soc.*, 1981, **103**, 7007; (b) W. G. Dauben and W. R. Baker, *Tetrahedron Lett.*, 1982, **23**, 2611.
18. R. Girotti, A. Marrocchi, L. Minuti, O. Piermatti, F. Pizzo and L. Vaccaro, *J. Org. Chem.*, 2006, **71**, 70.
19. (a) L. Minuti, A. Taticchi, E. Gacs-Baitz and E. Wenkert, *Tetrahedron*, 1995, **51**, 10033; (b) L. Minuti, R. Selvaggi, A. Taticchi, M. Guo and E. Wenkert, *Can. J. Chem.*, 1922, **70**, 1481.
20. E. Peredes, B. Biolatto, M. Kneeteman and P. Mancini, *Tetrahedron Lett.*, 2000, **41**, 8079.
21. L. Minuti, A. Taticchi, E. Gacs-Baitz and A. Marrocchi, *Tetrahedron*, 1998, **54**, 10891.
22. F. B. Mallory, K. E. Buttler and A. C. Evans, *Tetrahedron Lett.*, 1996, **37**, 7173.
23. (a) F. Fringuelli, A. Taticchi and E. Wenkert, *Org. Prep. Proced. Int.*, 1990, **22**, 131; (b) F. Fringuelli, L. Minuti, F. Pizzo and A. Taticchi, *Acta Chem. Scand.*, 1993, **47**, 255.

24. L. Minuti, A. Taticchi, A. Marrocchi, L. Costantini and E. Gacs-Baitz, *Tetrahedron: Asymmetry*, 2001, **12**, 1179.
25. L. Minuti, A. Taticchi, A. Marrocchi, A. Broggi and E. Gacs-Baitz, *Polycyclic Aromatic Compounds*, 2003, **23**, 483.
26. (a) F. A. L. Anet, *Tetrahedron Lett.*, 1962, 1219; (b) H. Kwart and I. Burchuk, *J. Am. Chem. Soc.*, 1952, **74**, 1219.
27. (a) N. Kawamura, Y.-M. Li, J. L. Engel, W. G. Dauben and J. E. Caside, *Chem. Res. Toxicol.*, 1990, **3**, 318; (b) W. G. Dauben, J. Y. L. Lam and Z. R. Guo, *J. Org. Chem.*, 1996, **61**, 4816.
28. (a) K. Tamura and T. Imoto, *Bull. Chem. Soc. Jpn*, 1975, **48**, 369; (b) H. Kotsuki, S. Kitogawa, H. Nishizawa and T. Tokoroyama, *J. Org. Chem.*, 1978, **43**, 1471; (c) H. Kotsuki, H. Nishizawa, S. Kitagawa, M. Ochi, N. Yamasaki, K. Matsuoka and T. Tokoroyama, *Bull. Chem. Soc. Jpn*, 1979, **52**, 544.
29. K. Kumamoto, I. Fukada and H. Kotsuki, *Angew. Chem. Int. Ed.*, 2004, **43**, 2015.
30. H. S. Prakash Rao, R. Murali, A. Taticchi and H. W. Scheeren, *Eur. J. Org. Chem.*, 2001, 2869.
31. (a) F. Fringuelli, F. Pizzo, A. Taticchi, E. Wenkert and T. D. J. Halls, *J. Org. Chem.*, 1982, **47**, 5056; (b) F. Fringuelli, F. Pizzo, A. Taticchi and E. Wenkert, *J. Org. Chem.*, 1983, **48**, 2802; (c) L. Minuti, A. Taticchi, R. Selvaggi and P. Sandor, *Tetrahedron*, 1993, **49**, 1071.
32. A. Marrocchi, L. Minuti, A. Taticchi and H. W. Scheeren, *Tetrahedron*, 2001, **57**, 4959.
33. L. Minuti, A. Taticchi, *Seminars in Organic Synthesis*, 22nd Summer School A. Corbella, Italian Chemical Society, 1997, 51.
34. J. Jurczak, M. Chemielewski and S. Filipek, *Synthesis*, 1979, 41.
35. M. Chemielewski and J. Jurczak, *J. Org. Chem.*, 1981, **46**, 2230.
36. M. Malinowska, P. Kwiatkowski and J. Jurczak, *Tetrahedron Lett.*, 2004, **45**, 7693.
37. D. L. Boger and K. D. Roberge, *J. Org. Chem.*, 1988, **53**, 3373.
38. D. L. Boger and K. D. Robarge, *J. Org. Chem.*, 1988, **53**, 5793.
39. A. de Meijere, A. Leonov, T. Heiner, M. Noltemeyer, M. Terosobes, *Eur. J. Org. Chem.*, 2003, 472. For previous examples of Diels Alder reactions of methylenecyclopropane derivatives under normal or high pressure conditions, see ref 7 of the paper.
40. B. Augustyns, N. Maulide and I. E. Markò, *Tetrahedron Lett.*, 2005, **46**, 3895.
41. S. Jarosz, E. Kolowska and A. Jezewski, *Tetrahedron*, 1997, **53**, 10775.
42. Y. Araki and T. Konoike, *J. Org. Chem.*, 1997, **62**, 5299.
43. T. Heiner, S. Michalski, G. Gerke, G. Kutchta, M. Buback and A. de Mejere, *Synlett*, 1995, 355.
44. M. Brunnere, G. Maas and F.-G. Klärner, *Helv. Chim. Acta*, 2005, **88**, 1813.
45. S. E. Deumark and A. Thorareusen, *Chem. Rev.*, 1996, **96**, 137.
46. I. F. Tietze and U. Beifuss, *Angew. Chem. Int. Ed.*, 1993, **32**, 131.

47. G. J. T. Kuster, H. W. Scheeren, Ch. 9, 284-304, of High Pressure Chemistry, Ed. Rudi van Eldik and F.-G. Klärner.

48. R. M. Uittenbogaard, J.-P. G. Seerden and H. W. Scheeren, *Tetrahedron*, 1997, **53**, 11929.

49. A. Barco, S. Benetti, G. Pollini, G. Spallato and V. Zanirato, *Tetrahedron Lett.*, 1991, **32**, 2517.

50. A. Papchikhin, P. Agback, J. Plavec and J. Chattopadhyaya, *J. Org. Chem.*, 1993, **58**, 2874.

51. G. J. Kuster, F. Kalmona, R. de Gelder and H. W. Scheeren, *Chem. Commun.*, 1999, 855.

52. L. W. A. van Berkom, G. J. T. Kuster, F. Kalmona, R. de Gelder and H. W. Scheeren, *Tetrahedron Lett.*, 2003, **44**, 5091.

53. C. Cameron, P. P. S. Saluja, M. A. Floriano and E. Whalley, *J. Phys. Chem.*, 1988, **92**, 3417.

54. H. Kotsuki, M. Kataoka and H. Nishizawa, *Tetrahedron Lett.*, 1993, **34**, 4031.

55. (a) H. Kotsuki, M. Nishiuchi, S. Kobayashi and H. Nishizawa, *J. Org. Chem.*, 1990, **55**, 2969; (b) H. Kotsuki, K. Hayashida, T. Shimanouchi and H. Hayakawa, *J. Org. Chem.*, 1996, **61**, 984; (c) H. Kotsuki, M. Wakao, H. Nishizawa, T. Shimanouchi and M. Shiro, *J. Org. Chem.*, 1996, **61**, 8915; (d) H. Kotsuki, M. Teraguchi, N. Shimomoto and M. Ochi, *Tetrahedron Lett.*, 1996, **37**, 3727; (e) H. Kotsuki, H. Hayakawa, M. Wakao, T. Shimanouchi and M. Ochi, *Tetrahedron: Asymmetry*, 1995, **6**, 2665.

56. H. Kotsuki, T. Shimanouchi, M. Teraguchi, M. Kataoka, A. Tatsukawa and H. Nishizawa, *Chem. Lett.*, 1994, 2159.

57. (a) H. Kotsuki, K. Matsumoto and H. Nishizawa, *Tetrahedron Lett.*, 1991, **32**, 4155; (b) H. Kotsuki, T. Kusumi, M. Inoue, Y. Ushio and M. Ochi, *Tetrahedron Lett.*, 1991, **32**, 4159.

58. Y. Yamamoto, T. Furuta, J. Matsuo and T. Kurata, *J. Org. Chem.*, 1991, **56**, 5737.

59. (a) T. Ibata, Y. Isogami and J. Toyoda, *Bull. Chem. Soc. Jpn.*, 1991, **64**, 42; (b) H. Kotsuki, S. Kobayashi, H. Suenaga and H. Nishizawa, *Synthesis*, 1990, 1145; (c) H. Kotsuki, S. Kobayashi, K. Matsumoto, H. Suenaga and H. Nishizawa, *Synthesis*, 1990, 1147.

60. (a) T. Ibata, X. Zou and T. Demura, *Tetrahedron Lett.*, 1993, **34**, 5613; (b) T. Ibata, X. Zou and T. Demura, *Bull. Chem. Soc. Jpn.*, 1994, **67**, 196; (c) T. Ibata, X. Zou and T. Demura, *Bull. Chem. Soc. Jpn.*, 1995, **68**, 3227.

61. (a) T. Ibata, M.-H. Shang and T. Demura, *Chem. Lett.*, 1994, 359; (b) T. Ibata, M-H. Shang and T. Demura, *Bull. Chem. Soc. Jpn.*, 1995, **68**, 2717; (c) T. Ibata, M.-H. Shang and T. Demura, *Bull. Chem. Soc. Jpn.*, 1995, **68**, 2941; (d) T. Ibata, M.-H. Shang and T. Demura, *Heterocyclic Commun.*, 1995, **1**, 241.

62. C. L. Brown, I. W. Muderawan and D. J. Young, *Synthesis*, 2003, **16**, 2511.

63. (a) K. Matsumoto, S. Hashimoto and S. Otani, *J. Chem. Soc., Chem. Commun.*, 1991, 306; (b) K. Matsumoto, M. Toda and S. Hashimoto,

Chem. Lett., 1991, 1283; (c) K. Matsumoto, S. Hashimoto, M. Hashimoto and M. Toda, *Synth. Commun.*, 1992, **22**, 787; (d) H. Kotsuki, H. Sakai and T. Shinohara, *Synlett*, 2000, **1**, 116.

64. S.-A. Poulsen, D. J. Young and R. J. Quinn, *Bioorg. Med. Chem. Lett.*, 2001, **11**, 191.

65. I. C. Barrett and M. A. Kerr, *Tetrahedron Lett*, 1999, **40**, 2439.

66. (a) H. Tsukube, H. Minatogawa, M. Munataka, M. Toda and K. Matsumoto, *J. Org. Chem.*, 1992, **57**, 542; (b) K. Matsumoto, M. Hashimoto, M. Toda, H. Tsukube and T. Uchida, *Chem. Express*, 1993, **8**, 105; (c) K. Matsumoto, M. Hashimoto, M. Toda and H. Tsukube, *J. Chem. Soc. Perkin Trans., 1*, 1995, 2497; (d) M. Toda, H. Tsukube, H. Minatogawa, K. Hirotsu, I. Miyahara, T. Higuchi and K. Matsumoto, *Supramolecular Chemistry*, 1993, **2**, 289.

67. (a) G. Jenner, *Tetrahedron*, 2002, **58**, 5185; (b) M. Bellassoued and E. Reboul, *F. Dumas Tetrahedron Lett.*, 1997, **38**, 1233; (c) G. Jenner, *Tetrahedron Lett.*, 2001, **42**, 243.

68. Y. Sekiguchi, A. Sasaoka, A. Shimomoto, S. Fujioka and H. Kotsuki, *Synlett*, 2003, **11**, 1655.

69. (a) K. Matsumoto, *Angew. Chem. Int. Ed.*, 1984, **23**, 617; (b) Y. Misumi and K. Matsumoto, *Angew. Chem. Int., Ed.*, 2002, **41**, 1031.

70. (a) Y. Yamamoto and K. Maruyama, *J. Am. Chem. Soc.*, 1983, **105**, 6963; (b) Y. Yamamoto, K. Maruyama and K. Matsumoto, *Tetrahedron Lett.*, 1984, **25**, 1075.

71. M. Bellassoued, E. Reboul and F. Dumas, *Tetrahedron Lett.*, 1997, **32**, 5631.

72. (a) J. Boyer, R. Corriù, R. Perz and C. Reye, *Tetrahedron*, 1983, **39**, 117; (b) Y. Nishigaichi, A. Takuwa, Y. Naruta and K. Maruyama, *Tetrahedron*, 1993, **49**, 7395.

73. G. Jenner, *New J. Chem.*, 1999, **23**, 525.

74. F. Dumas, C. Fressignè, J. Langlet and C. Giessner-Prettre, *J. Org. Chem.*, 1999, **64**, 4725.

75. H. Kotsuki and K. Arimura, *Tetrahedron Lett.*, 1997, **38**, 7583.

76. A. Yu. Rulev, J. Maddaluno, G. Plè, J.-C. Plaquevent and L. Duhamel, *J. Chem. Soc.; Perkin Trans.*, 1998, **1**, 1397.

77. A. Yu. Rulev, N. Yenil, A. Pesquet, H. Oulyadi and J. Maddaluno, *Tetrahedron*, 2006, **62**, 5411.

78. P. Harrington and M. A. Kerr, *Can. J. Chem.*, 1998, **76**, 1256.

79. C. Camara, D. Joseph, F. Dumas, J. d'Angelo and A. Chiarori, *Tetrahedron Lett.*, 2002, **43**, 1445.

80. K. Kumamoto, Y. Ichikawa and H. Kotsuki, *Synlett*, 2005, **14**, 2254.

81. (a) K. Matsumoto, S. Hashimoto, T. Uchida, H. Iida and S. Otani, *Chem. Express*, 1993, **8**, 475; (b) K. Matsumoto, T. Uchida, S. Hashimoto, Y. Yonezawa, H. Iida, A. kakeni and S. Otani, *Heterocycles*, 1993, **36**, 2215.

82. (a) K. Voigt, U. Schick, F. E. Meyer and A. De Meijere, *Synlett*, 1994, 189; (b) T. Sugihara, M. Takebayashi and C. Kaneko, *Tetrahedron Lett.*, 1995, **36**, 5547.

83. (a) W. G. Dauben and R. T. Hendrcks, *Tetrahedron Lett.*, 1992, **33**, 603; (b) H. Remoto, Y. Kubota, N. Sasaki and Y. Yamamoto, *Synlett*, 1993, 465.

84. (a) D. W. Coillet and S. D. Hamann, *Trans. Faraday Soc.*, 1961, **57**, 2231; (b) T. Asano, *Bull. Chem. Soc. Jpn*, 1969, **42**, 2005; (c) D. W. Coilet and S. D. Hamann, *Nature*, 1963, **200**, 166; (d) B. S. Farah, E. E. Gilbert and J. P. Sibilia, *J. Org. Chem.*, 1965, **30**, 998; (e) D. W. Coilet, S. D. Hamann and E. F. McCoy, *Aust. J. Chem.*, 1965, **18**, 1911; (f) S. Jarosz, P. Salanski and M. Mach, *Tetrahedron*, 1998, **54**, 2583.

85. (a) E. Herbert, Z. Welvart, M. Ghelfenstein and H. Szwarc, *Tetrahedron Lett.*, 1983, **24**, 1381; (b) N. S. Isaacs and O. H. Oben, *Tetrahedron Lett.*, 1986, **27**, 995; (c) W. G. Dauben and J. J. Takasugi, *Tetrahedron Lett.*, 1987, **28**, 4377; (d) N. S. Isaacs and G. N. El-Din, *Tetrahedron Lett.*, 1987, **28**, 2191.

86. F. Mugnier, A. Harrison-Marchand, F. Igersheim and J. Maddaluno, *J. Phys. Org. Chem.*, 2005, **18**, 268.

87. P. Kwiatkowski, E. Wojaczynska and J. Jurczak, *Tetrahedron: Asymmetry*, 2003, **14**, 3643.

88. S. H. Chong, D. J. Young and T. S. Andy Hor, *J. Organomet. Chem.*, 2006, **691**, 349.

89. S. Hillers, S. Sartori and O. Reiser, *J. Am. Chem Soc.*, 1996, **118**, 2087.

90. S. Has-Becher, K. Bodmann, R. Kreuder, G. Santoni, T. Rein and O. Reiser, *Synlett*, 2001, **9**, 1395.

91. K. Matsumoto, J.C. Kim, H. Iida, H. Hamana, K. Kumamoto, H. Kotsuki, G. Jenner, *Helv. Chim. Acta*, 2005, **88**, 1734 and reference cited therein.

Environmentally Benign Chemical Synthesis via Mechanochemical Mixing and Microwave Irradiation

VIVEK POLSHETTIWAR AND RAJENDER S. VARMA

Sustainable Technology Division, National Risk Management Research Laboratory, U.S. Environmental Protection Agency, 26 W. Martin Luther King Dr., MS 443, Cincinnati, Ohio 45268, USA

8.1 Introduction

The demands for development of green and sustainable synthetic methods in the fields of healthcare and fine chemicals, combined with the pressure to produce these substances expeditiously and in an environmentally benign fashion, pose considerable challenges to the synthetic chemical community.[1,2] Consequently, modern chemists are driven to focus on the development of efficient and environmentally benign synthetic protocols. This goal can be achieved through the correct choice of starting materials and atom economic methodologies with minimal chemical steps, the proper use of greener solvents and reagents, and well-organized strategies for product isolation and purification.[3,4]

Also, the use of non-conventional energy sources, *viz.* mechanochemical mixing and MW irradiation, has become a preferential choice for the chemist. The combination of green chemistry principles with MW heating and mechanochemical mixing techniques is dramatically reducing chemical waste and reaction

RSC Green Chemistry Book Series
Eco-Friendly Synthesis of Fine Chemicals
Edited by Roberto Ballini
© Royal Society of Chemistry 2009
Published by the Royal Society of Chemistry, www.rsc.org

times in several organic syntheses and chemical transformations and some of the representative important advancements in this area are reviewed here.

8.2 Environmentally Benign Synthesis *via* Mechanochemical Mixing

Mechanical mixing *via* high-energy ball-milling is a sustainable experimental technique, which is generally used for the preparation of metal hydrides and other organometallic materials.[5] Recently, it has also been used in organic synthesis and some important transformations are reviewed in this section.

8.2.1 Mechanically Induced Organic Transformations in a Ball-mill

A facile protocol for the synthesis of α-tosyloxy β-keto sulfones was developed by utilizing non-toxic [hydroxy(tosyloxy)iodo]benzene (HTIB) under solvent-free conditions at room temperature (Scheme 8.1). The simple procedure involves grinding together a neat mixture of α-benzenesulfonyl-acetophenone and HTIB at room temperature using a pestle and mortar. Within 5 min, the reaction afforded α-tosyloxybenzenesulfonylacetophenone in good yield.[6]

The synthesis of the triazole nucleus has gained great importance in organic synthesis. A solvent-free and expeditious synthesis of 1-aryl-4-methyl-1,2,4-triazolo[4,3-*a*]quinoxalines was developed that utilizes a mixing of a relatively benign non-metallic oxidant, iodobenzene diacetate (PhI(OAc)$_2$, Scheme 8.2).[7]

Nitrones have been synthesized in a ball-mill under mechanochemical and solvent-free conditions (Scheme 8.3). This protocol is clean and simple to perform, occurring under mild reaction conditions, with only water, NaCl and CO$_2$ as by-products. The methodology is highly efficient in terms of sustainable

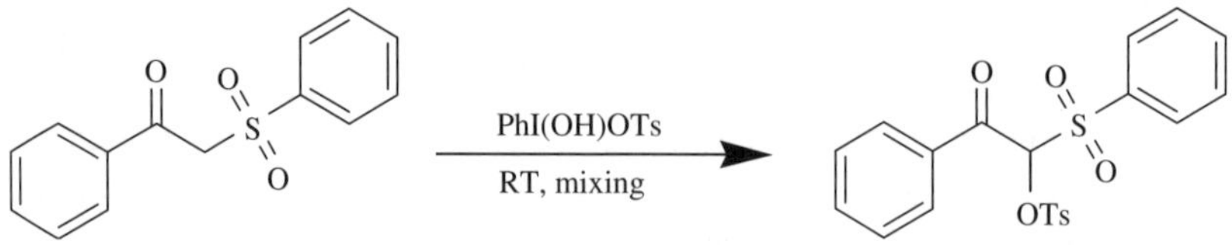

Scheme 8.1 Mechanochemical synthesis of α-tosyloxy β-keto sulfones.

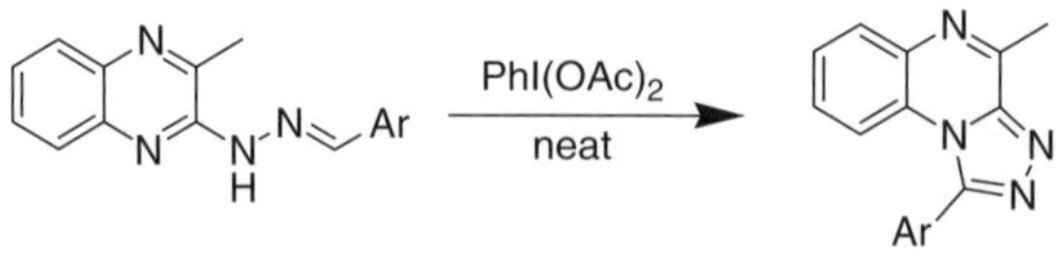

Scheme 8.2 PhI(OAc)$_2$-catalyzed solvent-free synthesis of triazoles.

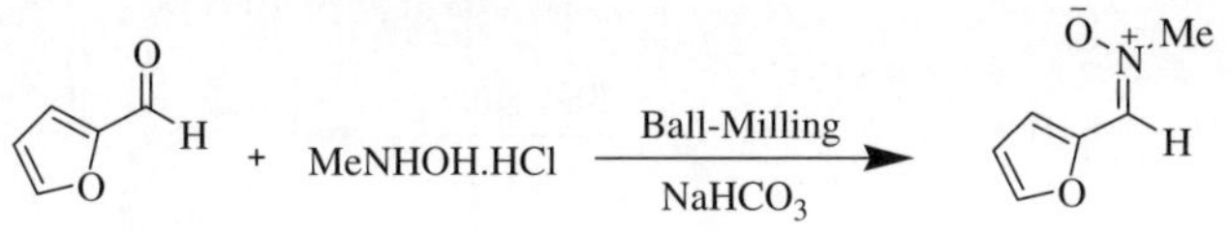

Scheme 8.3 Mechanochemical synthesis of nitrones in a ball-mill.

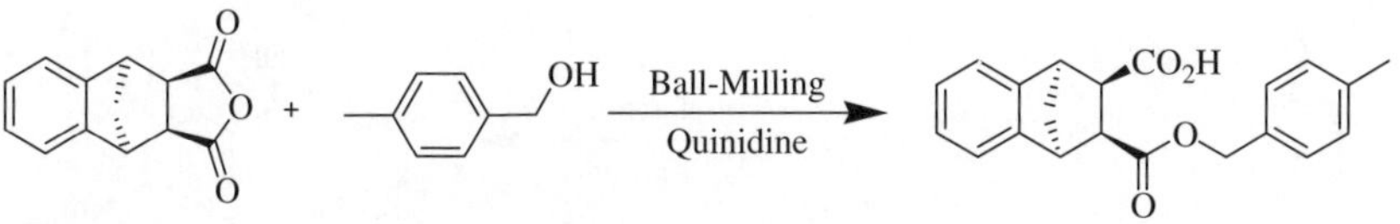

Scheme 8.4 Mechanochemical ring opening of anhydride in a ball-mill.

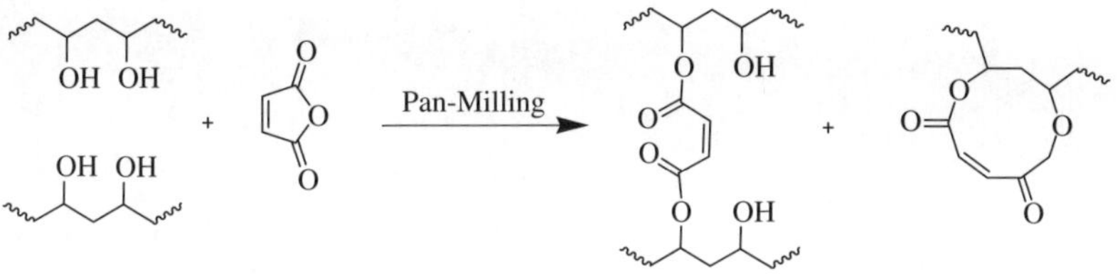

Scheme 8.5 Mechanochemical esterification of PVA in a pan-mill.

synthesis, affording products in high yield and in pure form that precludes further purification.[8]

The mechanochemical technique has also been applied to the asymmetric opening of *meso*-anhydrides, catalyzed by the cinchona alkaloid quinidine (Scheme 8.4). A simple procedure affords the optically active dicarboxylic acid monoesters in high yields. Even mixtures of purely solid substrates react under these conditions, and no solvent is required, which significantly simplifies the product isolation step.[9]

Mechanochemical processing has been used to conduct the esterification of poly(vinyl alcohol) (PVA) with maleic anhydride through stress-induced reaction by pan-milling (Scheme 8.5). In comparison with conventional methods for the esterification of PVA, this protocol is viable and environmentally friendly and can be an effective technique for the chemical modification of polymers.[10]

A one-pot protocol to oxidize aryl aldehydes and anilines directly to amides in good yields under mechanical milling conditions has been developed (Scheme 8.6). Compared with the traditional liquid-phase reaction, this solvent-free and metal catalyst-free procedure makes this protocol more efficient and eco-friendly.[11]

The proline-catalyzed aldol reaction can be conducted in a ball-mill, which yields *anti*-aldol products with up to >99% enantiomeric excess in excellent

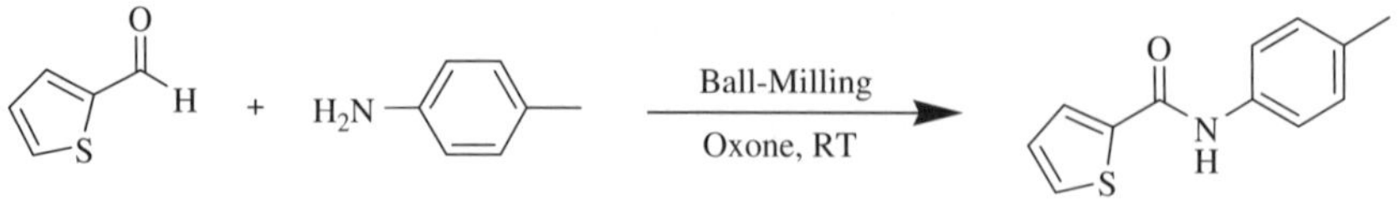

Scheme 8.6   Mechanochemical one-pot synthesis of amides in a ball-mill.

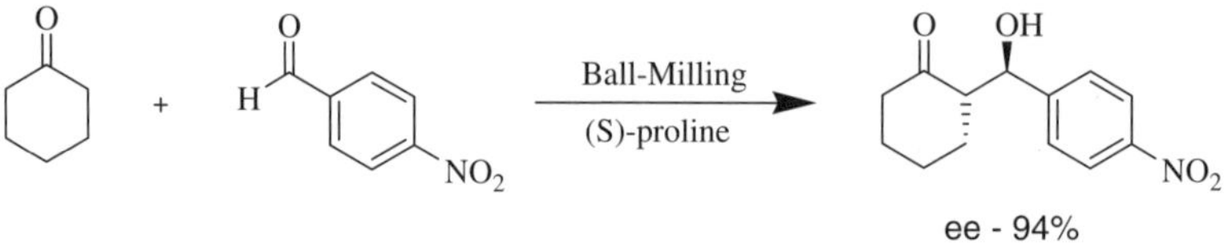

Scheme 8.7   Mechanochemical enantioselective aldol reaction in a ball-mill.

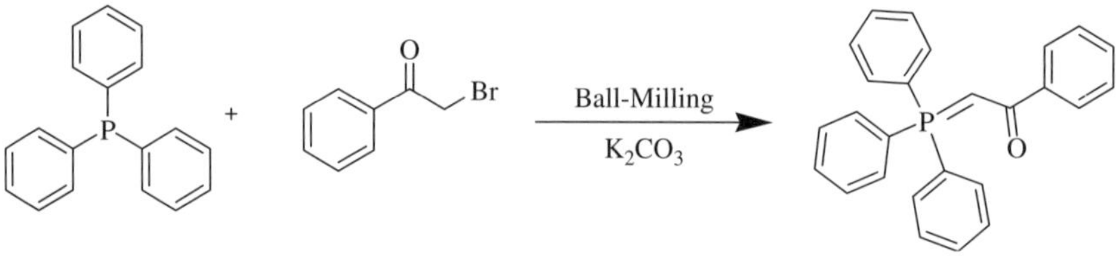

Scheme 8.8   Mechanochemical Wittig reaction in a ball-mill.

yields under solvent-free conditions (Scheme 8.7). Even mixtures of exclusively solid compounds react under these conditions, giving products through honey-like intermediates. Also, the equimolar reactant ratio makes the isolation step much easier, leading to higher product yields.[12]

Phosphorus ylides can also be prepared *via* solvent-free Wittig reaction in a ball-mill (Scheme 8.8). High-energy mechanical mixing played a critical role in this chemical transformation and was responsible for the amorphization of the reactants and mass transfer in a solid state.[13]

8.2.2   Mechanically Induced Fullerene and Carbon-nanotube Synthesis in a Ball-mill

The preparation and reactions of buckminsterfullerenes, as well as the carbon nanotubes, are currently attracting great interest. The [2 + 2] dimer of fullerene C_{60} has been synthesized by solid-state mechanochemical mixing of C_{60} with KCN (Scheme 8.9). The reaction can also be accomplished using potassium salts, metals and organic bases. Under optimum conditions, the reaction yielded only the dimer C_{120} and un-reacted C_{60} in a ratio of about 3 : 7.[14] As per *ab initio* and density functional calculations, under the mechanochemical mixing

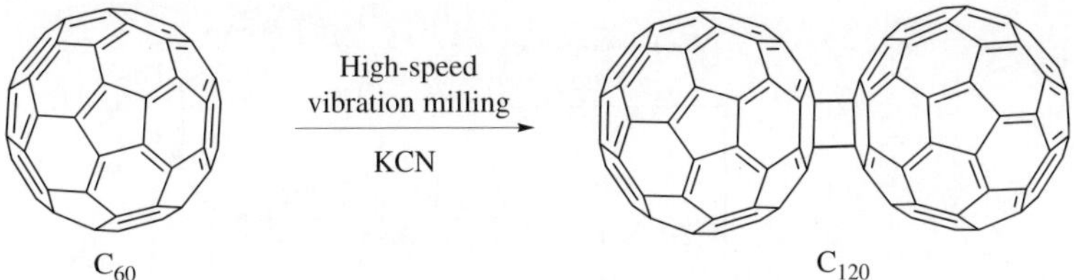

Scheme 8.9   Mechanochemical synthesis of fullerene C$_{120}$.

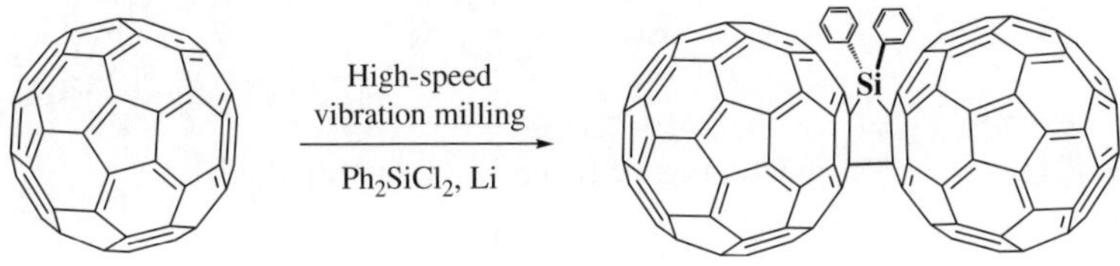

Scheme 8.10   Mechanochemical synthesis of a fullerene C$_{60}$ dimer.

conditions, the stability of C$_{120}$ appears to be comparable to that of two C$_{60}$ molecules.

Another C$_{60}$ dimer connected by a silicon bridge and a single bond was prepared by Fujiwara and Komatsu using the high-speed vibration milling technique (Scheme 8.10). They also proposed the mechanism for the formation of C$_{60}$ dimer, which involves the ionic reaction between dichlorodiphenylsilane and C$_{60}$ radical anion formed by one-electron transfer from lithium (Li) and radical coupling of the two C$_{60}$ cages.[15]

Chemical modification of single-walled carbon nanotubes (SWNTs) is a good tool to access the fundamental chemistry of SWNTs, to modify their chemical and physical properties, and to generate SWNTs with new characteristics. Pan and co-workers developed a new approach for functionalization of SWNTs with multiple hydroxyl groups *via* mechanochemical mixing with potassium hydroxide. The resulting SWNTols showed a robustly intertube interaction through hydrogen bonding. This mechanochemical protocol for chemical modification of SWNTs renders this methodology as a simple and sustainable approach towards large-scale production of functionalized SWNTs and self-assembled SWNT films.[16] Water-soluble SWNTs can be prepared by mechanochemical high-speed vibration milling; however, the solubility of SWNTs depends on the number of phosphate groups and the type of bases employed.[17] Interestingly, SWNTs can also be solubilized in organic solvents by mechanochemical mixing, which occurs by the formation of a complex between the SWNTs and polythiophene derivative.[18]

The [4 + 2] cycloaddition is one of the best-studied reactions of C$_{60}$. Komatsu *et al.* carried out this cycloaddition reaction to produce C$_{60}$ derivatives with new structural features. In particular, the reaction with phthalazine takes a completely different path depending on the reaction phase. In the solution-phase

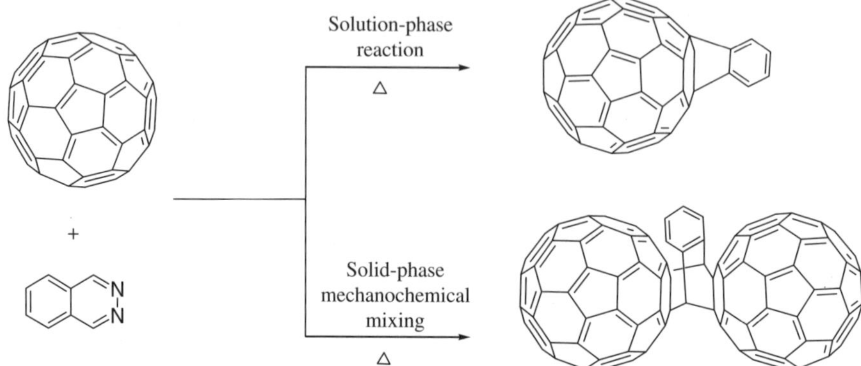

Scheme 8.11   [4 + 2] Cycloaddition of fullerene C_{60} with phthalazine.

Scheme 8.12   Solvent-free reactions of fullerenes and *N*-alkylglycines with aldehydes.

reaction, an open-cage fullerene derivative was formed and in the mechanochemical solid-state reaction the derivative with two C_{60} cages rigidly fixed in close proximity in a bicyclic framework was formed (Scheme 8.11).[19]

The solvent-free reactions of fullerenes and *N*-alkylglycines with aldehydes using high-speed vibration milling have been successfully accomplished by Wang and co-workers. Various fulleropyrrolidines were obtained in good yields by reactions of C_{60} with aldehydes and *N*-methylglycine (Scheme 8.12).[20]

The Bingel reaction of diethyl bromomalonate and C_{60}, under the mechanochemical high-speed vibration milling conditions, produced cyclopropanated C_{60} in good yield. However, the reactions of diethyl malonate and ethyl acetoacetate with C_{60} afforded 1,4-bisadduct and dihydrofuran-fused C_{60} derivative, respectively (Scheme 8.13).[21]

Wang and co-workers also evaluated the reactivity of 9-substituted anthracene in Diels–Alder reactions with fullerene. Corresponding adducts were prepared in high yields under solvent-free conditions using the high-speed vibration milling technique (Scheme 8.14).[22]

8.3   Environmentally Benign Synthesis *via* Microwave (MW) Irradiation

There are many examples of the booming application of MW heating in organic synthesis.[1–4] These include the use of benign reaction media such as

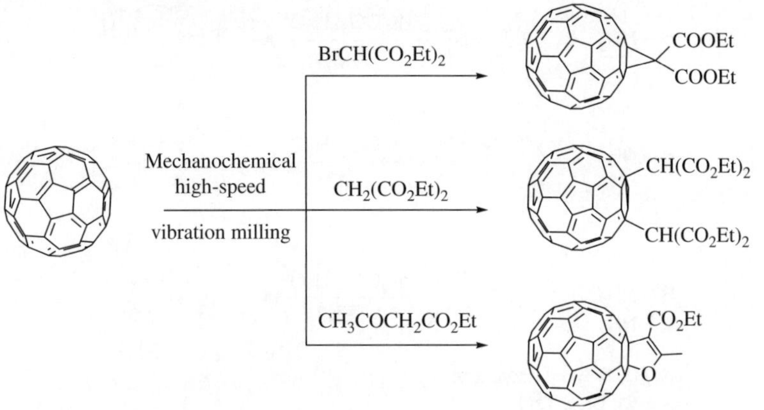

Scheme 8.13 The Bingel reaction of fullerenes under mechanochemical conditions.

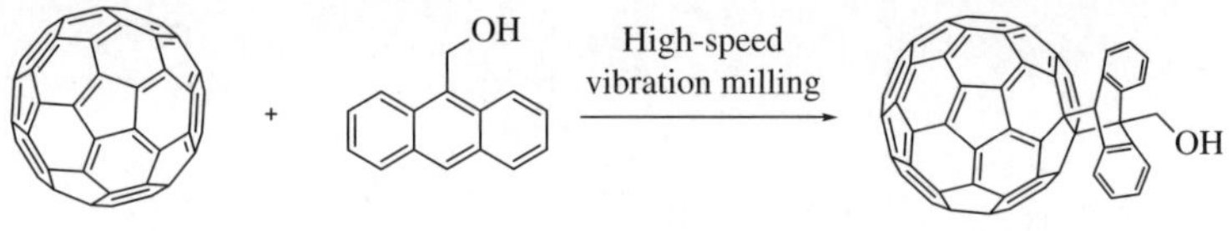

Scheme 8.14 Diels–Alder reactions of fullerene with a 9-substituted anthracene.

water[4] and ionic liquids,[3] reaction under solvent-free conditions[2] and the use of solid-supported catalysts.[23]

8.3.1 MW-assisted Synthesis of Heterocycles

Heterocyclic compounds hold a special place among pharmaceutically significant natural products and synthetic compounds. The synthesis of nitrogen-containing heterocycles such as substituted azetidines, pyrrolidines, piperidines, azepanes, *N*-substituted 2,3-dihydro-1*H*-isoindoles, 4,5-dihydropyrazoles, pyrazolidines and 1,2-dihydrophthalazines has been accomplished in an aqueous medium under the influence of MWs (Scheme 8.15).[24]

A range of nitrogen-containing heterocycles has also been synthesized by the condensation of hydrazine, hydrazide and diamines with diketones and β-keto esters, respectively, using polystyrene sulfonic acid (PSSA) in aqueous medium (Scheme 8.16).[25]

A Grignard type of addition of alkynes to *in situ* generated imines from aldehyde and amines, catalyzed by CuBr, provides an efficient solvent-free approach for the synthesis of substituted N-heterocycles such as propargylamines in excellent yields (Scheme 8.17).[26]

This N-alkylation of nitrogen heterocycles has been achieved in aqueous media under MW irradiation conditions (Scheme 8.18). Shorter reaction times

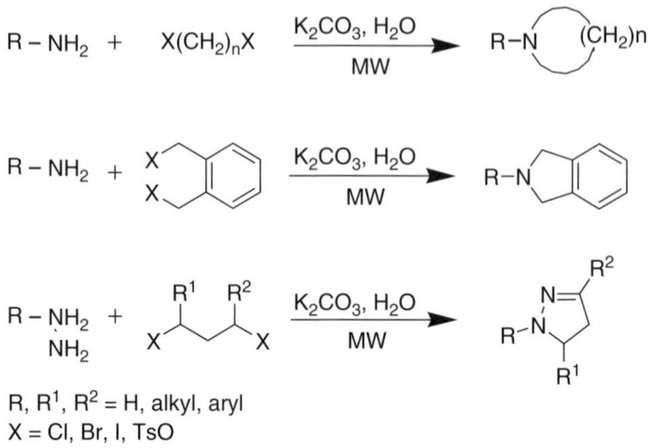
K2CO3, H2O
MW
R, R1, R2 = H, alkyl, aryl
X = Cl, Br, I, TsO

Scheme 8.15 Some examples of MW-assisted nitrogen-containing heterocycles synthesized in aqueous media.

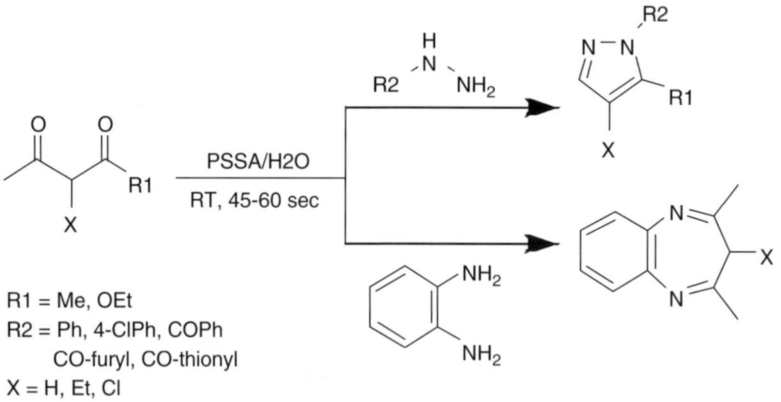
PSSA/H2O
RT, 45-60 sec
R1 = Me, OEt
R2 = Ph, 4-ClPh, COPh
CO-furyl, CO-thionyl
X = H, Et, Cl

Scheme 8.16 MW-assisted PSSA-catalyzed nitrogen-heterocycles synthesis.

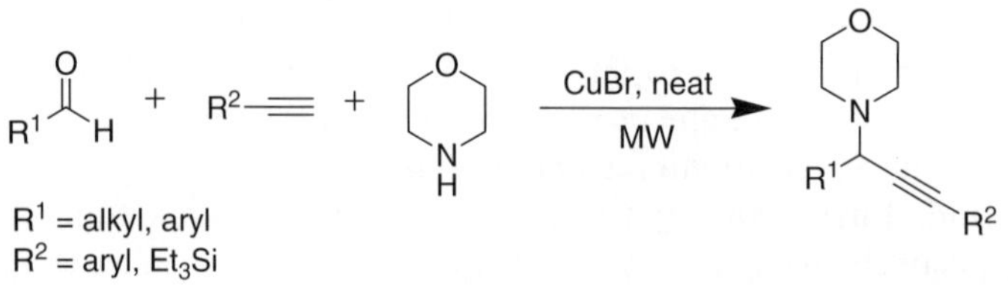
CuBr, neat
MW
R1 = alkyl, aryl
R2 = aryl, Et3Si

Scheme 8.17 MW-assisted CuBr-catalyzed solvent-free route to propargylamines.

and superior product yields are the advantages that make this procedure a benign alternative to conventional chemical synthesis.[27]

A MW-assisted protocol for the direct synthesis of cyclic ureas has been developed that proceeds expeditiously in the presence of ZnO (Scheme 8.19).

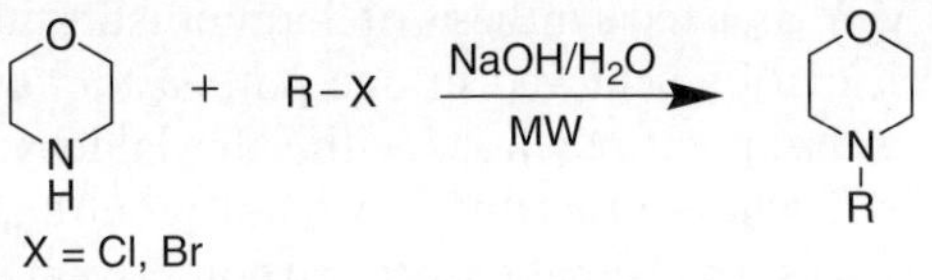

X = Cl, Br

Scheme 8.18 MW-assisted NaOH-catalyzed N-alkylation in water.

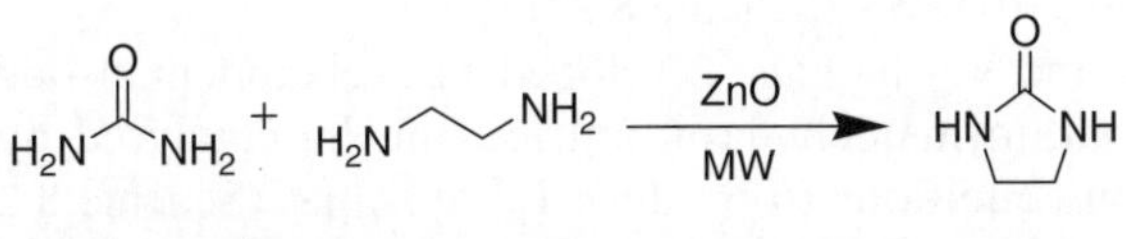

Scheme 8.19 MW-assisted synthesis of imidazolidine-2-one.

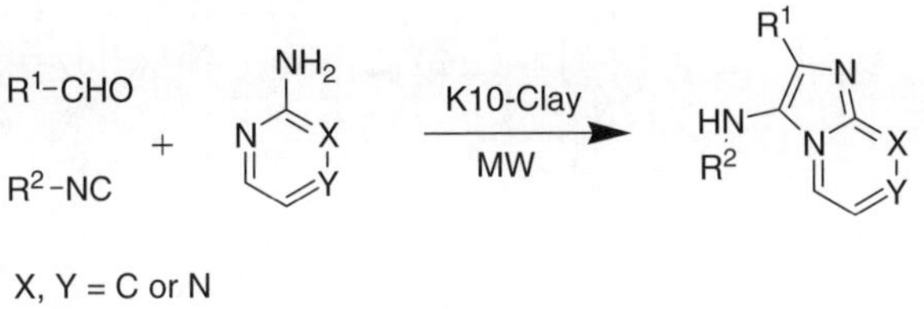

X, Y = C or N

Scheme 8.20 MW-assisted clay-catalyzed synthesis of nitrogen heterocycles.

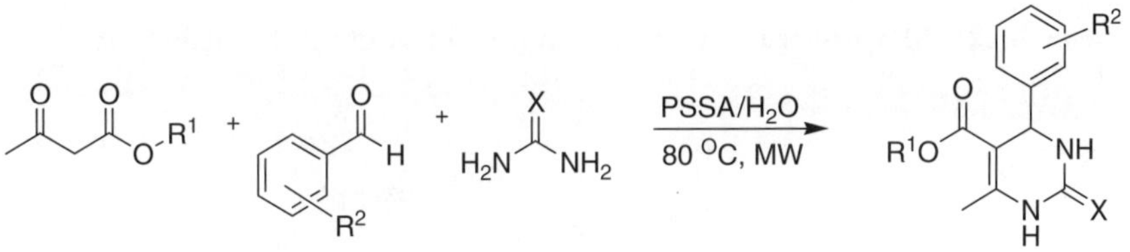

Scheme 8.21 MW-assisted Biginelli reaction in aqueous medium (PSSA = polystyrene sulfonic acid).

The reaction was not only accelerated upon exposure to MW irradiation but also the formation of by-products was eliminated when compared to the conventional heating methods.[28]

A one-pot MW protocol for the synthesis of imidazo[1,2-*a*]-annulated pyridines, pyrazines and pyrimidines has been developed (Scheme 8.20) that proceeded in the presence of solid catalyst montmorillonite K10 clay under solvent-free conditions.[29]

Dihydropyrimidinones, an important class of organic compounds that show prominent biological activity, were synthesized by a greener aqueous Biginelli protocol using PSSA as a catalyst. (Scheme 8.21).[30]

The solvent-free MW-assisted synthesis of 2-amino-substituted isoflav-3-enes has been reported; it can be achieved in one pot *via* the *in situ* generation of enamines and their subsequent reactions with salicylaldehydes (Scheme 8.22). This environmentally friendly procedure does not require azeotropic removal of water using large excess of aromatic hydrocarbon solvents for the generation of enamines.[31]

A rapid solvent-free synthesis of 2-aroylbenzo[*b*]furans has been developed, from readily accessible α-tosyloxyketones and mineral oxides, that is accelerated by exposure to MWs (Scheme 8.23).[32]

Polshettiwar and Varma have developed a novel tandem bis-aldol reaction of ketones with paraformaldehyde in aqueous media catalyzed by PSSA under MW irradiation conditions to produce 1,3-dioxanes (Scheme 8.24).[33]

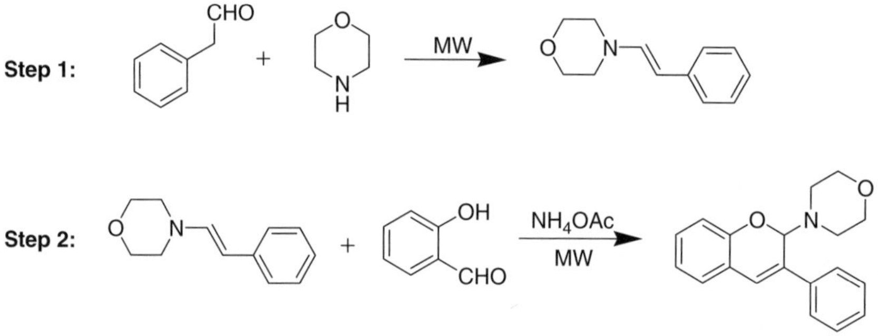

Scheme 8.22   MW-assisted solvent-free synthesis of substituted isoflav-3-enes.

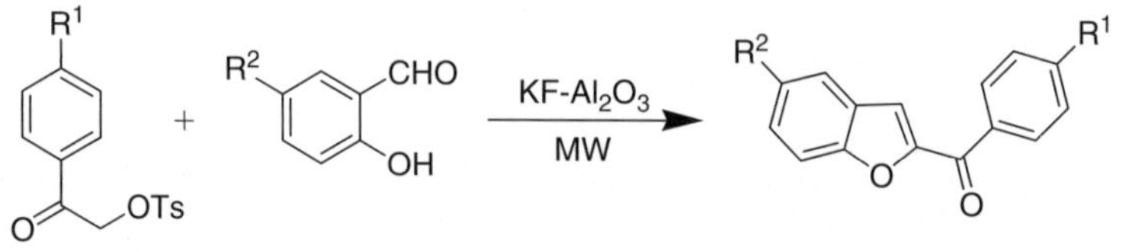

Scheme 8.23   MW-assisted synthesis of 2-aroylbenzo[*b*]furans.

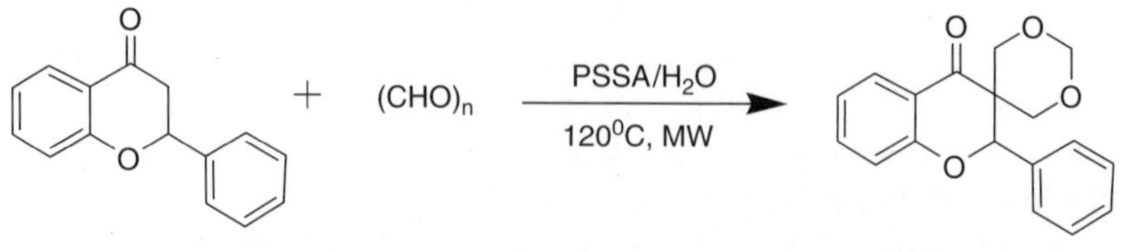

Scheme 8.24   MW-assisted PSSA-catalyzed synthesis of 1,3-dioxanes in aqueous media.

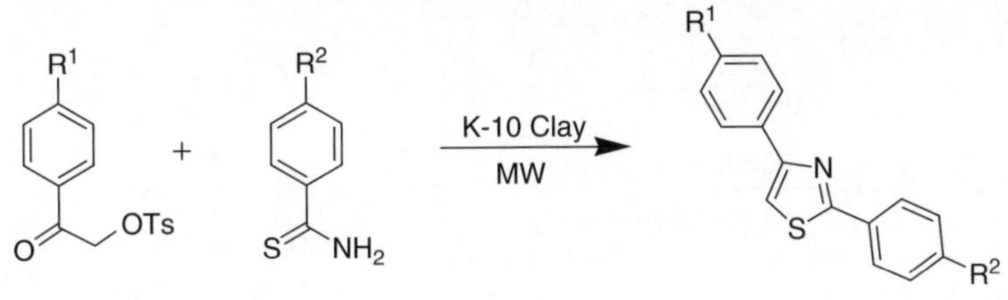

Scheme 8.25 MW-assisted solvent-free synthesis of 1,3-thiazoles.

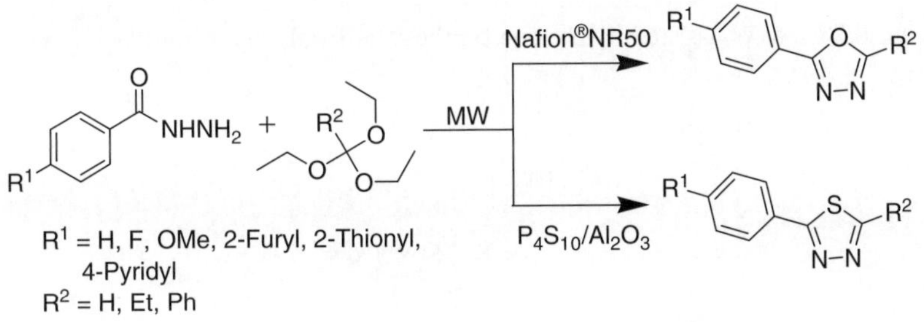

R^1 = H, F, OMe, 2-Furyl, 2-Thionyl, 4-Pyridyl
R^2 = H, Et, Ph

Scheme 8.26 MW-assisted synthesis of 1,3,4-oxadiazoles and 1,3,4-thiadiazoles.

1,3-Thiazoles have been readily obtained from thioamides and α-tosyloxy-ketones, catalyzed by montmorillonite K-10 clay under MW irradiation, in excellent yields. These compounds are very difficult to synthesize under conventional heating conditions (Scheme 8.25).[32]

Recently, a novel one-pot solvent-free synthesis of 1,3,4-oxadiazoles and 1,3,4-thiadiazoles has been developed (Scheme 8.26) by the condensation of acid hydrazide with triethyl orthoalkanates under MW irradiation conditions.[34]

An environmentally benign aqueous protocol for the synthesis of heterocyclic hydrazones using PSSA as a catalyst has been developed (Scheme 8.27). The simple reaction proceeds efficiently in water in the absence of any organic solvent under MW irradiation and involves basic filtration as the product isolation step.[35]

8.3.2 MW-assisted Synthesis of Ionic Liquids (ILs)

ILs are polar but consist of poorly coordinating ions and provide a polar alternative for biphasic systems. Other important attributes of these ILs include negligible vapor pressure, potential for recycling, compatibility with various organic compounds and organometallic catalysts, and ease of separation of products from reactions. ILs, being polar and ionic in character, couple with MW irradiation very efficiently and are, therefore, ideal MW-absorbing candidates for expediting chemical reactions. The first efficient preparation of 1,3-dialkylimidazolium halides *via* MW irradiation was developed by Varma and co-workers (Scheme 8.28). The reaction time was reduced from several hours to

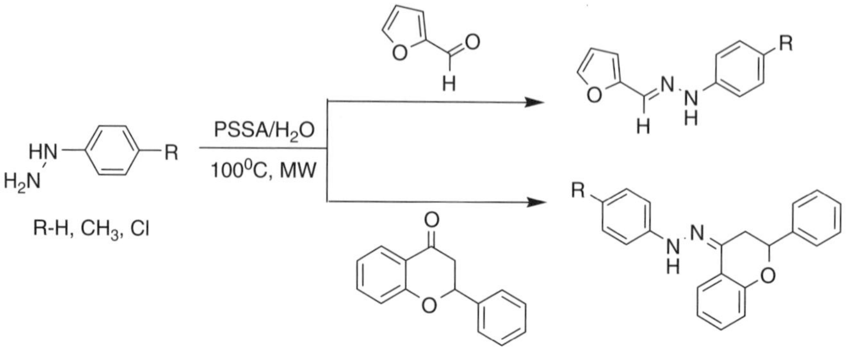

Scheme 8.27 MW-assisted hydrazone synthesis of furaldehyde and flavanone.

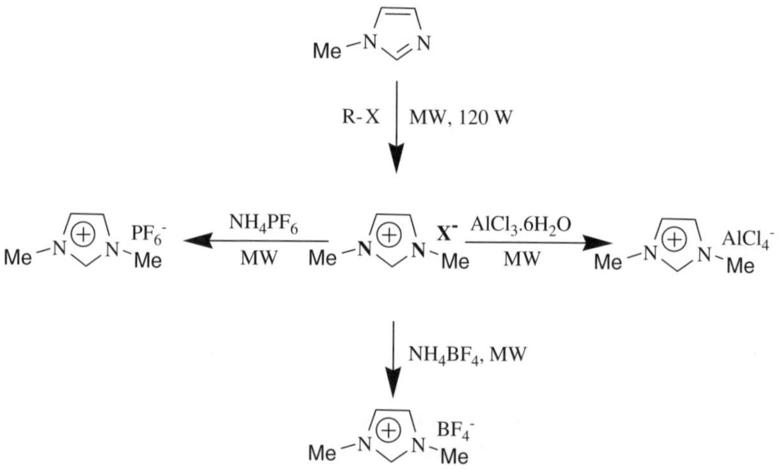

Scheme 8.28 MW-assisted synthesis of ILs.

minutes and it avoids the use of a large excess of organic solvents as the reaction medium.[36]

Tetrahaloindate(III)-based ILs, prepared using a similar MW protocol, have been used as catalysts for the synthesis of cyclic carbonates. CO_2 reacted with various epoxides in the presence of catalytic amounts of tetrahaloindate(III)-based ILs, producing important fine chemicals in good to excellent yields in benign fashion (Scheme 8.29).[37]

8.3.3 MW-assisted Oxidation–Reduction Reactions

MW protocols for oxidation–reduction reactions using immobilized reagents on solid supports have been extensively explored. The oxidation of sulfides to sulfoxides and sulfones was also developed under MW irradiation with good

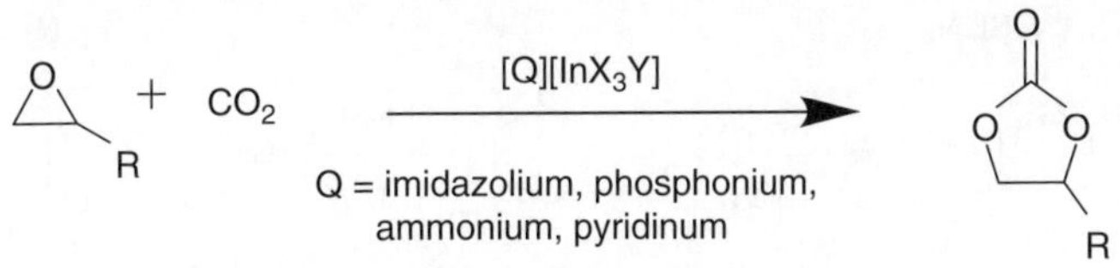

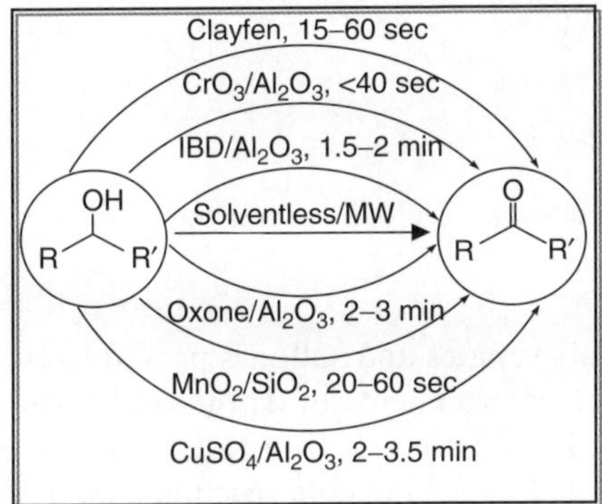

Scheme 8.29 Tetrahaloindate(III)-based IL-catalyzed synthesis of cyclic carbonates.

Scheme 8.30 MW-assisted solvent-free oxidation and reduction reactions.

selectivity to either sulfoxides or sulfones using sodium periodate ($NaIO_4$) on silica. A rapid and chemoselective reduction of aldehydes and ketones, which uses alumina-supported sodium borohydride ($NaBH_4$) and proceeds in the solid state, was also accelerated by MW irradiation. A solvent-free reductive amination protocol for carbonyl compounds using sodium borohydride supported on moist montmorillonite K10 clay was carried out under MW irradiation. Clay serves as a Lewis acid and provides water from its inner layers that enhances the reducing ability of $NaBH_4$ (Scheme 8.30).[38]

8.3.4 MW-assisted Nucleophilic Substitution Reactions

A rapid and sustainable MW-protocol for the synthesis of various azides, thiocyanates and sulfones has been developed in an aqueous medium (Scheme 8.31).

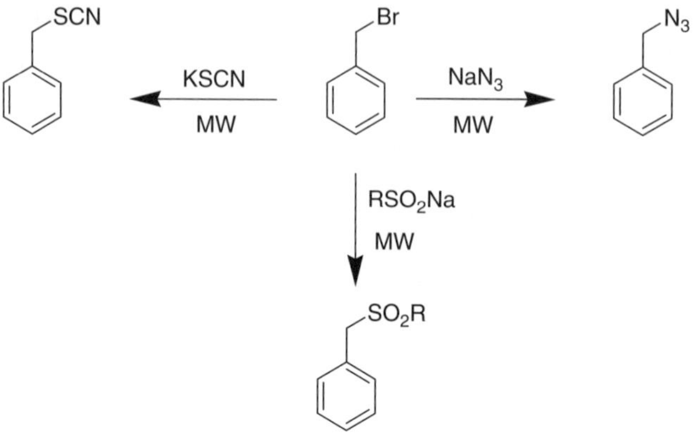

Scheme 8.31 Synthesis of azides, thiocyanates and sulfones in aqueous medium.

This MW-assisted synthesis of azides, thiocyanates and sulfones proved to be a useful alternative that avoids the use of environmentally detrimental volatile chlorinated hydrocarbons. Various functional groups such as ester, carboxylic acid, carbonyl and hydroxyl were unaffected under the mild reaction conditions employed.[39]

8.3.5 MW-assisted Suzuki and Heck Reactions using supported Pd-catalyst

The Heck and Suzuki reactions are among the most widely used reactions for the formation of carbon–carbon bonds. These reactions are generally catalyzed by soluble Pd complexes with various ligands. However, the efficient separation and subsequent recycling of homogeneous transition-metal catalysts remains a scientific challenge and an aspect of economical and ecological relevance.[40] Polshettiwar and Varma have developed the first Pd–N-heterocyclic carbene (NHC) complex in the form of organic silica that was prepared using the sol–gel method (Scheme 8.32).[41]

This catalyst was then used for Heck and Suzuki reactions under MW irradiation condition (Scheme 8.33). These C–C coupling reactions proceeded efficiently under the influence of MW, with excellent yield and high turnover frequency (TOF), without any change in catalytic activity for at least five reaction cycles.[41]

8.3.6 MW-assisted aza-Michael and Ritter Reaction

The aza-Michael addition is an important class of carbon–nitrogen bond-forming reactions, and has been demonstrated to be a powerful tool in organic synthesis. Efficient aza-Michael addition reactions of alkyl amine (Scheme 8.34)

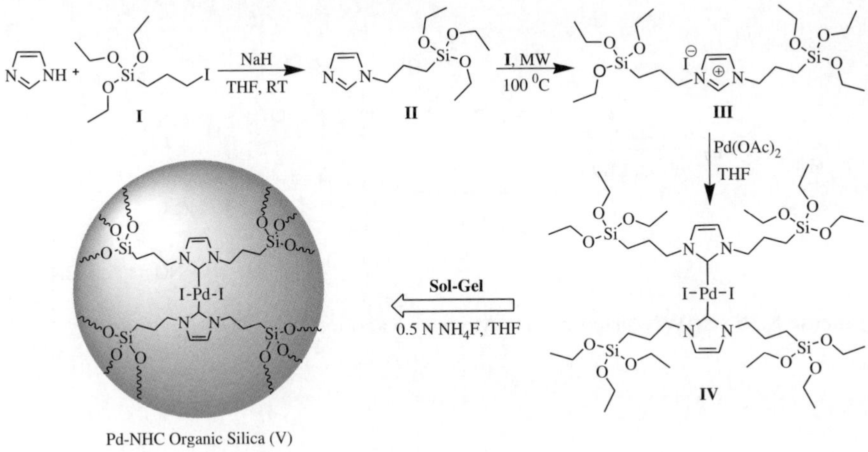

Scheme 8.32   Synthesis of Pd-NHC organic silica catalyst.

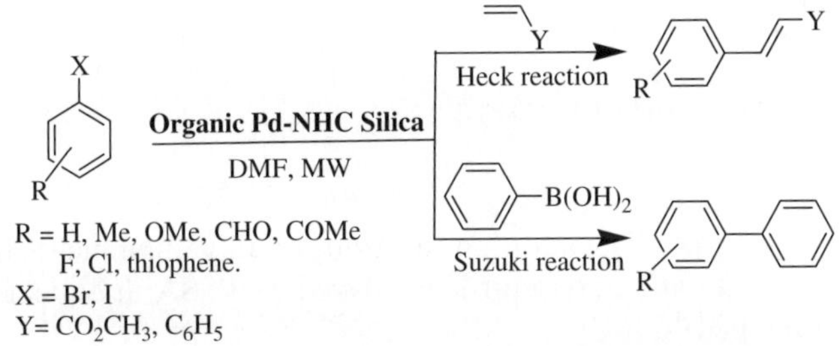

Scheme 8.33   Pd–NHC silica catalyzed Heck and Suzuki reactions.

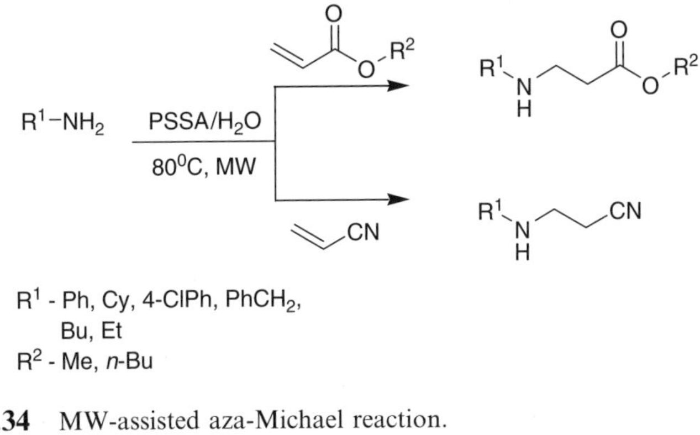

Scheme 8.34   MW-assisted aza-Michael reaction.

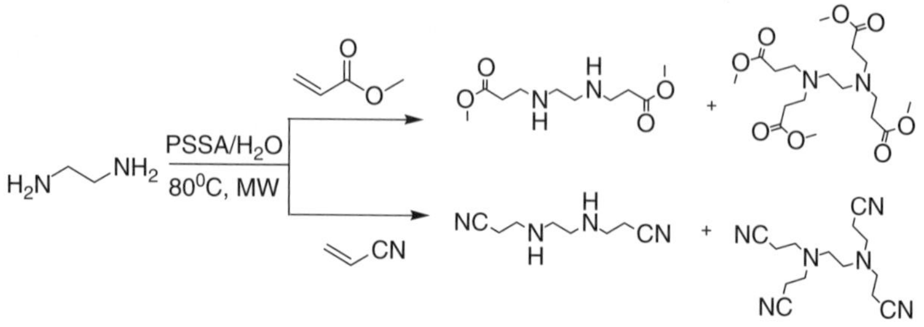
PSSA/H$_2$O
80^{0}C, MW

Scheme 8.35 MW-assisted bis-aza-Michael reaction.

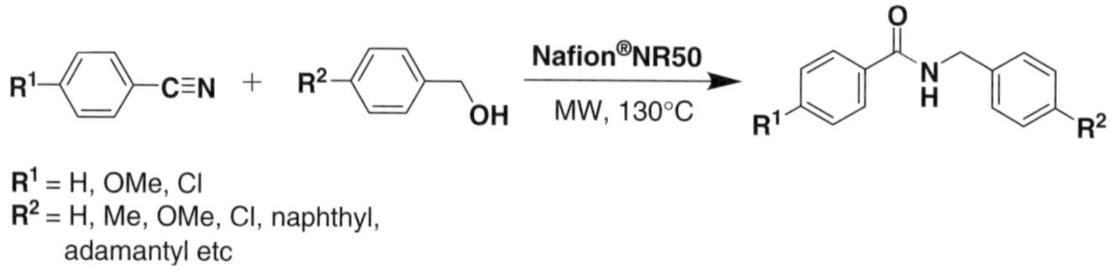
Nafion®NR50
MW, 130°C
R^1 = H, OMe, Cl
R^2 = H, Me, OMe, Cl, naphthyl, adamantyl etc

Scheme 8.36 MW-assisted Ritter reaction.

and tandem bis-aza-Michael addition reactions of alkyl diamine (Scheme 8.35) with methyl acrylate and acrylonitrile catalyzed by PSSA in aqueous medium have been developed.[42]

This operationally simple PSSA-catalyzed aza-Michael protocol proceeded efficiently in aqueous medium without use of organic solvent. Also, the use of polymer-supported, relatively low toxicity, and inexpensive PSSA as a catalyst renders this method greener and eco-friendly.

An efficient synthesis of amides by the Ritter reaction of alcohols and nitriles under MW irradiation was developed under solvent-free conditions (Scheme 8.36). This green protocol is catalyzed by solid-supported Nafion®NR50 with improved efficiency and reduced waste production.[43]

8.4 Conclusions

The most important goal of chemists is in ensuring that our next generation of synthetic protocols for drugs and fine chemicals are more sustainable and greener than the current generation. The uses of emerging MW-assisted chemistry and mechanochemical mixing in conjunction with the utility of greener reaction media or solvent-free conditions are effective and sustainable techniques to reduce chemical waste and reaction times in organic synthesis.

Acknowledgment

Vivek Polshettiwar was supported, in part, by the Postgraduate Research Program at the National Risk Management Research Laboratory administered by the Oak Ridge Institute for Science and Education through an interagency agreement between the U.S. Department of Energy and the U.S. Environmental Protection Agency.

References

1. (a) V. Polshettiwar and R. S. Varma, *Curr. Opin. Drug Discov. Devel.*, 2007, **10**, 723; (b) V. Polshettiwar and R. S. Varma, *Pure Appl. Chem.*, 2008, **80**, 777.
2. R. S. Varma, *Green. Chem.*, 1999, **1**, 43.
3. V. Polshettiwar and R. S. Varma, *Acc. Chem. Res.*, 2008, **41**, 629.
4. V. Polshettiwar, R. S. Varma, *Chem. Soc. Rev.*, 2008, **37**, 1546.
5. D. Braga, S. L. Giaffreda, F. Grepioni, A. Pettersen, L. Maini, M. Curzi and M. Polito, *Dalton Trans.*, 2006, 1249.
6. D. Kumar, M. S. Sundaree, G. Patel, V. S. Rao and R. S. Varma, *Tetrahedron Lett.*, 2006, **47**, 8239.
7. D. Kumar, K. V. G. Chandra Sekhar, H. Dhillon, V. S. Rao and R. S. Varma, *Green. Chem.*, 2004, **6**, 156.
8. E. Colacino, P. Nun, F. M. Maria Colacino, J. Martinez and F. Lamaty, *Tetrahedron*, 2008, **64**, 5569.
9. T. Rantanen, I. Schiffers and C. Bolm, *Org. Proc. Res. Devel.*, 2007, **11**, 592.
10. Bo. Zhao, C. H. Lu and M. Liang, *Chinese Chem. Lett.*, 2007, **18**, 1353.
11. J. Gao and G.-W. Wang, *J. Org. Chem.*, 2008, **73**, 2955.
12. B. Rodriguez, A. Bruckmann and C. Bolm, *Chem. Eur. J.*, 2007, **13**, 4710.
13. V. P. Balema, J. W. Wiench, M. Pruski and V. K. Pecharsky, *J. Am. Chem. Soc.*, 2002, **124**, 6244.
14. K. Komatsu, G.-W. Wang, Y. Murata, T. Tanaka and K. Fujiwara, *J. Org. Chem.*, 1998, **63**, 9358.
15. K. Fujiwara and K. Komatsu, *Org. Lett.*, 2002, **6**, 1039.
16. H. Pan, L. Liu, Z.-X. Guo, L. Dai, F. Zhang, D. Zhu, R. Czerw and D. L. Carroll, *Nano Lett.*, 2003, **3**, 29.
17. A. Ikeda, T. Hamano, K. Hayashi and J.-I. Kikuchi, *Org. Lett.*, 2006, **8**, 1153.
18. A. Ikeda, T. Hamano, K. Nobusawa, T. Hamano and J.-I. Kikuchi, *Org. Lett.*, 2006, **8**, 5489.
19. Y. Murata, N. Kato and K. Komatsu, *J. Org. Chem.*, 2001, **66**, 7235.
20. G.-W. Wang, T.-Z. Zhang, E.-H. Hao, L.-J. Jiao, Y. Muratab and K. Komatsu, *Tetrahedron*, 2003, **59**, 55.
21. G.-W. Wang, T.-Z. Zhang, Y.-J. Li, L. H. Zhan, Y.-C. Liu, Y. Murata and K. Komatsu, *Tetrahedron Lett.*, 2003, **44**, 4407.

22. G.-W. Wang, Z.-X. Chen, Y. Murata and K. Komatsu, *Tetrahedron*, 2005, **61**, 4851.
23. (a) R. S. Varma, *Tetrahedron*, 2002, **58**, 1235; (b) V. Polshettiwar and A. Molnar, *Tetrahedron*, 2007, **63**, 6949.
24. (a) Y. Ju and R. S. Varma, *Tetrahedron Lett.*, 2005, **46**, 6011; (b) Y. Ju and R. S. Varma, *Org. Lett.*, 2005, **7**, 2409; (c) Y. Ju and R. S. Varma, *J. Org. Chem.*, 2006, **71**, 135.
25. V. Polshettiwar and R. S. Varma, *Tetrahedron Lett.*, 2008, **49**, 397.
26. Y. Ju, C.-J. Li and R. S. Varma, *QSAR and Comb. Sci.*, 2004, **23**, 891.
27. Y. Ju and R. S. Varma, *Green. Chem.*, 2004, **6**, 219.
28. Y. J. Kim and R. S. Varma, *Tetrahedron Lett.*, 2004, **45**, 7205.
29. R. S. Varma and D. Kumar, *Tetrahedron Lett.*, 1999, **40**, 7665.
30. V. Polshettiwar and R. S. Varma, *Tetrahedron Lett.*, 2007, **48**, 7343.
31. R. S. Varma and R. Dahiya, *J. Org. Chem.*, 1998, **63**, 8038.
32. R. S. Varma and D. P. J. Liesen Kumar, *J. Chem. Soc. Perkin Trans. 1*, 1998, 4093.
33. V. Polshettiwar and R. S. Varma, *J. Org. Chem.*, 2007, **72**, 7420.
34. V. Polshettiwar and R. S. Varma, *Tetrahedron Lett.*, 2008, **49**, 879.
35. V. Polshettiwar and R. S. Varma, *Tetrahedron Lett.*, 2007, **48**, 5649.
36. (a) R. S. Varma and V. V. Namboodiri, *Chem. Commun.*, 2001, 643; (b) V. V. Namboodiri and R. S. Varma, *Tetrahedron Lett.*, 2002, **43**, 5381.
37. Y. J. Kim and R. S. Varma, *J. Org. Chem.*, 2005, **70**, 7882.
38. C. R. Strauss and R. S. Varma, *Top. Curr. Chem.*, 2006, **266**, 199.
39. Y. Ju, D. Kumar and R. S. Varma, *J. Org. Chem.*, 2006, **71**, 6697.
40. V. Polshettiwar and A. Molnar, *Tetrahedron*, 2007, **63**, 6949.
41. V. Polshettiwar and R. S. Varma, *Tetrahedron*, 2008, **64**, 4637.
42. V. Polshettiwar and R. S. Varma, *Tetrahedron Lett.*, 2007, **48**, 8735.
43. V. Polshettiwar and R. S. Varma, *Tetrahedron*, 2008, **64**, 2661.

Subject Index